The Stokes Phenomenon, Borel Summation and Mellin-Barnes Regularisation

Victor Kowalenko
University of Melbourne

CONTENTS

FOREWORD

Hyperasymptotics, known also as beyond-all-orders asymptotics and exponential asymptotics, has become important in the last couple of decades, but asymptotic/divergent series have been useful and confusing for centuries. Niels Henrik Abel, who died one hundred and eighty years ago, wrote: "Divergent series are the invention of the devil, and it is a shame to base on them any demonstration whatsoever. By using them, one may draw any conclusion one pleases and that is why these series have produced so many fallacies and so many paradoxes ...". Hyperasymptotics has been the struggle to remove these fallacies and paradoxes.

Sir George Stokes made a major advance in 1857 when he discovered the Stokes phenomenon [2]; the form of the asymptotic approximation must change discontinuously crossing certain sectors in the complex plane. In his own words, "the inferior term enters as it were into a mist, is hidden for a little from view, and comes out with its coefficient changed." Stokes' discontinuities have been successfully embedded into the special function software that evaluates Bessel functions, the exponential integral and many other transcendental functions in Matlab, Mathematica and FORTRAN. Thus, his coefficients from the "mist" have been reduced to black boxes, robust and invisible to the user. Yet Stokes himself was never entirely pleased with his "discontinuities". In a droll letter to his fiancee, he says that he stayed awake until 3 am, peering over his notes with only the feeble, guttering light of a candle, trying to understand them.

Robert Dingle offered an improved treatment in his 1973 book [3]. His student, Sir Michael Berry, wrote about "smoothing a Victorian discontinuity" in 1989 [9], and introduced "superasymptotic" and "hyperasymptotic" into the lexicon [47]. In this book, Victor Kowalenko boldly takes on these sages with his own interpretation of the Stokes phenomenon.

Who is right? We shall let the reader judge, but offer only two observations that are on sure ground. One is that, like Abel and Stokes, contemporary mathematicians are still struggling to interpret and conceptualize hyperasymptotics. In Enrico Fermi's words, "we are still confused, but at a higher level".

The second is that the reader may use the approximations and algorithms of Kowalenko's book without fear. One of his book's strengths is that it is full of very careful, high precision numerical experiments (as are the papers of many of his adversaries like Berry). There are no controversies about numerical accuracy;

the conflicts are about concepts, interpretation and the best strategies for extending hyperasymptotics to new realms where it has not yet been verified.

Kowalenko's book is thus an important step forward in a quest that has stretched over centuries and is still an active research frontier. His study captures only part of a very broad mathematical toolbox important in quantum mechanics, fluid mechanics, optics, quantum field theory and a dozen other fields. Refs. [27] and [48] give a perspective on this breadth, while Ref. [49] catalogues other books on hyperasymptotics.

Still, Kowalenko's book covers an important subset of this broad field. In particular, it reaches across the centuries to shed deeper light on the Stokes phenomenon and the Dingle-Berry theories. Borel summation, too, is more than a hundred years old, but Kowalenko shows that it is a powerful tool for regularizing divergent asymptotic series. The vagueness of the Landau gauge symbols, $O()$ and $o()$, which express only order-of-magnitude and limits, can be replaced, at least for the cases he analyzes, with precision. This book is a good addition to an ancient, but also modern, research frontier.

John P. Boyd

Ann Arbor, Michigan

USA

PREFACE

In 1993 Dr T. Taucher and I carried out a numerical study into the accuracy of the complete asymptotic expansion for the series $S_3(a) = \sum_{k=0}^{\infty} \exp(-ak^3)$. A complete asymptotic expansion is defined as an expansion that is composed of not only all the terms in the main or dominant asymptotic series, but also all the terms of each transcendental or exponentially subdominant asymptotic series, should they exist. Normally, the latter series, which are said to lie beyond all orders and whose derivation is the goal of the subject known as exponential asymptotics, are neglected by applied mathematicians when they seek asymptotic solutions to problems, while the dominant series is usually truncated after the first few terms.

There are two main reasons for adopting this relatively crude, but standard, approach in asymptotics. The first is that both the entire dominant and subdominant series are divergent and until now, there has been no adequate theory that can provide meaningful values to such series. The second is that truncating the dominant series after a few terms generally provides a very good approximation to the original function/integral provided one does not venture sufficiently far away from the limiting point of the expansion, which is generally zero or infinity. Hence, when an asymptotic solution is provided to a mathematical problem, it is often accompanied by the phrase "in the limit as such and such quantity tends to zero or infinity". Because of the second reason there has been no urgency on the part of mathematicians to develop a theory of divergent series so that exact values could be obtained from asymptotic expansions. Instead, the subject of asymptotics has developed by employing such vague symbols as $\sim$ or goes as and $O()$ or the order of, while in other instances, practitioners have sought to calculate bounds for the remaining terms neglected in the truncation process. Because of this imprecision, asymptotics as a mathematical discipline has often suffered from derisory remarks made in particular by pure mathematicians, who point out that mathematics is supposed to be an exact science.

In the case of $S_3(a)$, which represented a particular value of a series called the generalised Euler-Jacobi series, the complete asymptotic expansion consisted of a dominant algebraic series and one subdominant exponential series, whose coefficients resembled the dominant series in the asymptotic expansion for the Airy function. Moreover, the complete asymptotic expansion was described as a small a-expansion, which meant that its limiting point was zero, as opposed to a large a expansion, whose limiting point is infinity. At the time we were only interested iv

in real values of a. Therefore, provided a was less than 0.01, one could truncate the dominant series and neglect the subdominant series to obtain very accurate estimates for $S_3(a)$. Our aim, however, was to find the values of a, where the subdominant series could no longer be neglected. This meant that we needed to consider intermediate values of a lying between 0.01 and 5 and "large" values, where a was greater than 5.

An undertaking such as this is quite formidable because as one moves away from the limiting point, accurate results can no longer be obtained by truncating the dominant series. In fact, for large values of a, the first term may provide the closest result to the actual value, but it is generally very inaccurate. Furthermore, as a increases, the terms in the subdominant series become numerically important and cannot be neglected. This is because a subdominant term such as $\exp(-1/\varepsilon)$ is smaller than any power of ε as $\varepsilon \to 0^+$, but for $\varepsilon = 1/2$, $\exp(-1/\varepsilon) > \varepsilon^3$ and $10\exp(-1/\varepsilon) > 1$. Therefore, for $a > 0.5$ one must consider not only the entire dominant series, but also the entire subdominant series, to evaluate accurate values for $S_3(a)$. This means, in turn, that one must be able to determine meaningful results to divergent series, a feat that has eluded mathematicians since their discovery well over 200 years ago.

When we began our study of the complete asymptotic expansion for $S_3(a)$, the only summation technique for divergent series that we had at our disposal was Borel summation. The problem with this technique was that whilst it could yield the exact answers to both series, it was incapable of providing the precision we required for our analysis because the Borel-summed results were in the form of multi-dimensional integrals. These were found to be extremely difficult to evaluate numerically to a large number of decimal places within a short period of time on our computing system. So, out of necessity we developed a new numerical technique that we named the Mellin-Barnes (MB) regularisation of a divergent series. With this technique we were able to evaluate the entire dominant and subdominant asymptotic series in $S_3(a)$ to previously unattainable degrees of precision. As a result, exact values for $S_3(1)$ and $S_3(10)$ were calculated to at least 16 decimal figures by summing both divergent series in the asymptotic expansion. For smaller values of a, viz. $a < 0.1$, the degree of precision was even greater, due to the fact that the contribution from the subdominant series was extremely small and thus, hard to detect. Nevertheless, not only did we tame the divergence of each component series in the complete asymptotic expansion for $S_3(a)$, we had discovered for the first time ever and contrary to conventional ideas or theory about asymptotic expansions that a complete asymptotic expansion can yield the exact values of the original function, even far away from its limiting point.

All the results from this study were documented and discussed extensively in a London Mathematical Society (LMS) Lecture Note, No. 214 [13]. In the conclusion to this work it was stated that as a result of the remarkable success of the numerical study into $S_3(a)$, a theory of divergent series could now be envisaged. With such a theory the subject of asymptotics could be liberated from the shackles of inaccuracy and limitation in range of application and be elevated to a proper mathematical discipline yielding precise results over previously unexplored regions. This is of great importance to theoretical physicists since the world of modern physics turns out to be more asymptotic than convergent. However, it was also stated that before such a theory could be developed, one would need to consider complex values of a, but as a moves about the complex plane, the divergent series in the asymptotic expansion for $S_3(a)$ acquire jump discontinuities due to the Stokes phenomenon. Hence, whilst the techniques of Borel summation and MB regularisation would feature prominently in the proposed theory of divergent series, the theory would only be complete if it possessed a proper explanation for the Stokes phenomenon.

One of the principal aims of this work is to present an entirely new, if not radical, explanation for the Stokes phenomenon that builds upon its initial discovery by G.G. Stokes in 1857. Consequently, Ch. 2 begins with the very first example of the phenomenon. Following that, the main developments regarding this arcane effect are described up to the present day. Ch. 3 explains how the phenomenon is incorporated into Dingle's theory of terminants, while Ch. 5 applies this theory to the asymptotic forms for the error function of complex argument in order to see whether discontinuous or abrupt jumps occur in the vicinity of Stokes lines or whether a smoothing as postulated by M.V. Berry takes place.

In order to resolve the differing views about phenomenon, it has been necessary to introduce the key concept of regularisation of a divergent series. This is carried out in Ch. 4 where it is defined as the removal of the infinity in the remainder of a divergent series so as to make the series summable. The limit values obtained in this process are referred to as regularised values. To familiarise the reader with this key concept, the regularised values are evaluated for fundamental series such as the geometric and binomial series. By itself, regularisation is a mathematical abstraction, but as explained in the chapter, it is necessary in asymptotics for correcting the improprieties of the various methods used to derive asymptotic expansions in the first place. When carried out properly, the sum of the regularised values for each component asymptotic series in a complete asymptotic expansion must yield the exact value of the original function from which the expansion is derived as we found in the study of the complete asymptotic expansion for $S_3(a)$.

Ch. 5 begins with the derivation of the asymptotic forms for the error function over the entire complex plane, which later serve as a test-bed example for the new explanation of the Stokes phenomenon. It is shown that Borel summation of these asymptotic forms yields regularised values of the error function. Then an extensive numerical study is carried out, especially near the Stokes lines within the principal branch of the complex plane. The results confirm that the Stokes phenomenon does result in discontinuous jumps and that there is no smoothing effect. As a result, an expression for the Stokes multiplier is given, which is shown in another extremely accurate study to behave as a step-function when the argument of the variable moves from below the positive real axis to above it, regardless of the variable's magnitude. Thus, the results in this chapter vindicate Stokes's view of the phenomenon that bears his name.

Since Ch. 5 shows conclusively that the Stokes phenomenon is discontinuous across specific lines or rays in the complex plane and has no dependence upon the magnitude of the variable, Ch. 6 re-examines the analysis leading to the smoothing effect. It is seen that this erroneous observation has been arrived at by neglecting the remainder of or truncating an asymptotic series. Later in the same chapter an alternative description of the Stokes phenomenon based on resurgence analysis is presented. Though the proponents of this complicated approach claim that resurgence analysis is a natural progression from Stokes's seminal work, this view conflicts with Stokes's original observation of jump discontinuities emerging at specific rays in the complex plane. In resurgence analysis Stokes lines, which are determined by maximising the exponential factors of (J)WKB solutions now become curves, merging into one another in the complex plane. Consequently, Stokes regions are no longer regarded as specific sectors in the complex plane, which is nothing like Stokes's description of the phenomenon or effect that he discovered in divergent series. Another problem with this approach is that no one has ever conducted a thorough numerical analysis which demonstrates conclusively that an asymptotic solution of the Schrödinger equation obtained *via* resurgence analysis can yield the same values as the exact solution. Instead, the proponents of the theory refer readers to Stokes smoothing and the hyperasymp-totics of Berry and Howls [47] for such numerical studies claiming that these, too, fall under the umbrella of resurgence analysis.

Prior to the advent of Stokes smoothing and resurgence analysis, the conventional view of the Stokes phenomenon was that the constants or multipliers associated with subdominant terms in a complete asymptotic expansion developed discontinuous jumps, often of unity, as the phase or argument of the variable moved across a Stokes line of discontinuity. Though more correct than the concept of smoothing, this view is very limited in that it does not apply to all branches in the complex plane and is virtually restricted to series with coefficients that possess gamma function growth. Such series are referred to as terminants, a terminology due to R.B. Dingle, who realised that the late terms of the asymptotic series of a multitude of functions in mathematical physics could be approximated by them.

As indicated earlier, MB regularisation was first introduced in the analysis of the complete asymptotic expansion for $S_3(a)$, but that study was limited to positive real values of a. In Ch. 7 this alternative technique for regularising a divergent series is developed further into a proper theory so that it can be applied over the entire complex plane. Two general types of divergent series are considered: one where the coefficients alternate in sign initially and the other where they do not. In this analysis the regularised values of the two types of asymptotic series are expressed in terms of MB integrals with domains of convergence rather than as Cauchy integrals defined over Stokes sectors as in the case of Borel summation. Furthermore, since the domains of convergence for the MB integrals overlap, the regularised value can be expressed by two different MB-regularised forms for specific sectors of the complex plane that encompass the Stokes lines. This is in stark contrast to the Borel-summed forms for the regularised values, which are different once the argument of the variable moves from a Stokes sector to a Stokes line and then from the Stokes line to the adjacent Stokes sector.

After presentation of the general theory for MB regularisation, the various MB-regularised forms for the regularised value of the asymptotic expansion for the error function $\text{erf}(z)$ are derived in Ch. 8 over all Stokes sectors in the complex plane. Because the asymptotic series in the expansion for $\text{erf}(z)$ is a basic terminant, all these results can be expressed in terms of a Stokes multiplier. Then we describe how the specific MB-regularised forms over the principal branch of the complex plane can be implemented in the software package Mathematica. The numerical results for large and small values of $|z|$ and for different values of the truncation parameter N not only give the exact values for $\text{erf}(z)$ within the extremely high accuracy and precision goals set in the numerical integration routine, but also confirm that the Stokes multiplier alternates between $-1/2$ and $1/2$ across all the Stokes sectors. That is, there is no Stokes smoothing.

Ch. 9 begins with two propositions, which give the MB-regularised forms for the regularised values of generalisations of the two types of terminants studied by Dingle. It is shown that only in special cases can the extra terms to the MB integrals in the regularised values be expressed in terms of a multiplier and that for the most part, this multiplier does not behave according to the conventional view of the Stokes phenomenon. Again, these regularised values are valid over the domains of convergence for the MB integrals, not Stokes sectors. Hence, there is virtually no Stokes phenomenon when it comes to evaluating regularised values obtained *via* MB regularisation.

Ch. 10 presents the Borel-summed regularised values for the two types of generalised terminants over all Stokes sectors in the complex plane. In order to accomplish this, a more sophisticated explanation of the Stokes phenomenon is introduced, which incorporates the limited theory developed by Dingle. In the case of the first type of generalised terminant we develop the regularised value by stipulating a primary Stokes sector initially, while for the second type of generalised terminant we need to stipulate a primary Stokes line initially. In both cases the regularised values are dependent upon how the singularities in the Cauchy integrals that arise out of the Borel summation of the generalised terminants are interpreted. These singularities are very much affected by whether the traversal to another Stokes sector occurs in a clockwise or anti-clockwise direction, but at all times the overall phase of the singularity is conserved. Because of this radical view of the Stokes phenomenon we are no longer restricted to Borel-summable series when determining the regularised values of divergent series.

As a result of this observation, two propositions are presented in Ch. 11 that give the regularised values for more general series than the generalised terminants of Chs. 9 and 10. Specifically, the coefficients in the more general series need only be expressed as a Mellin transform in order to obtain the regularised values. Because of the connection to Borel summation, the regularised values obtained in this process are referred to as extended Borel-summed forms. In addition, the basic characteristics of the Stokes phenomenon apply to these more general regularised values. That is, they remain uniform over specific sectors in the complex plane until encountering lines of discontinuity. Furthermore, the regularised values are composed of Cauchy integrals, whose principal values must be evaluated along the lines of discontinuity, while the jump discontinuities that emanate from the singularities in the Cauchy integrals can only in special cases be expressed in terms of a Stokes multiplier times the same function that is applicable over all Stokes sectors.

In this book there are many numerical examples presented to astonishing accuracy. Since they are exact, they are far more accurate than the alternative approach of developing strategies for truncating asymptotic series beyond the optimal point of truncation which is the more familiar form of hyperasymptotics [47-49]. In fact, all the examples presented here consider situations where there is no opti-mal point of truncation, i.e. values of the power series variable where truncation breaks down completely and the asymptotic series are extremely divergent. More importantly, the numerical examples serve not only to verify the theoretical re-sults given in the numerous propositions, but also reveal many interesting prop-erties that arise out of the new approach to divergent series, particularly due to the concept of regularisation. In addition, with such symbols as $\sim$, $\simeq$, $O()$, $o()$, $+\ldots$, $\leq$ and $\geq$ proliferating throughout the entire discipline, one can virtually prove anything in standard asymptotics, which, as mentioned earlier, has resulted in the subject being regarded as vague and limited in range of applicability and being ridiculed by pure mathematicians. On the other hand, numbers do not lie, a fact that seems to have been overlooked by a great number of practitioners in the field who have tended to rely on "proving theorems" employing the aforemen-tioned symbols rather than verifying their results numerically. In fact, much of the theory discussed in the later chapters of this book only came into realisation as a result of the numerical examples in the earlier chapters. Furthermore, the numeri-cal examples demonstrate how the theoretical material in this book can be applied to wide-ranging problems in applied mathematics and theoretical physics. After all, there is little point in developing a new theory if it only provides an alternative view of explaining known problems without possessing the capacity to solve new problems.

Finally, I wish to express my gratitude to Gonzalo Medina of the Department of Mathematics and Statistics, Universidad Nacional de Colombia-Sede Manizales, for writing the LaTeX style file that enabled the manuscript to be produced in the Bentham e-book format.

Victor Kowalenko

Melbourne, Australia

September, 2009

CHAPTER 1

Introduction

Abstract. This introduction outlines the main issues that are to be analysed extensively in the following chapters. It begins with a general description of the Stokes phenomenon in asymptotics. Despite being discovered in 1857, it is pointed out that a complete theory giving the exact size of the jump discontinuities is still not available. Before such a theory can be realised, however, the developments up to the present day need to be discussed. After describing Stokes's seminal work, the introduction proceeds to the work of Heading and Dingle, a century later. Out of the observations of these three figures a conventional view of the Stokes phenomenon evolved whereby the multipliers of subdominant asymptotic series experience jumps of unity across Stokes lines. Then in 1989 Berry argued that the multipliers experience a rapid smoothing at Stokes lines, although there has never been a definitive demonstration of this behaviour. Such a demonstration can only be performed if methods exist for obtaining meaningful values from divergent series, a process referred to here as regularisation. The introduction concludes by discussing the two methods for regularising a divergent series used throughout this book, namely Borel summation and Mellin-Barnes regularisation.

The Stokes phenomenon [1] refers to the emergence of jump discontinuities as the argument or phase of the variable in an asymptotic expansion for a function is varied continuously. Ever since its discovery by Stokes in 1857 [2], it has tantalised mathematicians by crying out for an explanation or theory. Such a theory would be required: 1) to indicate precisely where the phenomenon occurs over all branches in the complex plane and 2) to yield the exact size of the discontinuities whenever they occur. As a consequence, an asymptotic expansion would take on different forms over all branches in the complex plane. Yet, as will be seen in this book, despite various attempts to explain the phenomenon over the past 150 years, a complete theory remains elusive to this day.

The first attempt to explain the phenomenon was the seminal paper by Stokes himself, which began with the observation of the phenomenon in the asymptotic expansion of a related function to the error function, $\mathrm{erf}(z)$. Stokes found that the asymptotic expansion for the function, which will be denoted by $u(a)$ in this book, developed jump discontinuities as soon as the argument of the variable moved above and below the positive real axis. In this case the jump discontinuity consisted of a single term, which ranked it as one of the simplest examples of the phenomenon. Furthermore, Stokes pointed out that the two different asymptotic forms remained valid over the upper and lower halves of the complex plane, i.e. until the argument of the variable reached the negative real axis where $\arg a = \pm \pi$. Hence, the positive and negative real axes formed the boundaries of sectors in the complex plane over which the forms for the asymptotic expansion remained uniform. Today, these sectors of uniformity are known as Stokes sectors, while the boundaries are referred to as Stokes lines. Then using the results of the first example, Stokes considered the more complicated case of the asymptotic expansion for the Airy function. In so doing, he had made the assumption that the phenomenon continued to behave in a similar manner to the first example, namely that jump discontinuities developed at specific lines in the complex plane. This was despite the fact that the jump discontinuities in the latter example are considerably more complex since they are composed of another asymptotic series. As a consequence of these successes, he was convinced that he had discovered how asymptotic expansions altered over the complex plane.

As described in Ch. 1 of Dingle's book [3], it appears for the next century or so that mathematicians were content to determine asymptotic expansions for the special functions of mathematical physics without relying on a general theory. This meant that the behaviour at the boundaries of Stokes sectors, where an asymptotic expansion experienced a sudden transition, often remained obscure. Then in a pair of companion papers in 1957, Heading [4, 5] was able to find a recognizable pattern in his investigation of the Stokes phenomenon on the asymptotic solutions to certain n-th order differential equations. In various cases the jump discontinuities could be expressed in terms of constants multiplying a subdominant asymptotic series to the dominant series of a complete asymptotic expansion. In this work an asymptotic series is defined as a power series with zero radius of absolute convergence. As we shall see, it need not be divergent according to the standard definition given on p. 15 of Ref. [6] or on p. 19 of Ref. [7]. More importantly, Heading was able to account for how the subdominant terms arising from a Stokes transition eventually became the dominant contribution in an asymptotic expansion as the argument of the variable moved further in a Stokes sector and that the previously dominant terms also experienced a Stokes transition. Later, he developed rules for continuing Liouville-Green approximations [8], which formed the basis of the rules presented in Ch. 1 of Ref. [3].

As discussed by Berry in the introduction to Ref. [9], a "conventional view" of the Stokes phenomenon emerged in which the "constants" associated with subdominant asymptotic series of complete asymptotic expansions, which are known today as Stokes multipliers, developed jump discontinuities as the phase of the variable was varied in the complex plane. Typically, when a Stokes line is encountered, on one side of it, the Stokes multiplier is given by a value of S_-, while on the other side of it, the multiplier equals $S_- + 1$. Frequently, S_- is equal -1/2 or 0. On the line itself, the Stokes multiplier takes on a value of $S_- + 1/2$. According to p. 237 of the book by Paris and Kaminski [10], the "hide-and-seek" nature of these multipliers and the vagueness in determining the precise location of the jump discontinuities continued to make the Stokes phenomenon appear just as mysterious or arcane as when Stokes discovered it. In this book we shall see that the concept of a Stokes multiplier is extremely simplistic, since such a quantity can only be isolated in simple asymptotic expansions such as the terminants studied by Dingle in Ch. 21 of Ref. [3]. When one considers higher level Stokes sectors of more sophisticated asymptotic expansions, the situation is markedly different and it becomes a struggle to isolate such a quantity.

In a radical approach to the Stokes phenomenon Berry then argued in the same reference that the coefficient of the subdominant terms in a complete asymptotic expansion should be regarded as a continuous function for fixed values of the variable and not discontinuous as put forward in the conventional view. Specifically, he claimed that on a suitably magnified scale the change in the Stokes multiplier in the vicinity of a Stokes line was continuous. Despite the fact that this claim is based on asymptotic methods and hence, by definition can only be approximate, it seems to have been embraced over the last decade and a half by the asymptotics community, presumably due to the attempts of others such as Olver [11] to provide a "rigorous mathematical foundation" for it. So much so, this radical approach is now referred to as Stokes smoothing, despite the fact that Stokes never held such a view. Moreover, as stated earlier, there has not been a definitive treatment or demonstration to this day where the claim has been substantiated.

The reason why a definitive demonstration of the Stokes smoothing has not been forthcoming is that previous attempts have been based on standard asymptotics procedure, where one truncates the dominant asymptotic series in a complete asymptotic expansion and bounds the remainder of the expansion. By doing this, the truncated asymptotic expansion is not only approximate, but is also valid over a limited range of applicability. For an asymptotic series in powers of z, it means considering values of $|z| \ll 1$ or what mathematicians refer to as the limit as $z \to 0$, the latter being referred to as the limit or limiting point. For such values of z, the dominant asymptotic series will in specific sectors of the complex plane yield the most accurate approximation to the original function at an optimal point of truncation N_T as discussed on p. 434 of Ref. [1]. The value of N_T increases as $|z| \to 0$. On the other hand, the contribution of the subdominant terms, which are said to lie beyond all orders [12], are masked by the bounded remainder estimates for the dominant asymptotic series. Hence, their contribution cannot be isolated and they are simply neglected according to the Poincaré definition for an asymptotic expansion given on p. 151 of Ref. [6].

What is really required is a method for summing all series in a complete asymptotic expansion regardless of the magnitude of the variable rather than truncating the dominant series and neglecting the subdominant series. Such a notion would appear to be nonsense since we are dealing with divergent series. However, as explained later, the divergence in an asymptotic expansion arises because the method used to derive it is improper. If this impropriety can be corrected, then we should be able to obtain exact values for both the dominant and subdominant series in a complete asymptotic expansion, thereby yielding the values of the original function when summing these values. To correct the impropriety in an asymptotic method, we need to introduce the concept of regularisation as discussed in Refs. [13-15]. As we shall see, regularisation of a divergent series is analogous to evaluating the Hadamard finite part of a divergent integral in the theory of generalised functions [16].

In this book we shall describe two methods for regularising divergent series, Borel summation and Mellin-Barnes (MB) regularisation. The first method is often employed when evaluating asymptotic series whose coefficients are directly related to the gamma function. As mentioned previously, these series of which there are basically two types are referred to as terminants by Dingle in Ref. [3]. He then proceeds to explain how the Stokes phenomenon affects one of the types for the two Stokes sectors covering the principal branch of the complex plane, but does not indicate how the other Stokes sectors for this terminant are affected by the Stokes phenomenon. Nor does he explain how the phenomenon affects the other type of terminant. Thus, his theory of terminants is unable provide a general form or expression for the regularised value of either type of terminant for an arbitrary Stokes sector. In the present work we aim to address this issue by presenting propositions with the regularised values for generalisations of the two types of terminants over arbitrary Stokes sectors. In the final chapter we shall modify or adapt Borel summation to handle series that are more complicated than these generalised terminants. In the process we shall develop a very profound understanding of the Stokes phenomenon that has never been envisaged before.

Although Borel summation can be extended to evaluate more complicated divergent series such as those arising in the asymptotic expansion of the Airy function by using Eq. (7.10) in Ref. [13], it can result in awkward regularised values composed of multi-dimensional integrals, which are not very amenable to numerical computation. For these situations it is better to employ MB regularisation, which was first introduced in Ref. [13], but has since been employed in various applications [14, 15, 17, 18]. In actual fact, the latter method is far more superior to the former because it can yield the regularised value of a divergent series when it is not possible to do so even by modifying Borel summation. Previously, nearly all the applications of this powerful method of regularisation were concerned with the behaviour of asymptotic series away from Stokes lines and hence, there was no need to consider Stokes phenomenon. In addition, because MB regularisation is vastly different from Borel summation, up to now it has not been clear whether the Stokes phenomenon has any effect on the expressions for the regularised value of a complete asymptotic expansion obtained *via* MB regularisation. It should be borne in mind that even though the expressions for the regularised value obtained *via* Borel summation develop jump discontinuities across Stokes lines, the regularised value will still be continuous if the original function from which the asymptotic expansion was derived is continuous there. So, an alternative method of regularisation does not necessarily mean that the expressions for the regularised value of a complete asymptotic expansion will encounter jump discontinuities at Stokes lines. In the present work we shall observe that the Stokes phenomenon affects only one type of general asymptotic series in one branch of the complex plane when the regularised value is derived *via* MB regularisation, but for other branches of the complex plane, the regularised value will be dependent upon the domains of convergence for the resultant MB integrals. These domains of convergence not only envelop the Stokes sectors for an asymptotic series, but also overlap each another. As a consequence, we shall see that the regularised values obtained *via* MB regularisation will not give rise to discontinuous jumps at Stokes lines except at what will be referred to as the primary Stokes line.

The principal tenet of regularisation is that the regularised value of a divergent series must remain invariant irrespective of the method used to derive it. As we develop the theory for the regularised value of a divergent series *via* the two methods of regularisation, we shall present several examples describing how to carry out numerical calculations of the regularised value from the general expressions in the various propositions appearing in this book. We shall see that not only do the expressions for the regularised values obtained by both methods yield identical values, but that the regularised values are also identical to the values of the functions from which the asymptotic expansions were derived originally. By identical in a computing environment we mean to the limit of the machine precision of the computing system. In our case this represents a maximum of 16 significant figures since the computing system used for the most of the numerical work conducted in this book consists of the software package Mathematica 4.1 [19] and a Pentium computer. Here it should be mentioned that with the advent of Versions 6.0 and 7.0 of this software package the restriction in machine precision may no longer apply since these more recent versions possess the capacity to perform numerical integration to unlimited accuracy. In fact, where the improved precision is deemed important for clarifying an example, we shall refer to the results obtained from using these later versions of the software. However, before we can embark on our ambitious journey into the exciting world of divergent mathematics, we need to survey and study in detail the main developments in the theory of the Stokes phenomenon over the past 150 years, beginning, of course, with Stokes's seminal paper in the following chapter.

CHAPTER 2

Stokes's First Example

Abstract. Ch. 2 reviews the derivation of the complete asymptotic expansion for a related function of the error function, which was the first example ever of the Stokes phenomenon. It is shown that the asymptotic method used to derive the expansion, namely the well-known method of iteration that is applied to differential equations, introduces an infinity into the remainder and hence, is improper. As a consequence, the concept of an equivalence statement is introduced, while regularisation is defined as the removal of the infinity in an asymptotic series in order to make its remainder summable. The analysis continues by demonstrating that the asymptotic expansion for the variant of the error function changes form at special rays of discontinuity and across specific sectors of the complex plane, which are now called Stokes lines and sectors respectively.

As indicated in the previous chapter Stokes's first observation of the phenomenon that now bears his name occurred with a function that is related to the error function. Specifically, Stokes studied the following function/integral:

$$u(a) = 2 \int_0^\infty dx \, e^{-x^2} \sin(2ax) = i\sqrt{\pi} \, e^{-a^2} \mathrm{erf}(-ia) \; , \qquad (2.1)$$

where the last form has been obtained via No. 2.5.36.1 from Ref. [20]. By expanding the sine function as a power series series, one obtains a convergent power series for $u(a)$, which is

$$u(a) = \sqrt{\pi} \sum_{k=0}^\infty \frac{(-1)^k a^{2k+1}}{\Gamma(k+3/2)} = 2a - \frac{4a^3}{3} + \frac{8a^5}{15} + \cdots \; . \qquad (2.2)$$

In obtaining Eq. (2.2) we have used the duplication formula for the gamma function, which is given as No. 8.335(1) in Ref. [21]. Because the powers of a are increasing, Stokes referred to the above result as the ascending series. He also noted that from a numerical perspective, for large values of $|a|$ the series begins by diverging rapidly, but ultimately it is convergent. Because he had observed that Eq. (2.2) was slowly convergent for large values of $|a|$, he sought to find a descending power series expansion, i.e. a series expansion in powers of $1/a$. Today, such a series is referred to as an asymptotic series, although what this actually means is rather unclear. For example, when such a series is derived, we do not know if it is always divergent or whether it is convergent for certain values of the variable. Stokes was also aware that the descending power series he derived was at first rapidly convergent, but eventually diverged with increasing rapidity. Therefore, in today's terminology he was aware that the asymptotic series for $u(a)$ reached an optimal point of truncation before it began to diverge, although he was unaware that he had effectively employed an asymptotic method to obtain the series in the first place.

As stated in the previous chapter, whenever an asymptotic method is employed, an impropriety has occurred, which is responsible for an infinity in the remainder of the expansion. The situation is no different with the method Stokes used to derive the asymptotic series for $u(a)$. Instead of applying an integrating method such as Laplace's method or the method of steepest descent to the integral in Eq. (2.1), which is perhaps the more common approach to obtaining an asymptotic series for a function, Stokes solved the differential equation for $u(a)$ by using the method of iteration. That is, he essentially replaced $u(a)$ in the differential equation of

$$\frac{du}{da} + 2au = 2 \ , \tag{2.3}$$

by $\sum_{k=0} \beta_k / a^{2k+1}$. The reason why $u(a)$ has not been set equal to the series directly will become apparent soon. Note also the upper limit of the series is indeterminate, which will also be explained shortly. In any case, by carrying out the replacement of $u(a)$ by the series Stokes found that $\beta_0 = 1$, $\beta_1 = 1/2$, $\beta_2 = 3/4$, $\beta_3 = 15/8$, etc. More generally, one finds that replacing $u(a)$ by the series results in the following recursion relation:

$$\beta_{k+1} = (k + 1/2)\beta_k \ . \tag{2.4}$$

This equation can be solved easily and yields $\beta_k = \Gamma(k + 3/2)/\Gamma(1/2)$ for $k \geq 1$.

If the asymptotic series for $u(a)$ is truncated at $k=N$ and introduced into Eq. (2.3), then we obtain

$$\frac{du}{da} + 2au = 2 - 2\frac{\Gamma(N+3/2)}{\Gamma(1/2)a^{2N+2}} \ . \tag{2.5}$$

That is, by truncating the series one obtains an error or correction term. For large values of $|a|$ and provided N is not too large, one can neglect the error term. As a consequence, the truncated series becomes an accurate approximation to the actual solution of Eq. (2.3) or $u(a)$, but it can never equal $u(a)$. When N becomes sufficiently large, irrespective of the magnitude of $|a|$, the correction term begins to diverge. Then the truncated series no longer represents an effective approximation to $u(a)$. The value of N where the neglected error or correction term begins to diverge represents the optimal point of truncation [1]. Nevertheless, in order to be able to remove the correction term in Eq. (2.5) so that we end up with Eq. (2.3), rather neglecting it, we must let $N \to \infty$, even though the correction term for N equal to infinity becomes infinite. Thus, we have effectively included an infinity in determining the entire asymptotic series for $u(a)$. However, $u(a)$ is finite, which can be observed by bounding the integral in Eq. (2.1). Therefore, like all asymptotic methods the method of iteration is improper and the infinity must be removed from the series in order to obtain finite values for $u(a)$. In this book the process of removing an infinity in the remainder of a divergent asymptotic series so that the series becomes summable is referred to as regularisation. Furthermore, because of this infinity, we express $u(a)$ in terms of the divergent asymptotic series as

$$u(a) \equiv \sum_{k=0}^{\infty} \frac{\Gamma(k+1/2)}{\Gamma(1/2)\,a^{2k+1}} \; . \tag{2.6}$$

That is, since the rhs is infinite for certain values of a such as $\arg a = 0$, the above result cannot possibly be an equation, but only an equivalence statement. We shall discuss this issue extensively in the following chapters for it is crucial if we wish to obtain the limiting values or limits to divergent series. For now, however, we note that in standard asymptotics jargon the series on the rhs of Equivalence (2.6) represents a large $|a|$ asymptotic series with the limit point situated at infinity.

The series on the rhs of Equivalence (2.6) represents the particular solution of the original differential equation. To obtain the complete solution, however, we need to include the solution to the homogeneous differential equation, which is

$$\frac{du}{da} + 2au = 0 \; . \tag{2.7}$$

The solution to this equation can be determined by using the integrating factor method [22]. Then we find that the complete solution to Eq. (2.3) can be written as

$$u(a) \equiv Ce^{-a^2} + \sum_{k=0}^{\infty} \frac{\Gamma(k+1/2)}{\Gamma(1/2)\,a^{2k+1}} \; . \tag{2.8}$$

In order to determine C, which is not directly dependent upon a, Stokes noted that $u(a)$ is an odd function in a. However, the first term on the rhs of Equivalence (2.8) is even, which means that when a is altered to $-a$, C must also change sign. Therefore, if $C = C'$ for $0 < \arg a < \pi$, then it will equal $-C'$ for $-\pi < \arg a < 0$ and for $\pi < \arg a < 2\pi$. That is, C is discontinuous across different sectors of the complex plane.

Now all that remains is to determine C', which can be accomplished by putting $a = i\rho$, where ρ is positively real. Then we can write Equivalence (2.8) as

$$2ie^{\rho^2} \int_0^\rho dx\, e^{-x^2} \equiv C' e^{\rho^2} - i \sum_{k=0}^\infty \frac{\Gamma(k+1/2)}{\Gamma(1/2)} \frac{(-1)^k}{\rho^{2k+1}} \; , \tag{2.9}$$

where we have introduced the integral representation for $u(a)$ given by Eq. (2.1). By dividing throughout by $\exp(\rho^2)$ and then taking the limit as $\rho \to \infty$, we find that the term involving the asymptotic series vanishes and we are left with

$$C' = 2i \int_0^\infty dx\, e^{-x^2} = i\Gamma(1/2) = i\sqrt{\pi} \; . \tag{2.10}$$

Hence, for $0 < \arg a < \pi$ and $-\pi < \arg a < 0$, C in Equivalence (2.8) equals $i\sqrt{\pi}$ and $-i\sqrt{\pi}$ respectively. This jump discontinuous behaviour of a complete asymptotic expansion, which occurs at specific lines or rays in the complex plane, is known today as the Stokes phenomenon, . The rays at which it occurs, e.g. $\arg a = 0$ in the above example, are called Stokes lines, while the regions of the complex plane over which each asymptotic result is valid are called Stokes sectors.

In actual fact, in the above example both the positive and negative real axes represent Stokes lines. That is, whenever $\arg a$ equals $2k\pi$ or and $(2k+1)\pi$, where k is any integer, C vanishes, thereby ensuring that $u(a)$ remains odd. Consequently, we can write the asymptotic forms for $u(a)$ covering the principal branch of the complex plane for a as

$$u(a) \equiv \begin{cases} -i\sqrt{\pi}\, e^{-a^2} + \sum_{k=0}^\infty \frac{\Gamma(k+1/2)}{\Gamma(1/2)} a^{-2k-1}, & -\pi < \arg a < 0 \; , \\ \sum_{k=0}^\infty \frac{\Gamma(k+1/2)}{\Gamma(1/2)} a^{-2k-1}, & \arg a = 0 \; , \\ i\sqrt{\pi}\, e^{-a^2} + \sum_{k=0}^\infty \frac{\Gamma(k+1/2)}{\Gamma(1/2)} a^{-2k-1}, & 0 < \arg a < \pi \; , \\ \sum_{k=0}^\infty \frac{\Gamma(k+1/2)}{\Gamma(1/2)} a^{-2k-1}, & \arg a = \pi \; , \\ -i\sqrt{\pi}\, e^{-a^2} + \sum_{k=0}^\infty \frac{\Gamma(k+1/2)}{\Gamma(1/2)} a^{-2k-1}, & \pi < \arg a < 2\pi \; . \end{cases} \tag{2.11}$$

By presenting the above results we are now in a position to give a proper definition for an asymptotic form in this book, which is defined simply as the complete asymptotic expansion for a function plus the Stokes sector or Stokes line for which it is valid. Therefore, in the above equivalence we have presented the five asymptotic forms for $u(a)$ that cover the principal branch of the complex plane for a. In other words, whenever we specify or derive a complete asymptotic expansion, we must also include the corresponding Stokes sector or Stokes line for the expansion. As we shall see, this will be crucial if we wish to obtain the exact values of a function via its the asymptotic forms. Furthermore, it is possible that the same complete asymptotic expansion can be derived from two different functions. Nevertheless, their asymptotic forms will not be identical because the Stokes sectors will be different in this situation.

The above asymptotic forms, however, can be expressed more succinctly with the introduction of a quantity S, known as a Stokes multiplier as discussed in Ref. [9]. Then the previous equivalence reduces to

$$u(a) \equiv \sum_{k=0}^{\infty} \frac{\Gamma(k+1/2)}{\Gamma(1/2)} a^{-2k-1} + 2i\sqrt{\pi} S e^{-a^2} , \qquad (2.12)$$

where

$$S = \begin{cases} 1/2 , & 2l\pi < \arg a < (2l+1)\pi , \\ 0 , & \arg a = l\pi \\ -1/2 , & (2l-1)\pi < \arg a < 2l\pi , \end{cases} \qquad (2.13)$$

and l is an arbitrary integer.

After the derivation of the asymptotic forms for $u(a)$ Stokes turned his attention to the asymptotic forms for the Airy function in the complex plane. Since the Airy function $\mathrm{Ai}(z)$ is related to the Macdonald function $K_\nu(z)$ via

$$\mathrm{Ai}(z) = \frac{1}{\pi} \sqrt{\frac{z}{3}} K_{1/3}\left(2z^{3/2}/3\right) , \qquad (2.14)$$

and the Macdonald function is, in turn, related to the modified Bessel function $I_\nu(z)$ by

$$K_\nu(z) = \frac{\pi}{2} \frac{I_\nu(z) - I_{-\nu}(z)}{\sin(\pi\nu)} , \qquad (2.15)$$

all we need to do is evaluate the asymptotic forms for the modified Bessel function to obtain the asymptotic forms for the Airy function. In this work, however, we shall be dealing with the role that regularisation of a divergent series plays in developing an understanding of the Stokes phenomenon. The problem of determining the asymptotic forms for the Bessel function family in the complex plane will be the subject of a future publication, which represents the sequel to Ref. [15]. In fact, it is envisaged that this work will investigate the asymptotic forms of the more general confluent hypergeometric function.

In concluding this chapter, the reader is reminded that Stokes did not study the behaviour of the various asymptotic forms for either $u(a)$ or the Airy function in the vicinity of the rays or lines of discontinuity in the complex plane that are named in his honour. That is, he was primarily interested in deriving asymptotic forms for specific sectors situated within the principal branch of the complex plane. Clearly, he believed that the asymptotic forms, which he derived, were discontinuous when crossing these sectors in the complex plane as is evidenced by the title of his seminal paper [2]. If there is smoothing of asymptotic forms near Stokes lines, then he was certainly not aware of it, mainly because he did not have the computing means at his disposal to investigate this issue. Therefore, it is peculiar that the asymptotics community refers to this relatively new phenomenon today as Stokes smoothing, especially when it is noted that the behaviour in the vicinity of Stokes lines was studied more than a century after Stokes's seminal paper by others such as Heading [5], Dingle [3] and Berry [9]. Only in the work of the last of these individuals does the purported Stokes smoothing appear. Therefore, in order to develop a complete understanding of the Stokes phenomenon, we shall in the following chapters examine the contributions that these individuals have made, particularly in the vicinity of Stokes lines.

CHAPTER 3

Dingle's Rules and Theory of Terminants

Abstract. Dingle's rules for deriving the changing forms of an asymptotic expansion as a result of the Stokes phenomenon are introduced. Of these rules only Nos. 1, 7 and 8 are necessary for the analysis in later chapters. Then with the aid of these rules his theory of terminants is presented, which is important because the asymptotic expansions of a multitude of mathematical functions can be approximated for large values of the summation index by these divergent series whose coefficients possess gamma function growth. There are basically two types of terminants, each possessing different properties according to Dingle's rules. Meaningful values for both types of terminants are obtained by Borel summation, which is shown to be a method of regularising asymptotic series.

From his study of the asymptotic solutions for certain n-th order differential equations, Heading [4, 5] was able to determine the behaviour of asymptotic expansions, in particular the coefficients of the subdominant terms, across Stokes lines. By using these observations with Stokes's seminal paper, Dingle was able to present a compact set of rules for locating Stokes lines and continuing asymptotic expansions across them in Ch. 1 of his book [3]. When presenting these rules he referred to the subdominant term(s) in a complete asymptotic expansion as the associated function or series since they become dominant as the argument of the variable continues to move throughout a Stokes sector. Generally in this work, we shall desist from investigating which parts of an asymptotic form are dominant and subdominant, but they are necessary for understanding Dingle's rules, which are, in turn, a vital part of our survey into the Stokes phenomenon.

The rules put forward by Dingle are:

1. Stokes lines for an asymptotic series are determined by those phases for which successive late terms are homogeneous in phase and all the same sign.

2. Stokes lines for an asymptotic series are determined by those phases for which the series attains peak exponential dominance over the associated term.

3. Relative to its associate, an asymptotic series is dominant where late terms

are uniform in sign, and recessed or maximally subdominant where the late terms alternate in sign.

4. The factor multiplying an asymptotic series is analytically continued across its own Stokes line.

5. In a sector lying between the Stokes line of an asymptotic series and the Stokes line of its associated series, the complete asymptotic expansion is the sum of the series dominant near one Stokes line and the dominant series near the other Stokes line.

6. On crossing a Stokes ray, an asymptotic series generates a discontinuity which is $\pi/2$ out of phase with the series and proportional to the associated function.

7. Half the discontinuity in form occurs on reaching the Stokes line and half on leaving it.

8. On the Stokes ray the factor multiplying the associated function is the mean of the factors to either side of the line.

The rule of primary importance to us is the first one, which states where the Stokes lines for an asymptotic series occur in the complex plane. Since it has already been mentioned that we shall not be concerned with which parts of an asymptotic expansion are dominant or subdominant within a Stokes sector, Rules 2 to 5 are superfluous for the material presented in this work. We shall see that Rule 6 follows naturally from our treatment of the Stokes phenomenon and so, it is not crucial for developing an understanding of the phenomenon. The last two rules, however, are of considerable interest as they are concerned with the behaviour of an asymptotic expansion at a Stokes line. Neither rule appears directly in the papers of Heading and Stokes, although one could argue that both rules are implied in the former. Rule 8 can also be paraphrased in the terminology of this book as:

8a. The asymptotic form on a Stokes line is the average of the asymptotic forms in the adjoining Stokes sectors to the line of discontinuity.
In employing this rule, we shall be required to make an additional qualification, which will be described in Ch. 5.

Remarkably, the above rules have never really been tested numerically by evaluating the asymptotic forms in the vicinity of Stokes lines. Because the associated series is maximally subdominant to the dominant asymptotic series on a Stokes line, which means that it is masked by the error term of the dominant series when the latter is truncated, it is often neglected as it is said to lie beyond all orders

[12, 13]. Since we shall not truncate any asymptotic series in this work, we shall also evaluate the associated/subdominant series whenever they emerge for only then will it be possible to obtain the exact values of the original function. This is essentially the reason why we need not be concerned with which parts of an asymptotic expansion are dominant and which are subdominant.

Dingle proceeds by showing how the rules can be used with ingenuity to derive the asymptotic forms for the parabolic cylinder function $D_p(z)$ once the dominant and associated asymptotic series have been determined. The question is, of course, what happens when one cannot determine either of these series. In this situation Dingle develops a theory of terminants, which will be discussed in detail later. Even if one knows the dominant and associated asymptotic series, then Dingle's rules are not really sufficient for deriving asymptotic forms. For example, by using his rules Dingle finds that the multiplier for the subdominant series of the asymptotic form for the parabolic cylinder function at the Stokes line of $\arg z = \pi$ can be expressed as $\exp(i\pi p) + i\alpha\beta/2$, but additional information is required to determine α and β. At this stage Dingle invokes a conjecture which he attributes to Zwaan [23]. This states that an initially real function cannot acquire an imaginary part if its derivatives are finite and continuous. This will be referred to as Zwaan-Dingle principle throughout this book. Hence, Dingle concludes that the multiplier must only be real, which yields one equation for the two factors. The other equation is obtained by the realisation that because replacing p by $-p-1$ in the subdominant series yields the dominant series and vice-versa, the same transformation applies to α and β. That is, $\beta(p) = \alpha(-p-1)$. Hence, the two equations yield

$$\alpha(-p)\alpha(-p-1) = -2\sin(\pi p) \ . \tag{3.1}$$

By relating the above equation to the reflection formula for the gamma function, which is given by

$$\Gamma(s)\Gamma(1-s) = \frac{\pi}{\sin(\pi s)} \ , \tag{3.2}$$

Dingle finds that $\alpha(p) = \sqrt{2\pi}/\Gamma(-p)$ and $\beta(p) = \sqrt{2\pi}/\Gamma(p+1)$.

The above heuristic approach is unsatisfactory for various reasons. First, it appears that we need additional information to determine the asymptotic forms, which may not be forthcoming. One wonders what would have happened if the final equation could not have been related to the reflection formula for the gamma function or whether the relation between p and $-p-1$ in the two asymptotic series did not apply. Another problem is that we have to keep applying the same arguments when we move to other Stokes sectors and lines in the complex plane. Furthermore, because the asymptotic series is real on a Stokes line, it does not necessarily mean that the value obtained for the series has always to be real, but we need the material in the next chapter before we can discuss this point further. What is really needed is a theory which allows one to obtain the asymptotic forms for the other Stokes sectors and lines based on the fact one has already determined the asymptotic form in one Stokes sector or on a particular Stokes line. That is, we should not have to rely on additional methods to derive the complete asymptotic expansions in Stokes sectors, now matter how ingenious they are.

To this end, in Ch. 21 of Ref. [3] Dingle shows how the above rules can be used to develop his theory of terminants. Basically, a terminant is a divergent power series summed from a finite value N to infinity, where the coefficients possess gamma function growth. The reason why such series are important to Dingle is that if N is sufficiently large, then the asymptotic series of a great number of functions in mathematical physics can be approximated by this form. Thus, as N gets larger, the terminant becomes a better approximation to the actual asymptotic series. Although this is the motivation behind the theory of terminants, Dingle does not discuss the issue of replacing the actual asymptotic series for a function by a truncated series to N and then approximating the remaining terms by a terminant. As demonstrated in Ref. [17], this is a very dangerous practice because as N passes the optimal point of truncation, the truncated part begins to diverge rapidly, while the regularised value of the remainder diverges in the opposite sense. However, if one approximates the remainder by a terminant for large values of N, then there will be a huge difference between the regularised values of the actual remainder series and its approximated counterpart. Therefore, when the regularised value of the terminant is added to the truncated part of the asymptotic series, it will lead to values that are nowhere near the values of the original function. Despite the fact that approximating the remainder of an asymptotic series by a terminant is wrought with danger, we shall continue with the presentation of Dingle's theory of terminants because such series are important for developing a general theory of divergent series anyway.

There are basically two types of terminants, one whose coefficients alternate in sign and the other whose coefficients are of the same sign. In Ch. 10 Dingle's original definition of terminants will be extended where we shall refer to asymptotic series as being either a Type I or a Type II generalised terminant. In order to obtain meaningful values to either type of terminant, Dingle employs the technique or method known as Borel summation. This method consists of introducing the integral representation for the gamma function into either type of series, interchanging the order of the summation and integration and then identifying the inner sum as the geometric series. For the first type of terminant Dingle finds that Borel summation yields

$$\sum_{k=N}^{\infty} \Gamma(k+\alpha+1)(-z)^k = \Gamma(N+\alpha+1)(-z)^N \Lambda_{N+\alpha}(z) \ , \tag{3.3}$$

which is valid for $|\arg z| < \pi$. In this result $\Lambda_s(z)$ is represented by the following integral:

$$\Lambda_s(z) = \frac{1}{\Gamma(s+1)} \int_0^{\infty} dy \, \frac{y^s e^{-y}}{1+zy} \ . \tag{3.4}$$

In the case of Type II terminants Borel summation yields

$$\sum_{k=N}^{\infty} \Gamma(k+\alpha+1)z^k = \Gamma(N+\alpha+1)z^N \bar{\Lambda}_{N+\alpha}(-z) \ , \tag{3.5}$$

which is only valid for z real and positive. When carrying out Borel summation for this type of terminant, one is eventually confronted with a singularity lying on the line of integration. That is, the resultant integral can be viewed as a Cauchy integral with a singularity lying on the contour of integration. In order to evaluate the resultant integral, Dingle investigates the behaviour of the residue. However, application of the residue theorem to the Cauchy integral results in an imaginary contribution, whereas the series in Eq. (3.5) is real. As a consequence, Dingle argues that a modification is required, whereby only the Cauchy principal value should be evaluated for z situated on the positive real axis. Thus, $\bar{\Lambda}_{N+\alpha}(z)$ in Eq. (3.5) is given by Dingle as

$$\bar{\Lambda}_s(-z) = \frac{1}{\Gamma(s+1)} P \int_0^{\infty} dy \, \frac{y^s e^{-y}}{1-zy}$$
$$= I(s,z) = \frac{1}{\Gamma(s+1)} P \int_C dy \, \frac{y^s e^{-y}}{1-zy} \ , \tag{3.6}$$

where C is the line contour along the positive real axis.

It should be mentioned that in presenting the above results we have altered Dingle's definition for both types of terminants slightly. Instead of representing them in powers of $1/z$, they have been written in powers of z. This means that z in the above results corresponds to $1/z$ in Ch. 21 of Ref. [3]. This slight modification generally has no effect on the Stokes sectors, but specific Stokes sectors are now reflected across the real axis when we wish to relate them to the actual Stokes sectors discussed in Dingle's book. That is, if we find that there is a Stokes sector of $0 < \arg z < \pi$ for the terminants defined above, then this Stokes sector will correspond to $-\pi < \arg z < 0$ in Dingle's theory.

The next insight made by Dingle is truly remarkable and stems from his study of the singularity in the Cauchy integral. It also results in Rule 7 given above. Basically, Dingle states that the discontinuity in $\bar{\Lambda}(-z)$ across the positive real axis is directly related to the residue of the integrand of $I(s,z)$. This is given by

$$\operatorname{Res}\{I(s,z)\} = -\frac{z^{-s-1}e^{-1/z}}{\Gamma(s+1)} \ . \tag{3.7}$$

If the argument or phase of z is infinitesimally greater than zero, then the Cauchy integral is defined. Consequently, we need to deform the line contour along the positive real axis so that it avoids the singularity by taking a semi-circular diversion in a clockwise direction. If $\arg z$ is infinitesimally less than zero, then we deform the contour by taking a semi-circular diversion around the singularity in an anti-clockwise direction. This is reminiscent of the Plemelj formulas as discussed on p. 412 of Ref. [24]. Now comes the really important point. Because we have excluded the contribution of the singularity in the second type of terminant by evaluating the Cauchy principal value when z lies on the positive real axis, we must do likewise to $\bar{\Lambda}(-z)$ when z is complex. That is, $\bar{\Lambda}(-z)$ is not only defined when z is complex, but also includes the effect of the semi-residue. According to Dingle, this contribution has to be removed when evaluating $\bar{\Lambda}(-z)$. That is, we need to subtract the semi-residue contributions for $\arg z > 0$ and $\arg z < 0$. Thus, Dingle arrives at

$$\bar{\Lambda}_s(-z) = \begin{cases} \Lambda_s(-z) - i\pi z^{-s-1}e^{-1/z}/\Gamma(s+1) \ , & 0 < \arg z < 2\pi \ , \\ \Lambda_s(-z) + i\pi z^{-s-1}e^{-1/z}/\Gamma(s+1) \ , & -2\pi < \arg z < 0 \ . \end{cases} \tag{3.8}$$

In Eq. (3.8), $\Lambda_s(-z)$ is given by

$$\Lambda_s(-z) = \int_C dy \, \frac{y^s e^{-y}}{1 - zy} \ , \tag{3.9}$$

while $\bar{\Lambda}_s(-z)$ is given by the series on the lhs of Eq. (3.5). Eq. (3.8) is identical to Eq. (33) in Ch. 21 of Ref. [3] except that $1/z$ has replaced x. We shall discuss these results in more detail in Ch. 5 when studying the asymptotic forms for the error function of imaginary argument. In terms of a Stokes multiplier S, Eqs. (3.6) and (3.8) can be written as

$$\bar{\Lambda}_s(-z) = \int_C dy \, \frac{y^s e^{-y}}{1 - zy} + 2S \, \frac{i\pi z^{-s-1} e^{-1/z}}{\Gamma(s+1)} \, , \qquad (3.10)$$

where it is understood for $\arg z = 0$ that the principal value of the Cauchy integral should be evaluated and

$$S = \begin{cases} 1/2 \, , & 0 < \arg z < 2\pi \, , \\ 0 \, , & \arg z = 0 \, , \\ -1/2 \, , & -2\pi < \arg z < 0 \, . \end{cases} \qquad (3.11)$$

The above results represent the situation where S_- mentioned in the introductory chapter is equal to -1/2. Note that by subtracting both the results for the Stokes sectors from each other, one obtains the entire Stokes discontinuity, which is imaginary. Thus, it appears that we have developed the Stokes phenomenon for the second type of terminant.

There is, however, a major problem with Eqs. (3.6) and (3.8) or the more general form of Eq. (3.10). Since the integrals are convergent, the rhs of each "equation" is finite. Yet, the lhs of each result is composed of a positively increasing divergent series, which means that it equals infinity, not a finite quantity. Therefore, these results cannot possibly be equations. In fact, to avoid the confusion due to these results, we need to replace the equals sign by the equivalence symbol. That is, Eq. (3.10) should be written as

$$\bar{\Lambda}_s(-z) \equiv \int_C dy \, \frac{y^s e^{-y}}{1 - zy} + 2S \, \frac{i\pi z^{-s-1} e^{-1/z}}{\Gamma(s+1)} \, . \qquad (3.12)$$

where $\bar{\Lambda}_s(-z)$ now only takes the series representation, while the rhs is finite. Before we are allowed to write such a statement, however, we need to establish that the method used to obtain these results, viz. Borel summation, is actually a technique for regularising divergent series.

The reader may well ask that if an equivalence symbol is required to make sense of the above results, should not the same apply to the first type of terminant given by Eq. (3.3)? One could, however, argue that because the series alternates in sign, it may be conditionally convergent. This is fine, perhaps as long as z is situated in the right hand complex plane, but it is not valid when z is situated in the left hand complex plane. E.g., for $\arg z = (2j+1)\pi$, where $j \in Z$, the first type of terminant becomes the second terminant, which is definitely divergent. As a consequence, the equals sign in Eq. (3.4) should also be replaced by the less stringent equivalence symbol, especially when $\arg z < 0$. Thus, for $|\arg z| < \pi$, Eq. (3.4) should be written as

$$\Lambda_s(z) \equiv \frac{1}{\Gamma(s+1)} \int_0^\infty dy \, \frac{y^s e^{-y}}{1+zy} \, , \qquad (3.13)$$

where the lhs side or $\Lambda_s(z)$ is given by the series representation. Furthermore, when $\arg z = (2j+1)\pi$, the above integral is undefined. Technically, we could replace $-z$ by z and apply the preceding analysis for the second type of terminant, but it may turn out that the evaluation of the Cauchy principal value is cumbersome. In addition, we shall see in later chapters that it is simply invalid to replace z by $-z$ in one type of terminant in order to obtain the regularised values of the other type of terminant. In any case, we should not be compelled to switch from one terminant to another because the argument of z has altered. What is really required, therefore, is an understanding of how the Stokes phenomenon affects the first type of terminant, which, unfortunately, is not presented in Ch. 21 of Ref. [3].

Though the material in this chapter has yet to be substantiated, it summarises the main developments in understanding the Stokes phenomenon up till the publication of Dingle's book in the mid-1970's. From the above results it is evident that the Stokes phenomenon was seen as a discontinuous effect, whereby the multiplier associated with a subdominant asymptotic series developed jump discontinuities as the argument of the variable in the dominant asymptotic series encountered a Stokes line. Moreover, the above results are entirely consistent with Stokes's view of the phenomenon. Dingle's theory of terminants represented an attempt to explain the origin of the jump discontinuities originally discovered by Stokes, but it was limited to asymptotic series where the coefficients only possessed gamma function growth, *i.e.* terminants. Although these series were Borel-summable, it led to confusion or worse still, inconsistency when both sides of the ensuing relations, *viz.* "Eqs." (3.3) and (3.10), were made equal to each other as described above. Nevertheless, by using his theory of terminants, Dingle was able to show how the rules given in Ch. 1 of Ref. [3] could be used to develop the asymptotic forms for the error function on p. 413 of the same reference. We shall substantiate these results by presenting a spectacular numerical study in a later chapter, but before this can be undertaken, we need to show that Borel summation is a method or technique for regularising a divergent series. Before this can be accomplished, however, we need to develop a better understanding of the key concept of regularisation. Consequently, we shall be able to develop a more general approach to the Stokes phenomenon that will enable us to go beyond terminants in addition to presenting forms for the first type of terminant as the argument or phase of its variable approaches a Stokes line. As stated above, this was never considered by Dingle. The presentation of this theory, however, will be postponed until we have discussed more modern interpretations of the Stokes phenomenon, in particular that proposed by Berry [9] in the late 1980s in which it is claimed that there is no jump discontinuity at a Stokes line, but a rapid smoothing. If this modern view of the Stokes phenomenon is indeed correct, then there would be no need for the generalised approach to the Stokes phenomenon presented in the later chapters of this book since this material is dependent upon interpreting the discontinuities in Cauchy integrals, which, in turn, emanates from Dingle's remarkable insight.

CHAPTER 4

Regularisation

Abstract.   Ch. 4 considers the regularisation of some basic divergent series. The first is the geometric series, which is shown to be conditionally convergent outside the unit circle of absolute convergence for $\Re z < 1$ and divergent elsewhere. Nevertheless, the regularised value is found to be identical to the limit of $1/(1-z)$ when the series is convergent. Then the regularisation of the binomial series is considered, where again, the regularised value is found to equal the limit when the series is convergent. Next the series denoted by $_2\mathcal{F}_1(a+1,b+1;a+b+2-x;1)$, which is divergent for $\Re x > 0$, is analysed. Here the regularised value is found to be different from the limit when the series is convergent or for $\Re x < 0$. Finally, the harmonic series is studied, whose regularised value equals Euler's constant. Unlike the previous examples, the last example involves logarithmic regularisation.

In the previous chapter when discussing "Eq." (3.5), it was mentioned that the series on the lhs was divergent when z was positively real and yet the result on the rhs was finite. Clearly, such a result cannot be an equation. Yet the asymptotic series was derived from a function that was finite, which means, in turn, that the method used to obtain it in the first place must be improper. Thus, a method of correcting this impropriety is required, which can be devised by introducing the concept of regularisation. In this book regularisation is defined as the removal of the infinity in the remainder of a divergent series so that the series becomes summable. The finite values produced when an asymptotic series is regularised must yield after multiplication by the specific factors outside the series the actual values of the original function from which the asymptotic series was derived. By itself, regularisation represents a mathematical abstraction, but in asymptotics, it serves to correct the impropriety in the method used to obtain the asymptotic series for a function or integral. This impropriety is responsible for the infinity in the remainder as we have already observed when we applied the method of iteration to the differential equation for $u(a)$ in Ch. 2. It also means that we cannot equate a function to an asymptotic series as in the case of "Eq." (3.5). Hence, we shall be required to replace the equals sign by the equivalence symbol when relating a divergent series to a finite quantity. It should also be noted that regularisation

Victor Kowalenko

does not affect the divergent behaviour of a truncated series once the optimal point of truncation has been exceeded. That is, the truncated part will continue to diverge past the optimal point of truncation. This behaviour is countered by the remainder, which diverges in the opposite sense, when it is regularised properly. Another important property of regularisation is that the points of $z = 0$ and $z = \infty$ in an asymptotic series will yield definite values provided the original function does not possess singularities at these points. In fact, the regularised value will be non-analytic only where the original function is non-analytic.

4.1 Geometric Series

We begin our introduction to regularisation by studying how the process affects the geometric series. Although this series is one of the simplest of all divergent series since its coefficients are all equal to unity, it will serve as the basis for regularising more complicated asymptotic series. The idea is that if a more complicated asymptotic series can be expressed in terms of the geometric series, then it will also be regularised once the regularised value of the geometric series is introduced. The geometric series can be written as

$$
{}_1\mathcal{F}_0(1;z) = \sum_{k-0}^{\infty} z^k \;, \tag{4.1}
$$

where

$$
{}_p\mathcal{F}_q(\alpha_1,\ldots,\alpha_p;\beta_1,\ldots\beta_q;z) = \sum_{k=0}^{\infty} \frac{\Gamma(k+\alpha_1)\cdots\Gamma(k+\alpha_p)}{\Gamma(\alpha_1)\cdots\Gamma(\alpha_p)}
$$
$$
\times \frac{\Gamma(\beta_1)\cdots\Gamma(\beta_q)}{\Gamma(k+\beta_1)\cdots\Gamma(k+\beta_q)} \frac{z^k}{k!} \;. \tag{4.2}
$$

The reader should note that the standard notation for a hypergeometric function, viz. ${}_pF_q$, has not been employed in the above definition. Instead the series has been defined according to a different notation ${}_p\mathcal{F}_q$. This is because the standard notation for a hypergeometric function is only employed when the series on the rhs is absolutely convergent, i.e. when $q \geq p$ or when $|z| < 1$ for $q = p - 1$. Since we are interested in the behaviour of divergent series in this book, we have introduced the above notation as a shorthand means of writing all series of the form given by the rhs of Eq. (4.2). That is, the lhs of Eq. (4.2) is merely an alternative means of expressing any series with the form on the rhs. It does not signify convergence as in the case of the notation for hypergeometric functions.

According to p. 19 of Ref. [6], the geometric series is absolutely convergent for $|z| < 1$ and divergent for $|z| \geq 1$. However, we shall now show that the series is actually conditionally convergent for $\Re z < 1$ and divergent for $\Re z > 1$ when $|z| > 1$. For the latter case, the series will need to be regularised in order to obtain a finite value. To see this more clearly, we elaborate on the explanation that first appeared in Ref. [14]. We begin by writing the series as

$$\sum_{k=0}^{\infty} z^k = \sum_{k=0}^{\infty} \Gamma(k+1) \frac{z^k}{k!} = \lim_{p \to \infty} \sum_{k=0}^{\infty} \frac{z^k}{k!} \int_0^p dt\, e^{-t} t^k \; . \qquad (4.3)$$

Since the integral in Eq. (4.3) is finite, technically, we can interchange the order of the summation and integration. In reality, an impropriety occurs when we do this, which will be discussed in more detail shortly. For now, interchanging the summation and integration yields

$$\sum_{k=0}^{\infty} z^k = \lim_{p \to \infty} \int_0^p dt\, e^{-t} \sum_{k=0}^{\infty} \frac{(zt)^k}{k!} = \lim_{p \to \infty} \int_0^p dt\, e^{-t(1-z)}$$

$$= \lim_{p \to \infty} \left[\frac{e^{-p(1-z)}}{z-1} + \frac{1}{1-z} \right]. \qquad (4.4)$$

For $\Re z < 1$, the first term of the last member of the above equation vanishes and the series yields a finite value of $1/(1-z)$. Therefore, we see that the same value is obtained for the series when $\Re z < 1$ as for when z lies in the circle of absolute convergence given by $|z| < 1$. According to the definition on p. 18 of Ref. [6], this means that the series is conditionally convergent for $\Re z < 1$ and $|z| > 1$. For $\Re z > 1$, however, the first term in the last member of Eq. (4.4) is infinite. Since we have defined regularisation as the process of removing the infinity that arises from employing an improper mathematical method as was the case when the method of iteration was employed in Ch. 2, we remove or neglect the first term of the last member of Eq. (4.4). Then we are left with a finite part that equals $1/(1-z)$, which we call the regularised value. Hence, for all complex values of z except $\Re z = 1$, we have

$$\sum_{k=0}^{\infty} z^k \begin{cases} \equiv \frac{1}{1-z} \; , & \Re z > 1 \; , \\ = \frac{1}{1-z} \; , & \Re z < 1. \end{cases} \qquad (4.5)$$

Frequently, we will not know for which values of z an asymptotic series is convergent and for which it is divergent. So, in these cases we shall replace the equals sign by the less stringent equivalence symbol on the understanding that we may be dealing with a series that is absolutely or conditionally convergent for some values of z. As a result, we adopt the shorthand notation of

$$\sum_{k=N}^{\infty} z^k = z^N \sum_{k=0}^{\infty} z^k \equiv \frac{z^N}{1-z} . \tag{4.6}$$

We shall refer to such results as equivalence statements or simply equivalences since they cannot be regarded strictly as equations. In fact, it is simply incorrect to refer to the above as an equation because we have seen for $\Re z > 1$ that the lhs is infinite. Furthermore, the above notation is only applicable when the form for the regularised value of the divergent series is identical to the form of the limiting value of the convergent series. As we shall see later, this is not always the case and when they are different, we need to specify the two different values in a similar form to Equivalence (4.5).

At the barrier of $\Re z = 1$, the situation appears to be unclear. For $z = 1$ the last member of Eq. (4.4) vanishes, which is consistent with removing the infinity from $1/(1-z)$. For other values of $\Re z = 1$, the last member of Eq. (4.4) is clearly undefined. This is to be expected as this line forms the border between the domains of convergence and divergence for the series. Because the finite value remains the same to the right and to the left of the barrier at $\Re z = 1$ and in keeping with the fact that regularisation is effectively the removal of the first term on the rhs of Eq. (4.4), we take $1/(1-z)$ to be the finite or regularised value when $\Re z = 1$. Hence, Equivalence (4.5) becomes

$$\sum_{k=0}^{\infty} z^k \begin{cases} \equiv \frac{1}{1-z} , & \Re z \geq 1 , \\ = \frac{1}{1-z} , & \Re z < 1. \end{cases} \tag{4.7}$$

As discussed in Ref. [14] the regularised value of a divergent series is analogous to the Hadamard finite part that arises in the regularisation of divergent integrals in the theory of generalised functions [16, 25]. For example, consider the divergent integral of

$$I = \int_0^{\infty} dx \, e^{ax} = \lim_{p \to \infty} \int_0^p dx \, e^{ax} = \lim_{p \to \infty} \left[\frac{e^{ap} - 1}{a} \right] . \tag{4.8}$$

For $\Re a > 0$ this integral is divergent, but removing the first term in the last member yields a finite part of $-1/a$, which is the same result one obtains when $\Re a < 0$. To show the connection with regularisation of a divergent series, the above integral can be written in terms of an arbitrary positive real parameter, say b, as

$$I = \int_0^\infty dx\, e^{-bx} e^{(a+b)x} = \int_0^\infty dx\, e^{-bx} \sum_{k=0}^\infty \frac{(a+b)^k x^k}{k!} \ . \tag{4.9}$$

In obtaining this result we have used the asymptotic method of expanding most of the exponential as described on p. 113 of Ref. [3]. Now we interchange the order of the summation and integration. Because most of the exponential has been expanded, an impropriety has occurred, which means that evaluating the integral can result in a divergent series depending on the values of a and b. As long as the series is divergent when I is divergent, this does not pose a problem and we can keep the equals sign. Then the integral I becomes

$$I = \sum_{k=0}^\infty \frac{(a+b)^k}{k!} \int_0^\infty dx\, e^{-bx} x^k = \frac{1}{a+b} \sum_{k=1}^\infty \left(1 + \frac{a}{b}\right)^k \ . \tag{4.10}$$

Since the geometric series has appeared, we know that the rhs is divergent for $\Re (a/b) > 0$. Furthermore, because b is an arbitrary positive real parameter, the rhs is divergent for $\Re a > 0$, which is when I is divergent. So the above result is consistent, which would not have been the case if we had applied the method of expanding most of the exponential to a convergent integral as is done usually. Now if we introduce the regularised value of the geometric series, viz. Equivalence (4.5), into the above equation, then we obtain a value of $-1/a$ for the finite part of I. That is, by regularising the series, we find that $I \equiv -1/a$, which is identical to evaluating the divergent integral directly and removing the infinity or the first term in the last member of Eq. (4.8). Hence, evaluating the Hadamard finite part of a divergent integral is equivalent to regularisation of a divergent series.

In Ref. [26] Farassat discusses the issue of whether the appearance of divergent integrals in applications constitutes a breakdown in physics or mathematics. He concludes that divergent integrals arise as a result of incorrect mathematics because an ordinary derivative has been wrongly evaluated inside an improper integral. Therefore, he regards regularisation of a divergent integral or taking the finite part as a necessary corrective measure. The same applies to the divergent series in asymptotic expansions. By itself a divergent series yields infinity, but when regularised, one obtains a finite value or part. When an asymptotic method is used to obtain an expansion, there is an impropriety or flaw associated with the method. The method of iteration used to obtain the expansion from the differential equation in Ch. 2 led to the inclusion of an infinity in the solution, while deriving asymptotic expansions from integral representations invariably involves integrating over a range that is outside the circle of absolute convergence of the expanded function. The observation that integrating outside the radius of absolute convergence results in divergent power series expansions was first presented in the forerunner to this work, *i.e.* Ref. [14] and has since been discussed at length in Ref. [27]. Therefore, while regularisation represents a mathematical abstraction for obtaining the finite value of a divergent series, it is necessary in asymptotics for correcting the impropriety in the method used to obtain an asymptotic expansion. After all, the original function or integral from which an asymptotic expansion is derived is finite, even though the latter is not.

From the above study of the geometric series we know that for certain values of z that the series given by Eq. (4.2) will be convergent in which case it can be expressed in terms of the standard notation for hypergeometric functions. For other values of z, however, the series will be divergent. For these values we need to define a regularised value in terms of the standard notation for hypergeometric functions. Because the regularised value of a series is also valid when the series is convergent, we write

$$\sum_{k=0}^{\infty} \frac{\Gamma(k+\alpha_1)\cdots\Gamma(k+\alpha_p)}{\Gamma(\alpha_1)\cdots\Gamma(\alpha_p)} \frac{\Gamma(\beta_1)\cdots\Gamma(\beta_q)}{\Gamma(k+\beta_1)\cdots\Gamma(k+\beta_q)} \frac{z^k}{k!}$$
$$\equiv {}_pF_q(\alpha_1,\ldots,\alpha_p;\beta_1,\ldots,\beta_q;z) \ . \tag{4.11}$$

Hence, when the series is absolutely convergent, i.e. for $q \geq p$ or $|z| < 1$ for $q = p-1$, the above becomes

$$_p\mathcal{F}_q(\alpha_1,\ldots,\alpha_p;\beta_1,\ldots\beta_q;z) = {}_pF_q(\alpha_1,\ldots,\alpha_p;\beta_1,\ldots,\beta_q;z) \ . \tag{4.12}$$

For the other values of q, p and z, the series can be either conditionally convergent or divergent. Then Equivalence (4.11) becomes

$$_p\mathcal{F}_q(\alpha_1,\ldots,\alpha_p;\beta_1,\ldots\beta_q;z) \equiv {}_pF_q(\alpha_1,\ldots,\alpha_p;\beta_1,\ldots,\beta_q;z) \ . \tag{4.13}$$

In order for the reader to gain a better understanding of the regularisation process, we now consider some interesting examples before we turn our attention to the asymptotic forms of the error function. As our first example we consider the regularised value of the binomial series. This has already been given as a definition in Ref. [15], although it was stated there that the result would be proved elsewhere. We shall do so in the following section.

4.2   Binomial Series

As indicated earlier, we can use our study of the regularised value of the geometric series as the basis for regularising other more complicated divergent series. In the first instance we begin by deriving the regularised value of the binomial series, which represents a generalisation of the geometric series. Throughout this book we shall present new formulations of the regularised values of divergent series in the form of propositions, the first of which appears immediately below.

Proposition 1. For all real values of ρ, regularisation of the binomial series yields

$$
{}_1\mathcal{F}_0(\rho;z) = \sum_{k=0}^{\infty} \frac{\Gamma(k+\rho)}{\Gamma(\rho)\,k!}
\begin{cases}
= 1/(1-z)^\rho & ,\ \Re z < 1 \\
\equiv 1/(1-z)^\rho & ,\ \Re z \geq 1.
\end{cases}
$$

Proof. There is no need to prove the above proposition for $|z| < 1$ since this is the standard form of the binomial theorem, which is discussed on p. 95 of Ref. [6]. We now establish that the series is conditionally convergent for $\Re z < 1$ and $|z| > 1$. Without loss of generality, let us assume that $\Re \rho < 0$. Then there exists a positive integer N such that $\rho^* = \rho + N$, where $\Re \rho^* > 0$. As a result, the binomial series can be written as

$$
{}_1\mathcal{F}_0(\rho;z) = \sum_{k=0}^{N-1} \frac{\Gamma(k+\rho)}{\Gamma(\rho)\,k!} z^k + L_N(z) \sum_{k=0}^{\infty} \frac{\Gamma(k+\rho^*)}{\Gamma(\rho)\,k!} z_N^k \ , \tag{4.14}
$$

where the operator $L_N(z)$ is defined as

$$
L_N(z) = \int_0^z dz_1 \int_0^{z_1} dz_2 \cdots \int_0^{z_{N-1}} dz_N \ . \tag{4.15}
$$

Introducing the integral representation for the gamma function into the numerator of the infinite sum and interchanging the order of the summation and integration, we can write Eq. (4.14) as

$$
{}_1\mathcal{F}_0(\rho;z) = \sum_{k=0}^{N-1} \frac{\Gamma(k+\rho)}{\Gamma(\rho)\,k!} z^k + \frac{1}{\Gamma(\rho)} L_N(z) \lim_{p\to\infty} \int_0^p dt\, e^{-t(1-z_N)}\, t^{\rho^*-1}. \tag{4.16}
$$

As in the case of the geometric series, we have committed an impropriety, which means that the series will have to be regularised in order to yield a finite limit for certain values of z. Next we use No. 3.381(1) from Ref. [21] to express the integral in terms of the incomplete gamma function $\Gamma(\rho^*, p(1-z))$. Hence, Eq. (4.16) becomes

$$_1\mathcal{F}_0(\rho;z) = \sum_{k=0}^{N-1} \frac{\Gamma(k+\rho)}{\Gamma(\rho)\,k!} z^k + \frac{1}{\Gamma(\rho)} L_N(z) \left((1-z_N)^{\rho^*} \Gamma(\rho^*) \right.$$
$$\left. - \lim_{p\to\infty} (1-z_N)^{\rho^*} \Gamma\left(\rho^*, (1-z_N)\,p\right) \right) , \qquad (4.17)$$

which is valid for $\Re\rho^* > 0$. The last term in Eq. (4.17) can be evaluated by taking the leading order term of the large $|z|$ asymptotic expansion for the incomplete gamma function, which appears as No. 8.357 in Ref. [21]. Then we find that

$$_1\mathcal{F}_0(\rho;z) = \sum_{k=0}^{N-1} \frac{\Gamma(k+\rho)}{\Gamma(\rho)\,k!} z^k + \frac{1}{\Gamma(\rho)} L_N(z) \left((1-z_N)^{\rho^*} \Gamma(\rho^*) \right.$$
$$\left. - (1-z_N)^{2\rho^*-1} \lim_{p\to\infty} p^{\rho^*-1} e^{-(1-z_N)p} \right) . \qquad (4.18)$$

Now we apply the operator $L_N(z)$ to the last term in Eq. (4.18), which yields

$$-\frac{1}{\Gamma(\rho)} L_N(z) \left((1-z_N)^{2\rho^*-1} \lim_{p\to\infty} p^{\rho^*-1} e^{-(1-z_N)p} \right) = -\frac{1}{\Gamma(\rho)}$$
$$\times \lim_{p\to\infty} p^{\rho^*-1} L_{N-1} \int_0^{z_{N-1}} dz_N\, (1-z_N)^{2\rho^*-1} e^{-(1-z_N)p}. \qquad (4.19)$$

By evaluating the integral over z_N, we are left with a term amongst others that will possess the exponential factor of $\exp(-(1-z_{N-1})p)$, courtesy of the upper limit of integration. To see this more clearly, put $\rho^* = 1/2$, which yields in the limit as $p \to \infty$ a value of $\exp(-(1-z_{N-1})p)/p$. Therefore, by carrying out the other integrations in $L_{N-1}(z)$, there will ultimately be one term that consists of powers of p multiplied by $\exp(-(1-z)p)$. It does not even matter if the powers of p are positive. All that matters is the exponential term since its behaviour determines whether the series is convergent or divergent. If $\Re z < 1$, then it will vanish and we are left with a finite limit to the binomial series. On the other hand, if $\Re z > 1$, then the exponential factor yields infinity, which has to be removed to yield a finite value for the binomial series. Hence, we see that the binomial series is conditionally convergent for $\Re z < 1$ and $|z| > 1$ and divergent for $\Re z > 1$.

We now show that the finite limit or regularised value of the series, $_1\mathcal{F}_0(\rho;z)$, is the same regardless of whether $\Re z$ is greater than or less than unity and is given by the value in the proposition. Initially, we assume that $0 < \Re \rho < 1$, which allows us to write the series as

$$\sum_{k=0}^{\infty} \frac{\Gamma(k+\rho)}{\Gamma(\rho)k!} z^k = \frac{1}{B(\rho, 1-\rho)} \sum_{k=0}^{\infty} \int_0^1 dt \, z^k t^{k+\rho-1} (1-t)^{-\rho} \, , \qquad (4.20)$$

where $B(x,y)$ represents the beta function and we have introduced its integral representation given by No. 8.380(1) in Ref. [21] into the numerator. In this form we see the connection with the geometric series when the order of the summation and integration is interchanged. As a consequence, if we replace the geometric series by its regularised value, then we obtain the regularised value of the binomial series. Hence, we arrive at

$$\sum_{k=0}^{\infty} \frac{\Gamma(k+\rho)}{\Gamma(\rho)k!} z^k \equiv \frac{1}{B(\rho, 1-\rho)} \int_0^1 dt \, \frac{t^{\rho-1}(1-t)^{-\rho}}{1-zt} \, . \qquad (4.21)$$

According to No. 9.111 of Ref. [21], the integral on the rhs of the above equivalence represents the integral representation for $_2F_1(1,\rho;1;z)$. It may not look as if we have gained much especially since this Gauss hypergeometric function reduces to $_1F_0(\rho;z)$, but the major difference is that the Gauss hypergeometric function has been analytically continued beyond its series representation, which is only convergent for $|z| < 1$. Furthermore, according to No. 7.3.1.1 in Ref. [28], $_1F_0(\rho;z) = (1-z)^{-\rho}$ for all values of z.

So far, we have established that the regularised value for the binomial series is as given in the proposition for $0 < \Re \rho < 1$. Let us now assume that $\Re \rho > 1$. Then there exists a positive integer N such that $\rho = \rho^* + N$ and $0 < \Re \rho^* < 1$. As a consequence, we can write the binomial series as

$$\sum_{k=0}^{\infty} \frac{\Gamma(k+\rho)}{\Gamma(\rho)k!} z^k = \sum_{k=0}^{\infty} \frac{\Gamma(k+\rho^*+N)}{\Gamma(\rho^*+N)k!} z^k = \frac{z^{1-\rho^*}\Gamma(\rho^*)}{\Gamma(\rho^*+N)}$$

$$\times \frac{d^N}{dz^N} z^{\rho^*+N-1} \sum_{k=0}^{\infty} \frac{\Gamma(k+\rho^*)}{\Gamma(\rho^*)k!} z^k \, . \qquad (4.22)$$

The final series in Eq. (4.22) is the binomial series for ρ^* and we have already determined its regularised value via Equivalence (4.21). Hence, the regularised value of the above series becomes

$$\sum_{k=0}^{\infty} \frac{\Gamma(k+\rho)}{\Gamma(\rho)k!} z^k \equiv \frac{z^{1-\rho^*}\Gamma(\rho^*)}{\Gamma(\rho^*+N)} \frac{d^N}{dz^N} \frac{z^{\rho^*+N-1}}{(1-z)^{\rho^*}} \, . \qquad (4.23)$$

Now we establish that the rhs of Equivalence (4.23) yields the value as given in the proposition. We do this by induction. Putting $N=1$ into the rhs yields

$$\frac{z^{1-\rho^*}\Gamma(\rho^*)}{\Gamma(\rho^*+1)}\frac{d}{dz}\frac{z^{\rho^*}}{(1-z)^{\rho^*}} = \frac{z^{1-\rho^*}}{\rho^*}\left(\frac{\rho^* z^{\rho^*-1}}{(1-z)^{\rho^*}} - \frac{z^{\rho^*}(-\rho^*)}{(1-z)^{\rho^*+1}}\right)$$

$$= \frac{1}{(1-z)^{\rho^*+1}} = \frac{1}{(1-z)^{\rho}} \ . \tag{4.24}$$

We assume that the above result holds for $N=n$ and put $N=n+1$. Then the rhs of Equivalence (4.23) becomes

$$\frac{z^{1-\rho^*}\Gamma(\rho^*)}{\Gamma(\rho^*+n+1)}\frac{d^{n+1}}{dz^{n+1}}\frac{z^{n+\rho^*}}{(1-z)^{\rho^*}} = \frac{z^{1-\rho^*}\Gamma(\rho^*)}{\Gamma(n+\rho^*+1)}$$

$$\times \frac{d^n}{dz^n}\left(\frac{(n+\rho^*)z^{n+\rho^*-1}}{(1-z)^{\rho^*}} - \frac{z^{n+\rho^*}(-\rho^*)}{(1-z)^{\rho^*+1}}\right) \ . \tag{4.25}$$

Since we have assumed the $N=n$ case is valid, this means that

$$\frac{\Gamma(\rho^*)z^{1-\rho^*}}{\Gamma(n+\rho^*)}\frac{d^n}{dz^n}\frac{z^{n+\rho^*-1}}{(1-z)^{\rho^*}} = \frac{1}{(1-z)^{\rho^*+n}} \ . \tag{4.26}$$

Introducing Eq. (4.26) into Eq. (4.25) yields

$$\frac{z^{1-\rho^*}\Gamma(\rho^*)}{\Gamma(\rho^*+n+1)}\frac{d^{n+1}}{dz^{n+1}}\frac{z^{n+\rho^*}}{(1-z)^{\rho^*}} = \frac{1}{(1-z)^{n+\rho^*}}$$

$$-\frac{z^{1-\rho^*}\Gamma(\rho^*+1)}{\Gamma(n+\rho^*+1)}\frac{d^n}{dz^n}\frac{z^{n+\rho^*}}{(1-z)^{\rho^*+1}} \ . \tag{4.27}$$

The last derivative can be viewed as the ρ^*+1 case of the assumption given by Eq. (4.26). Hence, we find that Eq. (4.27) can be written as

$$\frac{z^{1-\rho^*}\Gamma(\rho^*)}{\Gamma(\rho^*+n+1)}\frac{d^{n+1}}{dz^{n+1}}\frac{z^{n+\rho^*}}{(1-z)^{\rho^*}} = \frac{1}{(1-z)^{n+\rho^*}} - \frac{z^{1-\rho^*}\Gamma(\rho^*+1)}{\Gamma(n+\rho^*+1)}$$

$$\times \frac{\Gamma(n+\rho^*+1)z^{\rho^*}}{\Gamma(\rho^*+1)(1-z)^{n+\rho^*+1}} = \frac{1}{(1-z)^{n+\rho^*+1}} = \frac{1}{(1-z)^{\rho}} \ . \tag{4.28}$$

We have already seen that the rhs of Equivalence (4.23) gives the regularised value stated in the proposition for $N=1$. According to Eq. (4.28), it also holds for $N=2,3,\ldots$, i.e. for all N. Therefore, for $\Re\rho>1$, the regularised value of the binomial series as given by Equivalence (4.23) reduces to the value given in the proposition.

To complete the proof, we need to establish that the regularised value in Proposition 1 is also valid for $\Re \rho < 0$. In this case we let $\rho = \rho^* - N$, where $0 < \Re \rho^* < 1$ and N is again a positive integer. Then we can write the binomial series as

$$\sum_{k=0}^{\infty} \frac{\Gamma(k+\rho)}{\Gamma(\rho)\,k!} z^k = \sum_{k=0}^{N-1} \frac{\Gamma(k+\rho)}{\Gamma(\rho)\,k!} z^k + \sum_{k=N}^{\infty} \frac{\Gamma(k+\rho^*-N)}{\Gamma(\rho)\,k!} z^k \ . \tag{4.29}$$

In the second sum on the rhs of Eq. (4.29) we replace $k-N$ by k. After a little algebraic manipulation we arrive at

$$\sum_{k=0}^{\infty} \frac{\Gamma(k+\rho)}{\Gamma(\rho)\,k!} z^k = \sum_{k=0}^{N-1} \frac{\Gamma(k+\rho)}{\Gamma(\rho)\,k!} z^k + \frac{z^N \Gamma(\rho^*)}{\Gamma(\rho)} \sum_{k=0}^{\infty} \frac{\Gamma(k+\rho^*)}{\Gamma(\rho^*)\,k!}$$
$$\times \ \frac{\Gamma(k+1)}{\Gamma(k+N+1)} z^k \ . \tag{4.30}$$

We now prove by induction that the regularised value of the rhs of Eq. (4.30) yields the value given in the proposition. Let us examine what happens when we put $N=1$ in the above result. Then we find that

$$\sum_{k=0}^{\infty} \frac{\Gamma(k+\rho)}{\Gamma(\rho)\,k!} z^k {}_o = 1 + \frac{z\Gamma(\rho^*)}{\Gamma(\rho)} \sum_{k=0}^{\infty} \frac{\Gamma(k+\rho^*)}{\Gamma(\rho^*)\,k!\,(k+1)}$$
$$= \ 1 + (\rho^* - 1) \int_0^z dt \sum_{k=0}^{\infty} \frac{\Gamma(k+\rho^*)}{\Gamma(\rho^*)\,k!} z^k \ . \tag{4.31}$$

The final sum is the binomial series for ρ^*, whose regularised value has been already been determined via Equivalence (4.21). Introducing the regularised value into Eq. (4.31) yields

$$\sum_{k=0}^{\infty} \frac{\Gamma(k+\rho)}{\Gamma(\rho)\,k!} z^k \equiv 1 + (\rho^* - 1) \int_0^z dt \, (1-t)^{-\rho^*} = \frac{1}{(1-z)^{\rho^*-1}} \ . \tag{4.32}$$

Next we assume that for $N=n$ the rhs of Eq. (4.30) yields the regularised value of $(1-z)^{-\rho}$. For $N=n+1$, Eq. (4.30) becomes

$$\sum_{k=0}^{\infty} \frac{\Gamma(k+\rho)}{\Gamma(\rho)\,k!} z^k = \sum_{k=0}^{n} \frac{\Gamma(k+\rho)}{\Gamma(\rho)\,k!} z^k + \frac{z^{n+1} \Gamma(\rho^*)}{\Gamma(\rho)} \sum_{k=0}^{\infty} \frac{\Gamma(k+\rho^*)}{\Gamma(\rho^*)\,k!}$$
$$\times \ \frac{\Gamma(k+1)}{\Gamma(k+n+2)} z^k \ . \tag{4.33}$$

Alternatively, we can write the above equation as

$$\sum_{k=0}^{\infty} \frac{\Gamma(k+\rho)}{\Gamma(\rho)\,k!}\, z^k = 1 + \sum_{k=0}^{n-1} \frac{\Gamma(k+1+\rho)}{\Gamma(\rho)\,(k+1)!}\, z^{k+1} + \frac{\Gamma(\rho^*)}{\Gamma(\rho^*-n-1)}$$
$$\times \sum_{k=0}^{\infty} \frac{\Gamma(k+\rho^*)}{\Gamma(\rho^*\,k!}\, \frac{\Gamma(k+1)}{\Gamma(k+n+1)} \int_0^z dt\, t^{k+n} \ . \tag{4.34}$$

In the truncated sum on the rhs of Eq. (4.34), we replace the factor of $z^{k+1}/(k+1)$ by $\int_0^z dt\, t^k$. Then after a little manipulation we obtain

$$\sum_{k=0}^{\infty} \frac{\Gamma(k+\rho)}{\Gamma(\rho)\,k!}\, z^k = 1 + \frac{1}{\Gamma(\rho^*-n-1)} \int_0^z dt \left[\sum_{k=0}^{N-1} \Gamma(k+\rho+1) \frac{t^k}{k!} \right.$$
$$\left. + \ \Gamma(\rho^*) \sum_{k=0}^{\infty} \frac{\Gamma(k+\rho^*)}{\Gamma(\rho^*)\,k!} \frac{\Gamma(k+1)}{\Gamma(k+n+1)}\, t^{k+N} \right] \ . \tag{4.35}$$

Both the truncated series and the infinite series on the rhs can be combined into one series, which can be identified as the binomial series for $\rho^* - n$. By induction this has been assumed to have a regularised value of $(1-z)^{-\rho^*+n}$. Therefore, we obtain

$$\sum_{k=0}^{\infty} \frac{\Gamma(k+\rho)}{\Gamma(\rho)\,k!}\, z^k \equiv 1 + \frac{\Gamma(\rho^*-n)}{\Gamma(\rho^*-n-1)} \int_0^z dt\, \frac{1}{(1-t)^{\rho^*-n}}$$
$$= \frac{1}{(1-z)^{\rho^*-n-1}} = \frac{1}{(1-z)^{\rho}} \ . \tag{4.36}$$

We have shown that the result in Proposition 1 is valid for $N=1$ or $\rho=\rho^*-1$ in obtaining Equivalence (4.32). The above implies that the result in Proposition 1 is valid for $N=2$ or ρ^*-2, and then for $N=3$ and so on for all positive integer values of N.

To complete the proof, we briefly consider the cases when (1) ρ is an integer and (2) $\Re z = 1$. For $\rho=1$, the result in Proposition 1 reduces to the geometric series, whose regularised value has already been evaluated. For $\rho=-N$, where $N \geq 0$, the series on the lhs becomes finite, yielding

$$\sum_{k=0}^{\infty} \frac{\Gamma(k+\rho)}{\Gamma(\rho)}\, \frac{z^k}{k!} = \sum_{k=0}^{N} \frac{\Gamma(k-N)}{\Gamma(-N)}\, \frac{z^k}{k!} \ . \tag{4.37}$$

By introducing the reflection formula for the gamma function, i.e. Eq. (3.2), we find that the series becomes $\sum_{k=0}^{N} \binom{N}{k} (-z)^k$, which, of course, equals $(1-z)^N$. For $\rho = N$, where $N > 1$, the series becomes

$$\sum_{k=0}^{\infty} \frac{\Gamma(k+N)}{\Gamma(N)} \frac{z^k}{k!} = \frac{1}{\Gamma(N)} \frac{d^{N-1}}{dz^{N-1}} \sum_{k=0}^{\infty} z^k \equiv \frac{1}{(N-1)!} \frac{d^{N-1}}{dz^{N-1}} \frac{z^{N-1}}{1-z} \ . \qquad (4.38)$$

The rhs of the above equivalence can be shown to equal $1/(1-z)^N$ by induction. In the second case the line $\Re z = 1$ represents the border between convergence and divergence as in the case of the geometric series, but because the regularised value of the binomial series is the same as when the series is convergent for all values of ρ, it is also set equal to $1/(1-z)^{\rho}$ for $\Re z = 1$.

One final remark about this proof is required before we consider the next example. The bulk of the proof has been concerned with establishing the proposition for $\Re \rho > 1$ and $\Re \rho < 1$ by avoiding the introduction of the integral representation for the beta function in Eq. (4.21), which is divergent for these values of ρ. If, however, we had stated for all values of ρ that

$$\int_0^{\infty} dt \ t^{\rho-1} (1-t)^{-\rho} \equiv \frac{\Gamma(\rho)\Gamma(1-\rho)}{\Gamma(k+1)} \ , \qquad (4.39)$$

then it would have simplified the proof of Proposition 1 dramatically. That is, the rhs of the above result represents the Hadamard finite part of a divergent integral [25]. As an aside, Ninham [29] has confirmed this result by applying his general formula for the finite of part of a divergent integral to the integral representation for $B(-1/2, -1/2)$. Specifically, he demonstrates that the finite part for these values of the beta function integral vanishes, which is consistent with the value of $\Gamma(-1/2)\Gamma(-1/2)/\Gamma(-1)$ obtained from the above result. In proving the proposition we chose not to extend the integral representation of the beta function to divergent regimes of ρ because the reader might then claim that the divergence in the binomial series had been masked or hidden by the divergence in the integral representation for the beta function. This is actually not the case since the regularisation process occurs in the next step after the integral representation for the beta function has been introduced. So, to avoid any objection in using Equivalence (4.39), we opted for the much longer and less controversial method of proving the proposition given here. This completes the proof of Proposition 1.

It should be noted that in order to derive new identities from a series with a finite radius of absolute convergence such as that in Proposition 1, one must generally appeal to those values of z, where the series is convergent, as exemplified by the following theorem.

Theorem 1. The following finite sums involving quotients of the beta and gamma functions,

$$\sum_{j=0}^{k} \frac{B(k-j+\mu, j+\nu)}{(k+1)B(k+1-j, j+1)} = B(\mu, \nu) \ , \qquad (4.40)$$

and for $k \geq 1$,

$$\sum_{j=1}^{k} \frac{(-1)^j B(j+\mu, 1-\mu)}{(j-1)!\,(k-j)!} = \frac{\pi}{k!\,B(-\mu, k)\,\sin(\pi\mu)} \ , \qquad (4.41)$$

follow from the convergent form of Proposition 1.

Proof. From Proposition 1 we have

$$\sum_{k=0}^{\infty} \frac{\Gamma(k+\mu)}{\Gamma(\mu)\,k!} (-z)^k \sum_{k=0}^{\infty} \frac{\Gamma(k+\nu)}{\Gamma(\nu)\,k!} (-z)^k \equiv \frac{1}{(1+z)^{\mu+\nu}} \ . \qquad (4.42)$$

Multiplying both series on the lhs of the equivalence yields

$$\sum_{k=0}^{\infty} (-z)^k C_k \equiv \frac{\Gamma(\mu)\Gamma(\nu)}{(1+z)^{\mu+\nu}} \ , \qquad (4.43)$$

where

$$C_k = \sum_{j=0}^{k} \Gamma(k-j+\mu)\Gamma(j+\nu)/\Gamma(k+1-j)\Gamma(j+1) \ . \qquad (4.44)$$

That is, the rhs of Equivalence (4.43) represents the regularised value of the power series on the lhs whose coefficients are given by C_k. Since we have used Proposition 1 to obtain the above equivalence, the series on the lhs of the above equivalence is conditionally convergent for $\Re z > -1$ and $|z| > 1$, while it is divergent for $\Re z < -1$. For the former regime of values of z the equivalence symbol can be replaced by an equals sign.

Now if we introduce Proposition 1 into the rhs of Equivalence (4.43), then we find that

$$\Gamma(\mu)\Gamma(\nu) \sum_{k=0}^{\infty} \frac{\Gamma(k+\mu+\nu)}{\Gamma(\mu+\nu)\,k!} (-z)^k \equiv \sum_{k=0}^{\infty} C_k (-z)^k \ . \qquad (4.45)$$

Whilst it is not permissible to do so for $\Re z < -1$, we can replace the equivalence symbol by an equals sign in the above result for $\Re z > -1$. In so doing, z remains fairly arbitrary, which means that we can equate like powers of z in the resulting equation. Then we obtain

$$C_k = \sum_{j=0}^{k} \frac{\Gamma(k-j+\mu)}{(k-j)!} \frac{\Gamma(j+\nu)}{j!} = \frac{\Gamma(\mu)\Gamma(\nu)}{\Gamma(\mu+\nu)} \frac{\Gamma(k+\mu+\nu)}{k!} \ . \qquad (4.46)$$

By using the fact that the beta function is defined as

$$B(\alpha,\beta) = \frac{\Gamma(\alpha)\Gamma(\beta)}{\Gamma(\alpha+\beta)} \ , \qquad (4.47)$$

we arrive at the first identity in the theorem. Again, it needs to be stressed that this result could only be obtained with the introduction of an equals sign or the convergent form of Proposition 1.

To obtain the second finite sum, we use the fact that

$$\left(1 + \frac{z}{1-z}\right)^{\mu} = (1-z)^{\mu} \ . \qquad (4.48)$$

By introducing Proposition 1 into the above result, we obtain

$$\sum_{j=1}^{\infty} \frac{\Gamma(j+\mu)}{\Gamma(\mu)\,j!} \left(-\frac{z}{1-z}\right)^{j} \equiv \sum_{j=1}^{\infty} \frac{\Gamma(j-\mu)}{\Gamma(-\mu)\,j!} z^{j} \ . \qquad (4.49)$$

For $|z| < 1/2$, we can replace the equivalence by an equals sign. Since z is still fairly arbitrary, we can equate like powers of z again, thereby obtaining

$$\sum_{j=1}^{k} \frac{(-1)^{j}(k-1)!}{(j-1)!(k-j)!} \frac{\Gamma(j+\mu)}{j!} = \frac{\Gamma(k-\mu)}{\Gamma(-\mu)} \frac{\Gamma(\mu)}{k!} \ . \qquad (4.50)$$

Finally, after a little manipulation with the reflection formula for the gamma function and the definition of the beta function, we arrive at the second result in the theorem. This completes the proof of the theorem.

4.3   Regularisation of a $_2\mathcal{F}_1$ Series

So far, we have presented two important examples of regularisation where the regularised value is identical to the limit of the absolutely convergent form of the series. However, this is not always the case as is demonstrated in the following proposition.

Proposition 2. For $\Re x < 0$, the series given by $_2\mathcal{F}_1(a+1, b+1; a+b+2-x; 1)$ has the limiting value of

$$
\begin{aligned}
_2\mathcal{F}_1(a+1, b+1; a+b+2-x; 1) &= \frac{\Gamma(x-a)\Gamma(x-b)}{\Gamma(x+1)\Gamma(x-a-b-1)} \\
&\times \frac{\sin(\pi(x-a))\sin(\pi(x-b))}{\sin(\pi x)\sin(\pi(x-a-b))} ,
\end{aligned}
\tag{4.51}
$$

while for $\Re x > 0$, its regularised value is given by

$$
_2\mathcal{F}_1(a+1, b+1; a+b+2-x; 1) \equiv \frac{\Gamma(x-a)\Gamma(x-b)}{\Gamma(x+1)\Gamma(x-a-b-1)} .
\tag{4.52}
$$

Proof. The first task of the proof is to establish that the series is divergent for $\Re x > 0$ and convergent for $\Re x < 0$. To resolve this issue, we investigate the large k behaviour of the terms in the series. With the aid of Stirling's formula or approximation for the gamma function, viz. No. 6.1.37 in Ref. [30] or No. 8.327 in Ref. [21], we find that the quotient of k-dependent gamma functions in $_2\mathcal{F}_1(a+1, b+1; a+b+2-x; 1)$ behaves as

$$
\begin{aligned}
\frac{\Gamma(k+a+1)}{\Gamma(k+1)} \frac{\Gamma(k+b+1)}{\Gamma(k+a+b+2-x)} &\overset{k\to\infty}{\sim} \frac{(k+a+1)^{k+a+1/2}}{(k+1)^{k+1/2}} \\
&\times \frac{(k+b+1)^{k+b+1/2}}{(k+a+b+2-x)^{k+a+b+3/2-x}} e^{x-1} \sim k^{x-1} .
\end{aligned}
\tag{4.53}
$$

A series of the form $\sum_{k=N}^{\infty} k^{x-1}$ is similar to the Dirichlet series form for the Riemann zeta function. Consequently, it is absolutely convergent for $\Re x < 0$, while it is divergent for $\Re x \geq 0$. Thus, for the series in the proposition to yield a finite value it must be regularised when $\Re x > 0$.

We continue the proof by writing the integral representation of the beta function as

$$
B(\alpha, \beta) = \frac{\Gamma(\alpha)\Gamma(\beta)}{\Gamma(\alpha+\beta)} = \int_0^1 dt\, t^{\alpha-1}(1-t)^{\beta+\gamma-1}(1-t)^{-\gamma} .
\tag{4.54}
$$

To ensure that the integral is convergent, we assume that $\Re \alpha$ and $\Re \beta > 0$. Since $t < 1$, we can treat $(1-t)^{-\gamma}$ as the limit for the convergent form of the binomial series. Then introducing the binomial series we find that

$$B(\alpha, \beta) = \frac{\Gamma(\alpha)\Gamma(\beta)}{\Gamma(\alpha+\beta)} = \int_0^1 dt\, t^{\alpha-1}(1-t)^{\beta+\gamma-1}\, {}_1\mathcal{F}_0(\gamma; t)$$
$$= \sum_{k=0}^{\infty} \frac{\Gamma(k+\gamma)}{\Gamma(\gamma)\, k!} \frac{\Gamma(k+\alpha)\,\Gamma(k+\beta)}{\Gamma(k+\alpha+\beta+\gamma)}\ . \tag{4.55}$$

Alternatively, the above equation can be written as

$$B(\alpha, \beta) = \frac{\Gamma(\alpha)\Gamma(\beta+\gamma)}{\Gamma(\alpha+\beta+\gamma)}\, {}_2F_1(\alpha, \gamma; \alpha+\beta+\gamma; 1)\ . \tag{4.56}$$

If we put $\gamma = a+1$, $\alpha = b+1$ and $\beta = -x$, then we arrive at

$${}_2\mathcal{F}_1(a+1, b+1; a+b+2-x; 1) = \frac{\Gamma(-x)\,\Gamma(a+b+2-x)}{\Gamma(a+1-x)\Gamma(b+1-x)}\ . \tag{4.57}$$

The first result in the proposition follows when the reflection formula for the gamma function, viz. Eq. (3.2), is introduced into Eq. (4.57).

Eq. (4.57) is a standard result in the theory of Gauss hypergeometric functions, but it is only valid for $\Re x < 0$. Now we consider $\Re x > 0$, where the series representation for the above ${}_2F_1$ function is divergent as shown earlier in the proof. Returning to Eq. (4.54), we could have written it also as

$$B(\alpha, \beta) = \frac{\Gamma(\alpha)\Gamma(\beta)}{\Gamma(\alpha+\beta)} = \int_0^1 dt\, t^{\alpha-1}(1-t)^{\beta-\gamma-1}\big(1+t/(1-t)\big)^{-\gamma}$$
$$= \int_0^1 dt\, t^{\alpha-1}(1-t)^{\beta-\gamma-1}\, {}_1\mathcal{F}_0\big(\gamma; -t/(1-t)\big)\ . \tag{4.58}$$

The variable $t/(1-t)$ in the binomial series ranges from 0 to ∞. However, because of the minus sign outside the variable, the binomial series is conditionally convergent as demonstrated by the proof for Proposition 1. Hence, the equals sign has been retained above. Introducing the binomial series yields

$$B(\alpha, \beta) = \frac{\Gamma(\alpha)\Gamma(\beta)}{\Gamma(\alpha+\beta)} \equiv \sum_{k=0}^{\infty} (-1)^k \frac{\Gamma(k+\gamma)}{\Gamma(\gamma)\, k!} \int_0^1 dt\, t^{\alpha+k-1}(1-t)^{\beta-\gamma-k-1}. \tag{4.59}$$

The equivalence symbol now appears instead of an equals sign to compensate for the impropriety of interchanging the integration and summation without ensuring absolute convergence. As a consequence, the resulting series must be divergent, which occurs when $\Re(\alpha+\beta) > 1$. Furthermore, the resulting integral, which represents the integral representation of the beta function, will, if not initially, become divergent since k ranges from zero to infinity. Because of the equivalence symbol, however, we can replace the integral by its finite part. Hence, we arrive at

$$B(\alpha,\beta) = \frac{\Gamma(\alpha)\Gamma(\beta)}{\Gamma(\alpha+\beta)} \equiv \sum_{k=0}^{\infty}(-1)^k\frac{\Gamma(k+\gamma)}{\Gamma(\gamma)\,k!}\,\frac{\Gamma(k+\alpha)\Gamma(\beta-k-\gamma)}{\Gamma(\alpha+\beta-\gamma)} \ . \qquad (4.60)$$

By introducing the reflection formula for the gamma function into the above equivalence, we find that

$$_2\mathcal{F}_1(\gamma,\alpha;1+\gamma-\beta;1) \equiv \frac{\Gamma(\beta)\Gamma(\alpha+\beta-\gamma)}{\Gamma(\alpha+\beta)\Gamma(\beta-\gamma)} \ . \qquad (4.61)$$

Finally, if we put $\gamma = b+1$, $\alpha = a+1$ and $\gamma-\beta = a+b+1-x$, then we obtain the result given in the proposition. In addition, the condition $\Re(\alpha+\beta) > 0$ becomes $\Re x > 0$, which completes the proof of this proposition.

The regularised value in Proposition 2 has an interesting history. It first appeared without proof on p. 15 of Dingle's book [3]. Because of its importance in the construction of exponentially improved asymptotic solutions of linear differential equations, see for example p. 53 of Ref. [10], there have been several attempts to prove this puzzling result [31, 33]. All these attempts have neglected the divergent remainder and thus, do not employ regularisation. They must surely be fallacious since the series is clearly divergent, yet the quotient of the gamma functions on the rhs is finite. In particular, Olver in Ref. [33] is aware that the series is convergent for $\Re x < 0$, but is unable to provide an explanation as to why the regularised value of the series is different from the limit of the convergent series. This conundrum has been resolved here by introducing the key concept of regularisation.

The quotient of gamma functions in the regularised value for Proposition 2 arises in the asymptotic expansions of the confluent hypergeometric function $U(a;a-b+1;z)$, which includes the family of Bessel functions. For example, the Macdonald function has the following integral representation:

$$K_\nu(z) = \sqrt{\frac{\pi}{2}}\,\frac{z^\nu e^{-z}}{\Gamma(\nu+1/2)}\int_0^\infty dy\,e^{-zy}y^{\nu-1/2}(1+y/2)^{\nu-1/2} \ , \qquad (4.62)$$

where from No. 8.432(3) of Ref. [21], $\Re v > -1/2$ and $|\arg z| < \pi/2$ or $\Re z = 0$ when $v = 0$. If we introduce Proposition 1 into Eq. (4.62), then we obtain

$$K_v(z) \equiv K_{1/2}(z) \sum_{k=0}^{\infty} \frac{\Gamma(k-v+1/2)}{\Gamma(1/2-v)\,k!} \frac{\Gamma(k+v+1/2)}{\Gamma(v+1/2)} \left(-\frac{1}{2z}\right)^k . \qquad (4.63)$$

According to Rule 1 in the previous chapter, the Stokes sectors for the asymptotic series are defined by $2l\pi < \arg(-1/z) < 2(l+1)\pi$. If we take $-1/z$ to be $1/ze^{i\pi}$, then the Stokes sectors are given by $-(2l+3)\pi < \arg z < -(2l+1)\pi$, where l is any integer. However, the above result has been obtained under the condition that $|\arg z| < \pi/2$. This means that the only Stokes sector over which it is valid is the $l = -1$ sector. If we had taken $-1/z$ to be $1/ze^{-i\pi}$, then the Stokes sectors would have been given by $-(2l+1)\pi < \arg z < -(2l-1)\pi$, which means that the $l = 0$ Stokes sector encompasses the sector of validity. Although the Stokes sectors and hence, the asymptotic forms are different, they, nevertheless, indicate that the above asymptotic expansion is valid for $|\arg z| < \pi$.

By putting $x = k$, $a = -v - 1/2$ and $b = v - 1/2$ in Proposition 2, we find that

$$\frac{\Gamma(k+v+1/2)}{\Gamma(k+1)} \Gamma(k+1/2-v) \equiv \Gamma(k)\,{}_2\mathscr{F}_1(1/2-v, 1/2+v; 1-k; 1)$$

$$= \Gamma(k) - \left(\frac{1}{4} - v^2\right)\Gamma(k-1) + \left(\frac{1}{4} - v^2\right)\left(\frac{9}{4} - v^2\right)\Gamma(k-2) + \cdots , \qquad (4.64)$$

where $k > 0$. For large values of k, the above result is often truncated after a few terms and introduced into the asymptotic expansions of Bessel functions and related functions such as the Airy function. The advantage of this is that the asymptotic expansions can then expressed in terms of Dingle's terminants. In Ref. [11] Olver truncates the rhs of Equivalence (4.64) at the first term and introduces the resulting expression into Equivalence (4.63). The resulting asymptotic form for $K_v(z)$ becomes the platform for his "mathematically rigorous" description of Stokes smoothing, which is discussed in Ch. 6.

There are problems that the practitioners of Equivalence (4.64) or its more general counterpart of the regularised value in Proposition 2 have overlooked. The first is that whenever a truncated form is introduced into an asymptotic expansion possessing the quotient of gamma functions, it can only yield an approximation to the original function. On p. 3 of Ref. [12] Berry refers to this approach as "the asymptotics of the asymptotics". Secondly, since k has to be large in order to make this replacement, $|z|$ also has to be large so as to ensure that an optimal point of truncation exists. As a consequence, this method of replacing the quotient of gamma functions by its corresponding divergent series in accordance with Proposition 2 is not valid for $|z| < 1$. On the other hand, if one includes all the terms on the rhs of Equivalence (4.64), then because k is continually being decremented, one will obtain gamma function values with negative integers, which yield infinity. That is, Equivalence (4.63) becomes

$$\frac{K_V(z)}{K_{1/2}(z)} - 1 \equiv \frac{\cos(\pi v)}{\pi} \left(\sum_{k=1}^{\infty} \left(-\frac{1}{2z} \right)^k \Gamma(k) \sum_{l=0}^{\infty} \frac{\Gamma(l+v+1/2)}{\Gamma(v+1/2)\,l!} \right.$$
$$\times \frac{\Gamma(l+1/2-v)}{\Gamma(1/2-v)} \left(\frac{1}{2z} \right)^l + \sum_{k=1}^{\infty} \left(-\frac{1}{2z} \right)^k \Gamma(1-k)$$
$$\times \left. \sum_{l=0}^{\infty} \frac{\Gamma(l+v+1/2)}{\Gamma(v+1/2)} \frac{\Gamma(l+1/2-v)}{\Gamma(1/2-v)\,l!} \left(\frac{1}{2z} \right)^l \right) . \qquad (4.65)$$

The second sum over k contains infinities due to the presence of $\Gamma(1-k)$, which are inconspicuous when Equivalence (4.64) is truncated after the first few terms. It might be argued, however, that these terms do not contribute since regularisation removes all infinities. If this is the case, then we can neglect the second k series and let $R_V(z)$ represent the regularised value of the rhs of Equivalence (4.63). Hence, we would expect $R_V(z)$ to yield the exact values for the quantity on the lhs of Equivalence (4.65). Thus, neglecting the second k series gives

$$R_V(z) \equiv -\frac{\cos(\pi v)}{2\pi z} \Lambda_0(1/2z) \sum_{k=0}^{\infty} \frac{\Gamma(k+v+1/2)}{k!\,\Gamma(v+1/2)} \frac{\Gamma(k+1/2-v)}{\Gamma(1/2-v)} \left(\frac{1}{2z} \right)^k ,$$
$$(4.66)$$

where according to Eq. (3.4), for $|\arg z| < \pi$, $\Lambda_0(1/2z)/2z$ is given by

$$(2z)^{-1} \Lambda_0 \left(\frac{1}{2z} \right) = \int_0^{\infty} dy \, \frac{e^{-y}}{y+2z} = e^{2z}\Gamma(0,2z) = -e^{2z}\text{Ei}(-2z) , \qquad (4.67)$$

and $\text{Ei}(z)$ is the exponential integral. The interesting feature of the above equivalence is that when we consider all the terms in the first k series, we return to the same type of series in Equivalence (4.63) except for a change of phase. Hence, we have not really gained much by employing the result in Proposition 2. If we write $1/z$ in the series as $-1/ze^{i\pi}$, then we can replace the series by its regularised value of $K_v(ze^{i\pi})/K_{1/2}(ze^{i\pi})$ subject to the condition that $|\arg(ze^{i\pi})| < \pi$. In addition, since $|\arg z| < \pi$, we find that

$$R_v(z) = -\frac{e^{2z}}{\pi} \Gamma(0, 2z) \cos(\pi v) \frac{K_v(ze^{i\pi})}{K_{1/2}(ze^{i\pi})} \ , \qquad (4.68)$$

for $-\pi < \arg z < 0$.

The question we now ask is whether the above expression gives identical values to those on the lhs of Equivalence (4.65). If we put $v = 3/8$ and $z = (3/7)\exp(i\pi/6)$, then we find that $R_v(z) = -0.074\,688\,622 + 0.037\,306\,804\,i$, whereas the lhs of Equivalence (4.65) equals $-0.067\,586\,459 + 0.023\,463\,75\,i$ to nine decimal places. Hence, we see that $R_v(z)$ does not give the same values as the quantity on the lhs of Equivalence (4.65), which means that the second k series on the rhs of Equivalence (4.65) needs to be included in the analysis if we wish to obtain the exact values for the quantity on the lhs. Our aim, however, has been to demonstrate that when one evaluates the regularised value of the first series, which goes well beyond truncating the first few terms as the practitioners of Equivalence (4.64) invariably do, one still does not obtain the exact values for the quantity on the lhs of Equivalence (4.65). Finding a regularisation technique that can be applied to the second k series represents a formidable challenge, which is beyond the scope of this book. In addition, we have seen that replacing the regularised value in the coefficient of a divergent power series by another divergent series is fraught with danger.

There is another problem that arises when one endeavours to analyse the asymptotics of $K_v(z)/K_{1/2}(z)$ in terms of the series in Equivalence (4.66). Because the coefficients are all of the same sign and homogeneous in phase, the series is similar in form to the second type of terminant. From the previous chapter we have seen that the second type of terminant has a different behaviour across the Stokes line of $\arg z = 0$ with the emergence of jump discontinuities determined from semi-residue contributions of the Cauchy integral for the series. This fact has also been overlooked by the practitioners of Proposition 2. By using the material in Ch. 10 of this book and adopting a similar approach to the derivation of Eq. (7.92), we find that the regularised value for the series in Equivalence (4.66), is given by

$$_2\mathcal{F}_0\left(v + \frac{1}{2}, \frac{1}{2} - v; \frac{1}{2z}\right) \equiv \frac{K_v(ze^{-i\pi})}{K_{1/2}(ze^{-i\pi})} + \frac{2iz\,K_v(z)K_{1/2}(z)}{\Gamma(1/2 + v)\Gamma(1/2 - v)} \ , \qquad (4.69)$$

for $0 < \arg z < \pi$, while for $-\pi < \arg z < 0$, we find that

$$
{}_2\mathcal{F}_0\left(v+\frac{1}{2},\frac{1}{2}-v;\frac{1}{2z}\right) \equiv \frac{K_v(ze^{i\pi})}{K_{1/2}(ze^{i\pi})} - \frac{2iz\,K_v(z)K_{1/2}(z)}{\Gamma(1/2+v)\,\Gamma(1/2-v)}. \qquad (4.70)
$$

Nevertheless, the inclusion of the extra terms in the above regularised values for the series will still not yield the values given by the lhs of Equivalence (4.65).

The power series appearing in Propositions 1 and 2 and the geometric series are different from those appearing in Equivalences (4.63) and (4.66) and in the asymptotic results of the previous chapters. This is because the former possess a finite radius radius of absolute convergence and hence, belong to the family of ${}_pF_{p-1}$ hypergeometric functions when convergent. As a consequence, they do not display asymptotic behaviour according to the Poincaré definition as described on p. 151 of Ref. [6]. In particular, the remainder only diverges when $|z| > 1$ and for these values of z, there is no optimal point of truncation. Although these series still need to be regularised for $\Re z > 1$, their regularised values are not subject to discontinuous jumps as we have seen in the asymptotic forms for the function $u(a)$ in Ch. 2. Thus, only divergent power series with zero radius of absolute convergence are affected by the Stokes phenomenon. Since the Stokes phenomenon means that the regularised value of an asymptotic series varies over different sectors in the complex plane, the regularisation process for asymptotic series is different from that applied to power series expansions for the functions belonging to the ${}_pF_{p-1}$ family of hypergeometric functions. In the following chapters our primary aim will be to develop an understanding of how asymptotic series are affected by the regularisation process, which will be considerably more complicated than the introductory examples that have been presented in this chapter.

4.4 Logarithmic Regularisation

To conclude this chapter, we discuss the regularisation of a series possessing a logarithmic infinity in its remainder, which like asymptotic series need to be regularised differently from the examples discussed in this chapter. A typical example of such a series is the harmonic series $S(N) = \sum_{k=N}^{\infty} 1/k$. In order to regularise this series, we shall follow the same approach of employing the regularised value of the geometric series as we did when regularising the binomial series. First, we replace z by $-z$ in Equivalence (4.7) and then integrate over z between zero and z, which gives

$$\sum_{k=0}^{\infty} \frac{(-1)^{k+1}}{k+1} z^{k+1} \begin{cases} \equiv \ln(1+z) \;, & \Re z \le -1 \;, \\ = \ln(1+z) \;, & \Re z > -1 \;. \end{cases} \tag{4.71}$$

From this result we see that for $z=1$ the series is conditionally convergent with a limit of $\ln 2$, a standard result in the mathematical literature as given on p. 36 of Ref. [6]. On the other hand, the series is not only divergent for $z=-1$, but also yields a regularised value with a logarithmic infinity, which, in turn, needs to be removed. Unlike the previous examples in this chapter such as the geometric series, we cannot just subtract or cancel the logarithmic infinity because we have effectively integrated a singularity.

The $z=-1$ case of Eq. (4.71) is a famous divergent series. It was first regularised by Euler, who discovered that the removal of the logarithmic infinity yielded a constant that bears his name today [34]. Specifically, he found that

$$\sum_{k=0}^{\infty} \frac{1}{k+1} - \ln\infty = \gamma = 0.577\,215\,664\,901\dots \;. \tag{4.72}$$

Because of this result we see that Euler's constant cannot be obtained by merely subtracting a logarithmic infinity from the $z=-1$ case of Eq. (4.71) since the finite value would simply vanish. However, the problem of obtaining the constant can be resolved via Borel summation. That is, we multiply the summand by $(k-1)!/(k-1)!$ and introduce the integral representation for the gamma function in the numerator. Then the harmonic series becomes

$$\sum_{k=1}^{\infty} \frac{1}{k} = \int_0^{\infty} dt\, e^{-t} \sum_{k=1}^{\infty} \frac{t^{k-1}}{k!} = \int_0^{\infty} \frac{dt}{t} \left(1 - e^{-t}\right) \;. \tag{4.73}$$

We know from Eq. (4.71) that the above integral yields a logarithmic infinity. If we express this as the integral $\int_1^{\infty} dt\, t^{-1}$ and subtract it from the integral in Eq. (4.73), then after a change of variable we come up with the following integral:

$$I = \int_0^1 dt \left(\frac{1}{\ln t} + \frac{1}{1-t}\right) \;, \tag{4.74}$$

which according to No. 8.367(6) in Ref. [21] is equal to γ.

Euler's regularisation formula provides us with a method or scheme for obtaining the regularised value of a logarithmically divergent series such as $S(N)$ given above. We simply subtract the harmonic series from the original series and add Euler's constant. Hence, the regularised value of $S(N)$ is simply equal to $\gamma - H_{N-1}$, where the harmonic number $H_{N-1} = \sum_{k=1}^{N-1} 1/k$.

Another fascinating property of Eq. (4.72) that does not apply to the other regularised values calculated in this chapter is that we can replace infinity in Eq. (4.72) by a large integer and still come up with an accurate approximation to γ. That is,

$$\sum_{k=0}^{N} \frac{1}{k+1} - \ln N \approx \gamma \ . \tag{4.75}$$

As N increases, the value evaluated on the lhs of the above result becomes a more accurate approximation to γ. In fact, this is the standard approach for determining numerical values of γ. In the other examples of this chapter we could not possibly expect to obtain an accurate approximation to the regularised value by truncating an infinite series to a large number N and then subtracting N instead of infinity.

Euler's constant is a unique example of a mathematical quantity that has been evaluated first by employing the concept of regularisation. In an interesting development, it has recently been discovered that γ can be expressed as a more rapidly converging series than the harmonic series involving an infinite set of numbers A_k referred to as the reciprocal logarithm numbers in Ref. [35], but whose magnitudes according to p. 137 of Ref. [36] are called either the Gregory or Cauchy numbers. This result for γ, which is known as Hurst's formula, is given by

$$\gamma = \sum_{k=1}^{\infty} \frac{(-1)^{k+1}}{k} A_k \ , \tag{4.76}$$

where $A_0 = 1$, $A_1 = 1/2$, $A_2 = -1/12$, and

$$A_k = \frac{(-1)^k}{k!} \int_0^1 dt\, \frac{\Gamma(k+t-1)}{\Gamma(t-1)} \ . \tag{4.77}$$

By using the properties of Volterra functions and the orthogonality of Laguerre polynomials, Apelblat finds on p. 138 of Ref. [36] that the A_k can also be expressed as

$$A_k = (-1)^k \int_0^{\infty} dx\, \frac{1}{(x+1)^k(\pi^2 + \ln^2 x)} \ , \tag{4.78}$$

for $k \geq 1$. If this result is introduced into Hurst's formula and the order of the integration and summation interchanged, then we obtain a new integral representation for Euler's constant by using the lower form of Equivalence (4.71). Hence, we arrive at

$$\gamma = -\int_0^\infty \frac{dx}{\ln^2 x + \pi^2} \, \ln\left(\frac{x}{x+1}\right) . \tag{4.79}$$

CHAPTER 5

Asymptotics for the Error Function

Abstract. In this chapter the asymptotic forms for the error function and the related function $u(a)$ over the principal branch of the complex plane are regularised via Borel summation. The resulting equations are expressed in terms of a Stokes multiplier, which toggles between -1/2 and 1/2 for the different Stokes sectors. Numerical studies are then conducted for large and small values of the magnitude of the variable, viz. $|z|$, over the entire principal branch. For the large values of $|z|$ the truncated series is the dominant contribution which is consistent with standard asymptotics. Although the truncated series dominates for small values of $|z|$, so does the regularised value of its remainder in the opposite sense. Hence, when both contributions are combined, the remaining contribution with the Stokes multiplier can become substantial. Nevertheless, in each case where all the contributions are summed, one always obtains the exact values of the error function. Then an expression for the Stokes multiplier is obtained. By carrying out an extensive numerical analysis in the vicinity of the Stokes line along the positive real axis, it is found that irrespective of the value of variable, the Stokes multiplier is discontinuous and not smooth as implied by the leading order term.

Although Dingle states in Ch. 1 of Ref. [3] that the asymptotics of the error function provided one of the earliest examples of the Stokes phenomenon, we have seen in Ch. 2 that Stokes actually studied the asymptotics of the related function $u(a)$ as given by Eq. (2.1) over various sectors of the complex plane for a. Since the aim of this chapter is to substantiate the theory of terminants and to resolve its deficiencies by demonstrating that Borel summation represents a regularisation technique, we shall study, as Dingle does, the asymptotic forms for the error function rather than those for $u(a)$. In order to accelerate this study, however, we shall use the latter's asymptotic forms to derive the asymptotic forms for the error function rather than derive them anew by employing an asymptotic method.

To obtain the asymptotic forms for the error function, we simply set $a = iz$ in Equivalence (2.13) and divide by $i\sqrt{\pi}\exp(z^2)$. This yields

$$\operatorname{erf}(z) \equiv -\frac{e^{-z^2}}{\pi z}\sum_{k=0}^{\infty}\Gamma(k+1/2)\left(-\frac{1}{z^2}\right)^k + 2S\ , \qquad (5.1)$$

where the Stokes multiplier is now given by

$$
S = \begin{cases} 1/2 \ , & (2l-1/2)\pi < \arg z < (2l+1/2)\pi \ , \\ 0 \ , & \arg z = (l-1/2)\pi \ , \\ -1/2 \ , & (2l-3/2)\pi < \arg z < (2l-1/2)\pi \ . \end{cases} \tag{5.2}
$$

The $l=0$ result of the first form and the $l=1$ results of the second and third forms of the above equivalence appear in Ch. 1 of Dingle's book [3] prior to the introduction of the rules for the Stokes phenomenon, which we have discussed in Ch. 3. The only difference between Dingle's results and the corresponding forms in Equivalence (5.1) is that the equivalence symbol has replaced the equals sign. Hence, the above forms give the regularised value as $\arg z$ moves across the Stokes sectors. In addition, the asymptotic series appearing in both sets of results represents a specific case of the first type of terminant given by Eq. (3.3), namely $\alpha = 1/2$ and $N = 0$. As discussed in Ch. 3, we can obtain a convergent integral representation for this terminant via Borel summation. Before this convergent integral representation can be related to the error function, however, we need to demonstrate that Borel summation represents a method of regularising a divergent series.

The first step in Borel summation is to replace the gamma function in the remaining terms of the truncated asymptotic series by its integral representation. In the second step one interchanges the order of the summation and integration. Therefore, the asymptotic series appearing in Equivalence (5.1) can be expressed as

$$
\sum_{k=0}^{\infty} \Gamma(k+1/2) \left(-\frac{1}{z^2}\right)^k = \sum_{k=0}^{N-1} \Gamma(k+1/2) \left(-\frac{1}{z^2}\right)^k
$$
$$
+ \left(-\frac{1}{z^2}\right)^N \int_0^{\infty} dt \, e^{-t} \, t^{N-1/2} \sum_{k=0}^{\infty} \left(-\frac{t}{z^2}\right)^k . \tag{5.3}
$$

According to p. 78 of Whittaker and Watson [6], interchanging the order of the operations is only valid if the series is uniformly convergent, which is not the case here. Nevertheless, since we began with a divergent asymptotic series on the lhs, it is valid to interchange the order of the operations provided we acknowledge that the final term on the rhs of Eq. (5.3) is still divergent. Furthermore, from the previous chapter we have seen that the geometric series is absolutely convergent provided that $|t/z^2| < 1$. Since t ranges from zero to infinity, this means that the radius of absolute convergence for the geometric series on the rhs of Eq. (5.3) is zero. This is the major difference between the series given above and those studied in the first two propositions of the previous chapter, which were absolutely convergent within the unit circle of the complex plane. However, the series on the rhs of Eq. (5.3) is conditionally convergent whenever $\Re(1/z^2) > 0$, although its limit is the same as the regularised value when the series is divergent for $\Re(1/z^2) < 0$. Hence, we can replace the series in the above result by its regularised value.

If we introduce the regularised value for the geometric series into the rhs of Eq. (5.3), then we obtain

$$\sum_{k=N}^{\infty} \Gamma(k+1/2) \left(-\frac{1}{z^2}\right)^k \equiv \left(-\frac{1}{z^2}\right)^{2N} \int_0^{\infty} dt\, \frac{e^{-t}\, t^{N-1/2}}{1+t/z^2} \ . \tag{5.4}$$

Note the appearance of the equivalence symbol in the above result, since the rhs is now finite, while the lhs can still be divergent. As stated already, Equivalence (5.4) represents the $\alpha = -1/2$ case for the first type of terminant discussed in Ch. 3. With the aid of No. 3.383(10) in Ref. [21], the integral on the rhs of the above result can be written as

$$\int_0^{\infty} dt\, \frac{e^{-t}\, t^{N-1/2}}{1+t/z^2} = z^{2N+1}\, e^{z^2}\, \Gamma(N+1/2)\Gamma(1/2-N,z^2) \ , \tag{5.5}$$

where $|\arg z^2| < \pi$ and $\Gamma(a,z)$ is one of the forms for the incomplete gamma function. It should be noted that although the above integral converges, it grows rapidly or diverges as N increases, particularly for values of $|z|$ far away from the limit point of infinity for the series, viz. for $|z| < 1$. For large values of $|z|$, the integral decreases until the optimal point of truncation is reached and once N exceeds this point, it begins to diverge again. As a consequence of Equivalence (5.4), we can re-write Equivalence (5.1) as either

$$\mathrm{erf}(z) = -\frac{e^{-z^2}}{\pi z} \left(\sum_{k=0}^{N-1} \Gamma(k+1/2) \left(-\frac{1}{z^2}\right)^k + \left(-\frac{1}{z^2}\right)^N \right.$$
$$\left. \times \int_0^{\infty} dt\, \frac{e^{-t}\, t^{N-1/2}}{1+t/z^2} \right) + 2S \ , \tag{5.6}$$

or

$$\mathrm{erf}(z) = -\frac{e^{-z^2}}{\pi z} \sum_{k=0}^{N-1} \Gamma(k+1/2) \left(-\frac{1}{z^2}\right)^k - \frac{(-1)^N}{\pi}$$
$$\times\ \Gamma(N+1/2)\Gamma(1/2-N,z^2) + 2S \ . \tag{5.7}$$

The above results have become equations, not equivalence statements, because we have undone the impropriety due to employing the method of iteration in Ch. 2. However, they may not necessarily be correct, which can be seen if we put $N=0$. Then the final version of Eq. (5.7) yields

$$\Gamma(1/2,z^2) = \sqrt{\pi}\left(2S - \mathrm{erf}(z)\right) \ . \tag{5.8}$$

This result agrees with No. 8.359(3) of Ref. [21] when $2S=1$, a result that was originally taken from Erdelyi et al [37] , where it also appears without proof. Unfortunately, when $\pi/2 < |\arg z| < 3\pi/2$, $2S=-1$ in Eq. (5.8), while for $\arg z = \pi/2$, $2S=0$. For these values of z, $2S$ still equals unity in the result given in Refs. [21] and [37]. If the version in the latter references is correct, then it appears to imply that the Stokes phenomenon does not exist at all! Furthermore, we shall soon demonstrate conclusively that S jumps from $-1/2$ to $1/2$ across a Stokes line when carrying out numerical studies of the asymptotic forms of a related function, viz. the error function of imaginary argument. The asymptotic series of this function has a Stokes line situated at $\arg z = 0$. Despite this, however, it does not necessarily mean that No. 8.359(3) in Ref. [21] is incorrect. The reason is that we have not established the sector of the complex plane over which Equivalence (5.4) is valid. For example, the result seems to be valid for the Stokes sector of $|\arg z| < \pi/2$ since for these values of z, $2S=1/2$, which is in accordance with No. 8.359(3) in Ref. [21]. However, outside this sector it may be that the regularised value of the series on the lhs of Equivalence (5.4) acquires extra terms in addition to the integral, which is also in accordance with the Stokes phenomenon. These extra terms could result in $2S$ being always equal to unity in Eq. (5.8), thereby validating No. 8.359(3) of Ref. [21]. Consequently, we require a theory of asymptotic series that will yield its regularised value via Borel summation over all the Stokes sectors in the complex plane. This issue is addressed in Ch. 9, where we derive Borel-summed regularised values for generalisations of the two types of terminant presented in Ch. 3.

It should also be mentioned that Eq. (5.5) is no longer valid when $|\arg z| > \pi/2$. To see this more clearly, we replace z by $-z$ in the integral, thereby obtaining

$$\int_0^\infty dt\ \frac{e^{-t}\,t^{N-1/2}}{1+t/(-z)^2} = (-z)^{2N+1}\,e^{z^2}\,\Gamma(N+1/2)\Gamma\left(1/2-N,(-z)^2\right) \ . \tag{5.9}$$

The condition for this result becomes $\pi/2 < |\arg z| < 3\pi/2$. If we put $N=0$ and treat the lhs of Eq. (5.9) as the regularised value of the asymptotic series in Equivalence (5.4), then we find that

$$\sum_{k=N}^{\infty} \Gamma(k+1/2) \left(-\frac{1}{z^2} \right)^k \equiv -ze^{z^2}\Gamma(1/2)\Gamma\left(1/2,(-z)^2\right) \ . \tag{5.10}$$

Since $\Gamma(1/2,(-z)^2) = \sqrt{\pi} - \sqrt{\pi}\mathrm{erf}(-z)$, Equivalence (5.10) becomes

$$\mathrm{erf}\, z \equiv -\frac{e^{-z^2}}{\pi z} \sum_{k=0}^{\infty} \Gamma(k+1/2) \left(-\frac{1}{z^2} \right)^k - 1 \ . \tag{5.11}$$

Equivalence (5.4) demonstrates that it is not Borel summation, which is responsible for yielding a convergent result to a divergent series as is often reported in the literature. That is, the procedure of introducing the integral representation for the gamma function and interchanging the order of summation and integration is not responsible for yielding a convergent value for a divergent series. Instead, it is the replacement of the geometric series by its regularised value that produces a convergent value for the asymptotic series of the error function. This point has been missed by Dingle in his theory of terminants. Therefore, Proposition 1, which presents a generalisation of the regularised value of the geometric series by analysing the binomial series, is a very important result since the regularised value of a divergent series can be obtained if we can express it in terms of either the geometric or binomial series. Because this step occurs when we carry out Borel summation of a terminant, the equals sign appearing in Eqs. (3.3) and (3.5) must be replaced by the equivalence symbol. Then the equations become what they should have been when they were presented in Ref. [3], equivalences. Thus, we have seen that Borel summation is a mathematical technique for regularising divergent series.

When $\arg z = \pm\pi/2$, the asymptotic series in Equivalence (5.1) becomes a single-sign series resembling the second type of terminant. Moreover, according to the first rule in Ch. 3, we have encountered Stokes lines. If the asymptotic series in Eq. (5.1) is Borel-summed for $z = |z|\exp(\pm i\pi/2)$, then one obtains

$$\sum_{k=N}^{\infty} \Gamma(k+1/2) \left(\frac{1}{|z|^2} \right)^k \equiv \frac{1}{|z|^{2N}} \int_0^{\infty} dt\, \frac{e^{-t}t^{N-1/2}}{1-t/|z|^2}$$

$$= -|z|^{2-2N} \int_C ds\, \frac{e^{-s}s^{N-1/2}}{s-|z|^2} \ , \tag{5.12}$$

where the final member is a Cauchy integral with the contour of integration C being the positive real axis. Denoting this integral as $I(|z|)$, we see that it is an improper integral due to the singularity at $s = |z|^2$ in the denominator. The problem with evaluating the contour integral is one does not know whether to evaluate the semi-residue contribution in a clockwise or anti-clockwise direction. Furthermore, regardless of which direction is taken, the contribution from the semi-residue is imaginary. Yet, the asymptotic series only possesses real terms, which according to p. 407 of Ref. [3] dictates that the Cauchy principal value should be taken. Hence, the corrected version of Equivalence (5.12) is

$$\sum_{k=0}^{\infty} \Gamma(k+1/2)|z|^{-2k} \equiv \sum_{k=0}^{N-1} \Gamma(k+1/2)|z|^{-2k}$$
$$+ \frac{1}{|z|^{2N}} P \int_0^{\infty} dt \, \frac{e^{-t} t^{N-1/2}}{1 - t/|z|^2} \, . \tag{5.13}$$

From Ch. 3 we have seen that when $|z|$ in the above series is replaced by the complex variable z, we need to include the contribution due to the residue on the Stokes line because the terminant is now of the second type. Whenever this type of asymptotic series is derived as an asymptotic expansion for a function, it is done under the condition that the variable is real and positive initially. If, however, we wish to extend the asymptotic form into the complex plane, then we need to be aware that we are starting from the outset on a Stokes line, whereas for the first type of terminant z is already situated within a Stokes sector. The difference is that for the second type of terminant we experience Stokes discontinuities as soon as the argument of the variable in the asymptotic series changes, but for the first type of terminant we can continue with the same asymptotic form until the argument of the variable in the asymptotic series encounters a Stokes line.

The residue of the Cauchy integral in Equivalence (5.12) when $|z|$ is replaced by z is given by

$$\mathrm{Res}\,\{I(z)\} = -z^{2N-1} e^{-z^2} \, . \tag{5.14}$$

For $\arg z \neq 0$, Dingle argues that the Cauchy integral in Equivalence (5.13) is not the sole contribution to the regularised value of the asymptotic series, but that one must include the appropriate contribution due to the residue at $t = z^2$. If $\arg z$ is an infinitesimal distance above the positive real axis, then the contour on the positive real axis will pass under the singularity that is now situated at $t = z^2$. That is, the semi-residue contribution is evaluated in an anti-clockwise direction. Therefore, by this slight movement the Cauchy integral has acquired a contribution, which was previously removed when the principal value was taken for $\arg z = 0$. To maintain consistency, Dingle argues that this semi-residue contribution must be subtracted from the integral representation, which gives the regularised value for this infinitesimal movement from the Stokes line. Then the regularised value will maintain this form until the next Stokes line is encountered at $\arg z^2 = 2\pi$. Hence, for $0 < \arg z^2 < 2\pi$, we find that

$$\sum_{k=N}^{\infty} \Gamma(k+1/2)\, z^{-2k} \equiv \frac{1}{z^{2N}} \int_0^{\infty} dt\, \frac{e^{-t}\, t^{N-1/2}}{1-t/z^2} + \pi i z\, e^{-z^2}\ , \qquad (5.15)$$

which is consistent with the sixth rule in Ch. 3. For the case of $-2\pi < \arg z^2 < 0$, the same argument applies except that the singularity is situated below the positive real axis. This means that the contour C will pass over the pole in a clockwise direction if it is situated an infinitesimal distance below the positive real axis. Although the semi-residue contribution is again subtracted from the integral representation, it will now be the negative of the final term in Equivalence (5.15). Therefore, summarising Dingle's peculiar, but insightful, argument, we arrive at

$$\sum_{k=N}^{\infty} \Gamma(k+1/2)\, z^{-2k} \equiv \begin{cases} z^{-2N} I(z,N) + \pi i z\, e^{-z^2}, & 0 < \arg z < \pi, \\ z^{-2N} P\, I(z,N)\ , & \arg z = 0, \\ z^{-2N} I(z,N) - \pi i z\, e^{-z^2}, & -\pi < \arg z < 0. \end{cases} \qquad (5.16)$$

where

$$I(z,N) = \int_0^{\infty} dt\, \frac{e^{-t}\, t^{N-1/2}}{1-t/z^2}\ , \qquad (5.17)$$

and P denotes that the Cauchy principal value of the integral should be evaluated.

As discussed in Ch. 3, Dingle considers the more general case of terminants, where the asymptotic series possesses coefficients $\Gamma(k+\alpha)$ and variable z instead of $\Gamma(k+1/2)$ and z^2 as in the above asymptotic series. It is a simple exercise to obtain the corresponding results for the more general case, which are given on p. 412 of Ref. [3] or in terms of $1/z$ by Equivalence (3.12). Again, it should be stressed that all these results should have the equivalence symbol relating both sides of the statement rather than an equals sign. Hence, the generalisation of

Equivalence (5.16) results in

$$\sum_{k=N}^{\infty} \Gamma(k+\alpha)z^{-k} \equiv \begin{cases} z^{-N}I(z,N,\alpha)+i\pi z^{\alpha}e^{-z} \;, & 0 < \arg z < 2\pi, \\ z^{-N}P\,I(z,N,\alpha) \;, & \arg z = 0, \\ z^{-N}I(z,N,\alpha)-i\pi z^{\alpha}e^{-z} \;, & -2\pi < \arg z < 0, \end{cases} \tag{5.18}$$

where now

$$I(z,N,\alpha) = \int_0^{\infty} dt\, \frac{e^{-t}\,t^{N+\alpha-1}}{1-t/z} \;. \tag{5.19}$$

From the above result we are able to observe how Rules 7 and 8 in Ch. 3 have evolved. It was also mentioned that the eighth rule could be modified so that the asymptotic form of a divergent series along a Stokes line is given by the average of the two asymptotic forms in the adjoining Stokes sectors. However, an extra qualification was required in order to make this modification. From Equivalence (5.19) we see that the average of the two results for the adjacent Stokes sectors does give the regularised value along the Stokes line of $\arg z = 0$ provided that we take the principal value of the resulting Cauchy integral, which is the qualification we need to make when invoking Rule 8a.

To convince the reader of the correctness of Equivalence (5.16) or its more general counterpart given by Equivalence (5.18), we consider a known function whose asymptotic forms are composed of the asymptotic series on the lhs of the equivalence. Then we can compare the values generated by the rhs of the equivalence with the corresponding values for the function. The function that we choose for this numercial study is the error function of imaginary argument $\mathrm{erfi}(z)$, which is defined as

$$\mathrm{erfi}(z) = -i\,\mathrm{erf}(iz) = i\,\mathrm{erf}(-iz) = \frac{2}{\sqrt{\pi}} \int_0^z dt\, e^{t^2} \;. \tag{5.20}$$

We can obtain the asymptotic form for this function along the Stokes line at $\arg z = 0$ by introducing the second form of Equivalence (5.16) into the above equation. This yields

$$\mathrm{erfi}(z) \equiv \frac{e^{z^2}}{\pi z} \sum_{k=0}^{\infty} \frac{\Gamma(k+1/2)}{z^{2k}} \equiv \frac{e^{z^2}}{\pi z} \sum_{k=0}^{N-1} \frac{\Gamma(k+1/2)}{z^{2k}}$$

$$+ \frac{e^{z^2}}{\pi z^{2N+1}} P \int_0^{\infty} dt\, \frac{e^{-t}\,t^{N-1/2}}{1-t/z^2} \;. \tag{5.21}$$

In this extraordinary result the series in the middle is divergent, while the expressions to either side are forms for the regularised value of the series. That is, if we remove the middle expression, then $\mathrm{erfi}(z)$ is equal to the expression on the rhs. Hence, the statement becomes an equation. By using the other forms of Equivalence (5.16) we obtain the asymptotic forms for $\mathrm{erfi}(z)$ above and below the positive real axis, which are

$$\mathrm{erfi}(z) = \frac{e^{z^2}}{\pi z} \sum_{k=0}^{N-1} \frac{\Gamma(k+1/2)}{z^{2k}} + \frac{e^{z^2}}{\pi z^{2N+1}} \int_0^\infty dt\, \frac{e^{-t} t^{N-1/2}}{1 - t/z^2} + 2iS \ . \qquad (5.22)$$

The Stokes multiplier S in Eq. (5.22) is given by

$$S = \begin{cases} 1/2 \ , & 0 < \arg z < \pi \ , \\ -1/2 \ , & -\pi < \arg z < 0 \ . \end{cases} \qquad (5.23)$$

For $\arg z = 0$, we have $S = 0$, but then we must only evaluate the Cauchy principal value of the integral as in Eq. (5.21). We shall refer to the last term in Eq. (5.22) as the Stokes discontinuous term or simply the Stokes discontinuity. Furthermore, since $u(a) = \sqrt{\pi}\exp(-a^2)\mathrm{erfi}(a)$, we also find that

$$u(a) = \frac{1}{\sqrt{\pi}\,a} \sum_{k=0}^{N-1} \frac{\Gamma(k+1/2)}{a^{2k}} + \frac{1}{\sqrt{\pi}\,a^{2N+1}} \int_0^\infty dt\, \frac{e^{-t} t^{N-1/2}}{1 - t/a^2}$$
$$+ \ 2\sqrt{\pi}\,i e^{-a^2} S \ . \qquad (5.24)$$

The result given above for $\mathrm{erfi}(z)$ ought to be compared with the standard asymptotics procedure where the asymptotic series in Equivalence (5.21) is truncated after a few terms in the large $|z|$ or $z \to \infty$ limit. Thus, one would normally write Eq. (5.22) as

$$\mathrm{erfi}(z) \sim \frac{e^{z^2}}{\sqrt{\pi}\,z} \left(1 + \frac{1}{2z^2} + \frac{3}{4z^4} + \cdots \right) \ , \qquad (5.25)$$

for $|\arg z| < \pi/2$. On other occasions, one might find $O(z^{-6})$, which is referred to as a Landau gauge symbol in Refs. [27] and [49], appearing in the result instead of $\cdots$. Comparing this result with the exact result given by Eq. (5.22) we see that the remainder of the series whose regularised value is represented by the integral in Eq. (5.22) has been neglected as is customary in standard asymptotics.

Nevertheless, for this result to yield good approximations to $\mathrm{erfi}(z)$, $|z|$ must be significantly greater than unity, ie. $|z| \gg 1$. In addition, we note that the Stokes discontinuity has been neglected even though according to Eq. (5.23) it is different for $0 < \arg z < \pi/2$ and $-\pi/2 < \arg z < 0$. Hence, it is simply not correct to state that above result is valid for $|\arg z| < \pi/2$. Of course, the reason why this can be done is that the Stokes discontinuity is very small or subdominant to the leading terms of the asymptotic series in these halves of the Stokes sectors and hence, can be neglected as is again customary in standard asymptotics. Neglecting the Stokes discontinuity yields the same asymptotic result over the first halves of the Stokes sectors as indicated above. Of course, when the Stokes discontinuity dominates in the other halves of the Stokes sectors, i.e. $\pi/2 < |\arg z| < \pi$, it has to be included and then the above result is no longer valid in standard asymptotics.

For those readers, who are not entirely convinced that Eq. (5.22) is a valid representation for the error function of imaginary argument or are still unable to appreciate the implication of the key concept of regularisation to asymptotic forms, let us consider two numerical examples. In the first example we shall assume that $|z|$ is "large". Therefore, we shall put $z = 3\exp(i\theta)$ and consider various values of θ over the principal branch of the complex plane. Such a value of $|z|$ is sufficiently large to ensure that an optimal point of truncation exists. In actual fact, the optimal point of truncation represents the first term in the asymptotic series of Equivalence (5.21) that is larger than its preceding term or the first value of N, which we denote as N_T, when

$$\left| \frac{\Gamma(N+3/2)}{\Gamma(N+1/2)\, z^2} \right| > 1 \ . \tag{5.26}$$

Therefore, N_T is the first integer greater than $|z|^2 - 1/2$. Alternatively, this means that a non-zero optimal point of truncation exists whenever $|z^2| > 3/2$. For $N < N_T$, the truncated series in Eq. (5.22) is expected to dominate the contribution due to the integral, which, as has already been stated, represents the regularised value for the remainder of the series.

Table 5.1 presents values of $\mathrm{erfi}(z)$ and the various contributions on the rhs of Eq. (5.22) for different values of θ with the truncation parameter N set equal to 5. In actual fact, far more values of θ as well as other values of N were considered. Nevertheless, the sample in the table are deemed sufficient to demonstrate the validity of Eq. (5.22) across the Stokes sectors. These results were obtained by using Mathematica 4.1 [19] on a Pentium IV computer. In particular, the values of

Table 5.1: Values of erfi(3 exp(iθ) and the various contributions on the rhs of Eq. (5.22) with N=5

θ		
	Truncated Series	$-119.237\,895\,578\,773 + 398.690\,937\,676\,295\,i$
	Remainder	$0.153\,297\,288\,569 - 0.256\,359\,541\,387\,i$
$-3\pi/34$	Stokes Discontinuity	$-i$
	Sum	$-119.084\,598\,290\,203 + 397.434\,578\,134\,907\,i$
	erfi(3 exp(-3πi/34))	$-119.084\,598\,290\,203 + 397.434\,578\,134\,906\,i$
	Truncated Series	$1\,628.106\,331\,997\,973$
	Remainder	$1.888\,290\,603\,591$
0	Stokes Discontinuity	0
	Sum	$1\,629.994\,622\,601\,564$
	erfi(3)	$1\,629.994\,622\,601\,565$
	Truncated Series	$-0.001\,850\,050\,647 + 0.000\,829\,800\,252\,i$
	Remainder	$4.419\,316\,466\,901 \times 10^{-7} - 6.031\,051\,604\,965 \times 10^{-7}\,i$
$2\pi/3$	Stokes Discontinuity	i
	Sum	$-0.001\,849\,608\,715 + 1.000\,830\,403\,357\,i$
	erfi(3exp(2πi/3))	$-0.001\,849\,608\,715 + 1.000\,830\,403\,357\,i$

the Borel-summed integral or remainder of the asymptotic series for non-zero values of θ were evaluated by using the numerical integration routine in the software package called NIntegrate. For $\theta=0$, the Cauchy principal value was evaluated by a special add-on routine called CauchyPrincipalValue. Both integration routines are limited by the machine precision of the computer and thus, 16 figure accuracy is the best that one can hope for with a Pentium computer. As stated in the introduction, the limitation in accuracy can be overcome by using the Mathematica 7.0, although the results can take much longer to compute. To avoid the possibility of expending too much CPU time, both AccuracyGoal and PrecisionGoal in the integration routines were set to a lower value of 14, while WorkingPrecision was set to the maximum value of 16. In addition, the options, MinRecursion and MaxRecursion, were set to 3 and 12 respectively. As a consequence, all the results in the table were evaluated in only a few CPU secs. Due to limited space the results in the table have been presented to 12 decimal places despite being for the most part more accurate than this number of decimal places.

Rows 2 to 6 of the table present the results of the various terms in Eq. (5.22) for

$z = 3\exp(-3\pi i/34)$, which lies in the Stokes sector of $-\pi < \arg z < 0$. According to Eq. (5.23), the Stokes multiplier is equal to -1/2. The second row in the table displays the value of the truncated series or the first term on the rhs of Eq. (5.22). The next value is the regularised value for the remainder of the series or the integral on the rhs of Eq. (5.22). As expected, for this value of z the truncated series dominates the contribution from the integral. In fact, the Stokes discontinuity, which appears in the next row, also dominates the contribution from the integral. The fifth row presents the value obtained by summing all three contributions, which should give the value of erfi(z) according to Eq. (5.22). The sixth row presents the actual value obtained by calling the function directly in Mathematica. As can be seen, the sum of the contributions agrees to 15 decimal places with the actual value of $\text{erfi}(3\exp(-3\pi i/34))$.

The corresponding results for $\theta = 0$ appear in the next five rows of the table. As stated previously, the regularised value of the remainder has been obtained by using the routine called CauchyPrincipalValue, which is identical to NIntegrate except that one must specify the pole at $t = 9$ in the routine. This routine has become redundant in Versions 6.0 and 7.0 of Mathematica, which is discussed in Ch. 11. As for the case of $\theta = -3\pi/34$, the truncated series dominates the regularised value for the remainder of the asymptotic series. However, for this value of $\arg z$, $S = 0$ and hence, the Stokes discontinuity vanishes. Thus we find that the sum of the truncated series and regularised value of the remainder yields the value of erfi(3) to 15 decimal places.

The last set of values in the table are those for $\theta = 2\pi/3$, which according to Eq. (5.23) means that the Stokes multiplier S is now equal to 1/2. Because this value of z lies across the positive imaginary axis or centreline of the Stokes sector, sometimes called an anti-Stokes line, the Stokes discontinuity will dominate the sum of the truncated series and the regularised value of the remainder. As expected, the regularised value of the remainder is significantly less than the value of the truncated series, but when all three contributions are summed, they yield a value that agrees with the actual value of $\text{erfi}(3\exp(2\pi i/3))$ to thirteen significant figures.

In the examples presented in Table 5.1, z was sufficiently large that the remainder was very small compared with the truncated series. This means that the first few terms of Approximation (5.25) will yield reasonably accurate results, although nothing like the accuracy of the results in Table 5.1. Now we consider a small value of $|z|$, where the contribution from the remainder is significant. Thus, we put $|z| = 1/4$. In this case there is no optimal point of truncation and the integral in

Eq. (5.22) will diverge rapidly as N increases from zero. In addition, since z is said to lie outside the region of applicability, Approximation (5.25) is no longer useful or valid. In actual fact, although the truncated series diverges, the integral in Eq. (5.22) diverges in the opposite direction so that when the two are summed with the Stokes discontinuity, they yield the original value of the function. However, there is a reduction in accuracy because a cancellation of decimal places occurs in the process of eliminating the divergence in the truncated sum and in the integral or regularised value of the remainder. This is disturbing because our computing system possesses limited precision, which may result in the value obtained for the rhs of Eq. (5.22) being significantly less accurate than the actual value of erfi(z). Therefore, to obtain the most accurate values for the original function when there is no optimal point of truncation in the asymptotic form, it is recommended that N be kept close to zero.

Table 5.2: Values of erfi($\exp(i\theta)/4$) and the various contributions on the rhs of Eq. (5.22) with N=2

θ		
	Truncated Series	$8.411\,088\,102\,650 - 12.470\,813\,891\,771\,i$
	Remainder	$-8.457\,217\,753\,552 + 13.198\,024\,093\,143\,i$
$-5\pi/9$	**Stokes Discontinuity**	$-i$
	Sum	$-0.046\,129\,650\,902 - 0.272\,789\,798\,628\,i$
	erfi(exp(-5πi/9)/4)	$-0.046\,129\,650\,902 - 0.272\,789\,798\,628\,i$
	Truncated Series	$21.620\,760\,676\,767$
	Remainder	$-21.332\,677\,056\,972$
0	**Stokes Discontinuity**	0
	Sum	$0.288\,083\,619\,794$
	erfi(1/4)	$0.288\,083\,619\,794$
	Truncated Series	$12.376\,250\,572\,370 - 16.728\,208\,121\,633\,i$
	Remainder	$-12.108\,250\,101\,855 + 15.829\,889\,405\,839\,i$
$\pi/9$	**Stokes Discontinuity**	i
	Sum	$0.268\,000\,470\,515 + 0.101\,681\,283\,705\,i$
	erfi(exp(πi/9)/4)	$0.268\,000\,470\,515 + 0.101\,681\,283\,705\,i$

In a similar manner to Table 5.1, Table 5.2 presents the various contributions on the rhs of Eq. (5.22) for erfi($\exp(i\theta)/4$) with a different set of values for θ. In view of the discussion in the previous paragraph the truncation parameter N was set equal to 2. The results confirm that the truncated series represents a very poor approximation to the values of $\mathrm{erfi}(\exp(i\theta)/4)$. Hence, it is critical in the region of the complex plane where Approximation (5.25) is no longer applicable that the regularised value for the remainder of the asymptotic series in the asymptotic forms for $\mathrm{erfi}(z)$ cannot be neglected and must be calculated as accurately as possible.

For $\theta = -5\pi/9$ the Stokes multiplier is equal to -1/2 according to Eq. (5.23). In addition, this value of z lies across the negative imaginary axis, another anti-Stokes line. Therefore, the Stokes discontinuity dominates the combined contribution of the truncated series and the regularised value of the remainder. For z on the Stokes line, viz. $\theta = 0$, the Stokes discontinuity is zero. Hence, we see that the sum of the truncated series and the Cauchy principal value for the regularised remainder yields the value of $\mathrm{erfi}(1/4)$ to 12 decimal places. This is one instance where 14 figure accuracy could not be maintained due to the loss of two decimal places in adding the separate contributions. The final value of $\theta = \pi/9$ was chosen to be close to the positive real axis, which is a Stokes line. According to Eq. (5.23) the Stokes multiplier is equal to 1/2. Once again, we see that Approximation (5.25) cannot yield an accurate value for $\mathrm{erfi}(\exp(i\pi/9)/4)$, but including all the terms on the rhs of Eq. (5.22) does. The interesting property in this example, however, is that the Stokes discontinuity is equally as important as the combined contribution of the truncated series and the regularised value of the remainder even though individually the contributions dominate the Stokes discontinuity.

The last example brings into question the approach adopted by Heading of bisecting Stokes sectors with anti-Stokes lines. According to Heading [5, 8], across these half-sectors the dominance of the component asymptotic terms is switched. That is, the dominant series becomes subdominant, while the previously subdominant series becomes the dominant contribution. This observation is based on the behaviour of the exponential factors outside the series. From the last example in Table 5.2 we have seen that the Stokes discontinuous term, which would be considered subdominant to the asymptotic series according to Heading and Dingle, can provide the dominant contribution to the value of $\mathrm{erfi}(z)$ in the half of the Stokes sector where the sum of the truncated asymptotic series and the regularised value of the remainder is expected to dominate. This is essentially why we have refrained as much as possible from discussing which parts of an asymptotic form are dominant or subdominant.

The results in Tables 5.1 and 5.2 demonstrate clearly that a function can be represented by a complete asymptotic expansion. According to p. 154 of Ref. [6], however, an asymptotic series can be the asymptotic expansion for several distinct functions. Because of this the subject of asymptotics has developed a bad reputation for being vague. This unfortunate conclusion is due mainly to the Poincaré definition, which is basically unable to handle exponentially subdominant terms. As indicated previously, these terms have traditionally been neglected because they are masked by the divergence in the dominant asymptotic series. As a result, this leads to the conclusion that different functions can possess the same asymptotic series. We have already seen an example of this with Approximation (5.25), which gives the asymptotic series for $\mathrm{erfi}(z)$ for $|\arg z| < \pi/4$ when according to Equivalence (5.22), $\mathrm{erfi}(z) - i$ is the regularised value of the asymptotic series for $0 < \arg z < \pi/2$ and $\mathrm{erfi}(z) + i$ is the regularised value of the series for $-\pi/2 < \arg z < 0$. In contrast, we have seen that once the dominant series is regularised, the subdominant terms provide a vital contribution in obtaining the exact values of the original function. Consequently, every function possesses its own complete asymptotic expansion. Furthermore, it is now valid to differentiate an asymptotic expansion provided it is complete and the original function is differentiable. This contradicts the statement made on p. 153 of Whittaker and Watson [6], which has been arrived at because of truncation. Once a complete asymptotic expansion for a function has been differentiated, meaningful values for the derivative of the function can be obtained by regularising the resulting expression.

The Stokes multiplier in Eq. (5.22) can be expressed as

$$ S = \frac{1}{2i} \left(\mathrm{erfi}(z) - \frac{e^{z^2}}{\pi z} \sum_{k=0}^{N-1} \frac{\Gamma(k+1/2)}{z^{2k}} - \frac{e^{z^2}}{\pi z^{2N+1}} \int_0^\infty dt \, \frac{e^{-t} t^{N-1/2}}{1 - t/z^2} \right). \quad (5.27) $$

With this result we can conduct a thorough investigation into the behaviour of a Stokes multiplier very close to a Stokes line, which as mentioned in Ch. 1 has never been accomplished before. Despite the appearance of z in the above result, we shall see that S will only depend upon where z is situated in the complex plane, i.e. it will depend upon the value of $\arg z$, not $|z|$. In conducting this investigation, we shall again consider a relatively "large" value of $|z|$, viz. $|z| = 10/3$, and a small one, viz. $|z| = 1/5$.

Table 5.3 presents a sample of the results obtained for the first value of $|z|$ with θ or $\arg z$ ranging from $\pm\pi/100000$ to $\pm\pi/10$. Here the truncation parameter N has been set equal to 5. Like the results in Tables 5.1 and 5.2 these results have also been obtained by using the NIntegrate routine in Mathematica except for the value on the Stokes line, i.e. $\theta = 0$, which has been evaluated by using the CauchyPrin-

Table 5.3: Values of the Stokes multiplier in Eq. (5.27) for $z=10\,\exp(i\theta)/3$ and N=5

θ	Stokes Multiplier
$-\pi/10$	$-0.500\,000\,000\,000\,085 + 1.136\,868\,377\,216\,160 \times 10^{-12}\,i$
$-\pi/100$	$-0.500\,000\,000\,008\,185 - 8.185\,452\,315\,956\,354 \times 10^{-12}\,i$
$-\pi/1000$	$-0.500\,000\,000\,002\,728 - 2.910\,383\,045\,673\,370 \times 10^{-11}\,i$
$-\pi/10000$	$-0.499\,999\,999\,999\,602 - 2.091\,837\,814\,077\,735 \times 10^{-11}\,i$
$-\pi/100000$	$-0.499\,999\,999\,999\,641 + 7.275\,957\,614\,183\,426 \times 10^{-12}\,i$
0	$0.0 - 1.273\,292\,582\,482\,099 \times 10^{-11}\,i$
$\pi/100000$	$0.499\,999\,999\,999\,641 + 7.275\,957\,614\,183\,426 \times 10^{-12}\,i$
$\pi/10000$	$0.499\,999\,999\,999\,602 - 2.091\,837\,814\,077\,735 \times 10^{-11}\,i$
$\pi/1000$	$0.500\,000\,000\,002\,728 - 2.910\,383\,045\,673\,370 \times 10^{-11}\,i$
$\pi/100$	$0.500\,000\,000\,008\,185 - 8.185\,452\,315\,956\,354 \times 10^{-12}\,i$
$\pi/10$	$0.500\,000\,000\,000\,085 + 1.136\,868\,377\,216\,160 \times 10^{-12}\,i$

cipalValue routine. One can gain an idea of the numerical errors involved in these calculations from the magnitudes of the complex parts in the values for the Stokes multiplier, which should be zero. For θ equal to $-\pi/1000$ and $-\pi/10000$, Mathematica gave an error warning that NIntegrate was converging too slowly due to a singularity or insufficient WorkingPrecision. This leads one to suspect that these values of θ were so small that the integration routine was being affected by the pole on the positive real axis. On other occasions, the NIntegrate routine was unable to converge after a set number of recursive bisections, but amazingly, still produced accurate results such as that for $\theta = -\pi/100000$. In these cases it appears that $|\theta|$ was so small, the integral was almost turning into an integral along the positive real axis. Hence, the singularity was again affecting the calculations.

Table 5.4 presents a sample of the results for $|z| = 1/5$ using the same values of θ as in the previous table. In this case the truncation parameter N was set equal to 2. In obtaining these results a greater number of error warnings occurred. Despite this, however, the complex parts of the values obtained for the Stokes multiplier were significantly lower than those than in the previous table.

As stated in the introduction, the accuracy of the results in Tables 5.3 and 5.4 can be greatly improved by using Mathematica 7.0. If WorkingPrecision is extended to 200 and AccuracyGoal and PrecisionGoal to 150 in the NIntegrate routine, then it is found that for θ equal to $-\pi/10$ to $-\pi/1000$, Mathematica 7.0 gives a value

Table 5.4: Values of the Stokes multiplier in Eq. (5.27) for $z=\exp(i\theta)/5$ and N=2

θ	Stokes Multiplier
$-\pi/10$	$-0.499\,999\,999\,999\,994 - 2.289\,834\,988\,289\,385 \times 10^{-15}\,i$
$-\pi/100$	$-0.499\,999\,999\,999\,998 - 1.047\,772\,979\,489\,991 \times 10^{-14}\,i$
$-\pi/1000$	$-0.499\,999\,999\,999\,999 - 1.200\,428\,645\,375\,950 \times 10^{-14}\,i$
$-\pi/10000$	$-0.500\,000\,000\,000\,000 + 1.203\,204\,202\,937\,513 \times 10^{-14}\,i$
$-\pi/100000$	$-0.500\,000\,000\,000\,105 - 2.038\,438\,862\,150\,826 \times 10^{-12}\,i$
0	$0.0 - 3.053\,113\,317\,719\,180 \times 10^{-16}\,i$
$\pi/100000$	$0.500\,000\,000\,000\,105 - 2.038\,438\,862\,150\,826 \times 10^{-12}\,i$
$\pi/10000$	$0.500\,000\,000\,000\,000 + 1.203\,204\,202\,937\,513 \times 10^{-14}\,i$
$\pi/1000$	$0.499\,999\,999\,999\,999 - 1.200\,428\,645\,375\,950 \times 10^{-14}\,i$
$\pi/100$	$0.499\,999\,999\,999\,998 - 1.047\,772\,979\,489\,991 \times 10^{-14}\,i$
$\pi/10$	$0.499\,999\,999\,999\,994 - 2.289\,834\,988\,289\,385 \times 10^{-15}\,i$

of -0.5 followed by 200 zeros and a complex term of the order of 10^{-200} for the Stokes multiplier. For $\theta = -\pi/1000$ it gives a value of -0.5 followed by over a hundred zeros and a complex value of the order of 10^{-104} for the Stokes multiplier. For the positive values of θ we obtain the same results except that the real part is positive beginning with 0.5. For $|\theta| < 10^{-6}\pi$ the Stokes multiplier is inaccurate as the routine is unable to converge after recursive bisections due to the presence of the singularity in the integral.

The results in Tables 5.3 and 5.4 plus the significantly more accurate results obtained from Mathematica 7.0 as described in the previous paragraph demonstrate clearly that the Stokes multiplier S for the asymptotic forms of the error function of imaginary argument is discontinuous across a Stokes line. For $-\pi < \theta < 0$, S equals -1/2, while for $0 < \theta < \pi$, it equals 1/2. For the Stokes line of $\theta = 0$, it simply vanishes. These results hold no matter what value $|\theta|$ takes within a Stokes sector. Nor are these results affected by how large $|z|$ is. Therefore, the results vindicate Eq. (5.23), which means that the behaviour of S is entirely consistent with the conventional view of the Stokes phenomenon discussed in the introductory chapter with $S_- = -1/2$. On the other hand, these results contradict the two main contemporary views of the Stokes phenomenon, but before this can be seen clearly, these rather different views need to be presented to the reader. This is the subject of the following chapter.

CHAPTER 6

Contemporary Views of the Stokes Phenomenon

Abstract. This chapter presents the two main contemporary views of the Stokes phenomenon. In the first view known as Stokes smoothing, it is claimed that rather than experiencing discontinuities at specific rays in the complex plane, an asymptotic expansion, when magnified on a suitable scale, is a rapidly smoothed function. Here we show that this fallacious conclusion has arisen because an asymptotic expansion for the Stokes multiplier has been truncated, thereby giving the misleading impression that it is equal to a term involving the error function when it is in fact only the leading term of a complicated expression that needs to be regularised. Based on resurgence analysis, the second view bears very little semblance to the original discovery made by Stokes. In this view the Stokes lines become analytic curves that are determined by setting the real part of the action in the one-dimensional Schrödinger equation to zero due to a strange interpretation of maximal dominance in a complete asymptotic expansion. Hence, the resulting asymptotic forms are no longer uniform over specific sectors of the complex plane in marked contrast to the conventional view of the Stokes phenomenon.

In this chapter the two main contemporary views of the Stokes phenomenon are described. The first of these has become known as Stokes smoothing. Although it is not in accordance with Stokes's original conception of the phenomenon, it will be analysed extensively because it appears to have been embraced by most of the asymptotics community ever since its introduction by Berry [9] in 1989. The second contemporary view of the Stokes phenomenon has evolved out of Ecalle's theory of resurgent functions [41, 43], but as we shall see, it bears even less semblance to Stokes's conception and the developments carried out later by Heading and Dingle. Surprisingly, its proponents claim their view of Stokes lines and regions is identical to that perceived by both Stokes and Dingle. Yet in this view Stokes sectors are no longer separated by specific rays emanating from the origin but are now bordered by complicated curves, which only in very simple cases reduce to sectors in the complex plane. Another drawback with the second view is that one cannot carry out a numerical investigation designed to verify the purported behaviour of the Stokes curves generated by it. As a consequence, we shall not delve too deeply into this esoteric view. Instead, interested readers will be directed to the extensive material associated with the view so that they can make a final decision.

6.1 Stokes Smoothing

So far, the survey presented in Chps. 1 to 3 has covered the conceptual understanding of the Stokes phenomenon up to the appearance of Dingle's book [3] in the mid-1970s. For the next decade or so, there was little new material on the phenomenon. Whilst it could be argued that the conventional view became entrenched in this period, it could also be argued that due to the lack of new material the phenomenon remained almost as arcane as when it was discovered by Stokes. Then came the revolutionary paper by Berry [9] in 1989, in which it was claimed that the conventional view of the Stokes phenomenon was fictitious. Instead of giving rise to jump discontinuities, an asymptotic expansion, it was said, undergoes a smooth, but rapid, transition across a Stokes line. Shortly afterwards, Olver [11] presented a "more rigorous mathematical approach" to substantiate the rapid smoothing postulated by Berry. Strangely, the smoothing near a Stokes line has become known as Stokes smoothing despite the fact that Stokes never held such a view. Largely as a result of these two papers, Stokes smoothing seems to have been embraced by the asymptotics community as discussed in Ch. 6 of Ref. [10] and in Ref. [44].

It should be pointed out that if smoothing does occur near Stokes lines, then it has major ramifications to Dingle's theory of terminants, which is based on the fact that the jump discontinuity occurring at a Stokes line emanates from the residue of a singular integral. It would imply that there are terms are missing in Equivalences (3.12) and (5.18). In order to determine these terms, we would require a new theory of terminants. Furthermore, it should be emphasised that despite being accepted whole-heartedly by the asymptotics community, Stokes smoothing has never really been tested or studied at the level of accuracy of the numerical results presented in the previous chapter. There, we observed that the Stokes multiplier given by Eq. (5.27) in the vicinity of a Stokes line was discontinuous not only to the high machine precision of our computing system, but also by employing Mathematica 7.0 with a WorkingPrecision of 200 and both PrecisionGoal and AccuracyGoal set to 150. For the selected values of $|z|$ in the second numerical example of the previous chapter, in particular the smaller value of 1/5, we would have expected to have observed smoothing by obtaining different values for the Stokes multiplier other than -1/2 and 1/2 for the values of θ or $\arg z$ closest to zero.

In his study of the two (J)WKB or phase-integral solutions of the one-dimensional Helmholtz equation given by

$$\partial_Z^2 y = k^2 Q(Z) y \ , \tag{6.1}$$

where $\Re Q > 0$, Berry found via Borel summation that the Stokes multiplier of the subdominant solution is given by

$$S_n(F) = \frac{1}{2} - \frac{i}{2\pi} \, P \int_0^\infty dt \, \frac{t^{n-\beta}}{1-t} e^{F(1-t)} \ . \tag{6.2}$$

In Eq. (6.2) F is known as a singulant, which is a terminology again due to Dingle [3]. It is defined as

$$F = k(\Phi_+ - \Phi_-) \ . \tag{6.3}$$

The $\Phi_\pm$ functions appear in the exponential factors of the asymptotic solutions of Eq. (6.1) and are given by

$$\Phi_\pm = \pm \int_a^Z dz' \sqrt{Q} \ . \tag{6.4}$$

In the above equation a represents a simple zero of $Q^2(Z)$. Furthermore, in obtaining the result for $S_n(F)$, Berry has assumed that the value of S_- mentioned in the introduction is equal to zero. In fact, he mentions that other choices of S_- only differ by real constants. For example, in the previous chapter we found that the Stokes multiplier for the asymptotic forms of the error function jumped or toggled from -1/2 to 1/2 across a Stokes line which, in turn, means that S_- is equal to $-1/2$ in this case. Hence, Berry would argue that the Stokes multiplier for the asymptotic forms of the error function would still be given by Eq. (6.2) except that the term of $1/2$ on the rhs would be removed.

Eq. (6.2), therefore, represents the main result upon which Stokes smoothing is based. Furthermore, by employing "more rigorous mathematics", Olver obtains the same result in Ref. [11] with $\beta = 1$ and $F = 2z$. Consequently, the derivation leading to this result cannot be questioned, which means that the analysis following its derivation in both the papers of Olver and Berry needs to be investigated in great depth. If we adopt Olver's form of Eq. (6.2) and change the variable from t to $1+\tau$, then the Stokes multiplier becomes

$$S_n = -\frac{1}{2\pi i} \int_{-1}^{\infty} \frac{d\tau}{\tau} e^{-A(\tau - \ln(1+\tau))} (1+\tau)^{\mu} e^{-iB\tau} , \tag{6.5}$$

where $2z = -A - iB$, $n - 1 = A + \mu$, A is large, real and positive, n is an integer and B and μ are real. Olver states that by applying Laplace's method, another well-known asymptotic method, one would expect as $A \to \infty$ that the major contribution to the above integral comes from the neighbourhood around $\tau = 0$. As a result, he expands the first exponential in the above integral in powers of τ and extends the lower limit of the integral to $-\infty$.

There are basically two issues concerned with this approach, which produces the same result for S_n as obtained by Berry in Ref. [9]. The first is that we can write Eq. (6.5) as

$$S_n = -\frac{1}{2\pi i} \int_{-\infty}^{\infty} d\tau \, \tau^{-1} e^{-A(\tau - \ln(1+\tau))} (1+\tau)^{\mu} e^{-iB\tau}$$
$$+ \frac{1}{2\pi i} \int_{-\infty}^{-1} d\tau \, \tau^{-1} e^{-A(\tau - \ln(1+\tau))} (1+\tau)^{\mu} e^{-iB\tau} . \tag{6.6}$$

Thus, if we extend the lower limit to $-\infty$, then we must consider the evaluation of a second integral along a branch cut. Because this integral is neglected by Berry and Olver, it means that even if the first integral can be evaluated exactly, the value for S_n is only an approximation. Therefore, Stokes smoothing is predicated upon an approximation, not exact mathematics as has been employed in our study of the Stokes multiplier in the asymptotic expansion for erfi(z) presented in the previous chapter. That is, at no stage did we neglect any term in a series or approximate an integral when evaluating the Stokes multiplier via Eq. (5.27).

If we consider the $A \to \infty$ limit, which is the type of vague statement that gives asymptotics a bad name, and neglect the second integral or even assume that it vanishes, then Eq. (6.6) can be written as

$$S_n \equiv -\frac{1}{2\pi i} \int_{-\infty}^{\infty} d\tau \, \tau^{-1} e^{-A\tau^2/2} \Big(\cos B\tau - i \sin B\tau \Big)$$
$$\times \sum_{k=0}^{\infty} \frac{\Gamma(k-\mu)}{\Gamma(-\mu)k!} (-\tau)^k \sum_{k=0}^{\infty} g_k(A) \tau^k . \tag{6.7}$$

The above result has been obtained by once again applying the asymptotic method of expanding most of the exponential as described on p. 113 of Ref. [3]. As a consequence, the special polynomials, $g_k(A)$, in Equivalence (6.7) are determined from the following equivalence statement:

$$\exp\left(-A\left(\tau - \tau^2/2 - \ln(1+\tau)\right)\right) \equiv \sum_{k=0}^{\infty} g_k(A)\,\tau^k \ . \tag{6.8}$$

The equivalence symbol appears here because there is a finite radius of absolute convergence that applies to the rhs. This issue is discussed in more detail in the appendix. Individually, the $g_k(A)$ can be calculated either by modifying the partition method for a power series expansion, which has been applied to various wide-ranging problems, e.g. see Refs. [18, 35, 38, 39, 40], or via the recursion relation given by

$$k g_k(A) = A \sum_{j=0}^{k-3} (-1)^{k+1-j} g_j(A) \ . \tag{6.9}$$

This result has been derived by differentiating Equivalence (6.8) and equating powers of τ after the power series expansions for the various quantities have been introduced into the lhs. This procedure of turning an equivalence statement into an equation is valid only for those values of τ lying within the radius of absolute convergence for Equivalence (6.8). From Eq. (6.9) we find that $g_0(A)=1$, $g_1(A)=g_2(A)=0$, $g_3(A)=A/3$, $g_4(A)=-A/4$, etc. General k-dependent formulae for the polynomials are derived in the appendix.

Because the rhs of Equivalence (6.8) has been introduced into the first integral in Eq. (6.6), which means that the range of integration is outside the radius of absolute convergence, the equivalence symbol now appears in Equivalence (6.7). However, the derivation of this result is based on assuming that the second integral in Eq. (6.6) vanishes. Since this integral is more than likely to be non-zero, we need to introduce an alternative symbol, perhaps the $\overset{\sim}{=}$ symbol, to indicate that not only does the result on the rhs of Equivalence (6.7) need to be regularised, but once this is accomplished, the resulting value will still only be an approximation to S_n. Furthermore, the $k=0$ terms in both series combine to yield an integral that is technically divergent, but Olver states that the integral is actually a contour integral with a semi-circular indentation of the path above the origin. Therefore, he arrives at

$$-\frac{1}{2\pi i} \int_{-\infty}^{\infty} d\tau\, \tau^{-1} e^{-A t^2/2} \cos B\tau = \frac{1}{2} \ . \tag{6.10}$$

On the other hand, Berry avoids the problem by evaluating the Cauchy principal value and providing the value of $1/2$ as in Eq. (6.2). This, of course, yields the same result obtained by Olver. As a consequence, both authors find that the Stokes multiplier is approximately equivalent to

$$S_n \cong \frac{1}{2} + \frac{1}{2}\operatorname{erf}\left(\frac{B}{\sqrt{2A}}\right) - i\,\frac{e^{-B^2/2A}}{\sqrt{2\pi A}} \sum_{k=0}^{\infty}\left(-\frac{1}{2A}\right)^k H_{2k}\left(\frac{B}{\sqrt{2A}}\right)$$

$$\times \sum_{j=0}^{2k+1}\frac{\Gamma(2k+1-j-\mu)}{\Gamma(-\mu)(2k+1-j)!}(-1)^j g_j(A) - \frac{1}{\sqrt{\pi}}e^{-B^2/2A}\sum_{k=1}^{\infty}\left(-\frac{1}{2A}\right)^k$$

$$\times\ H_{2k-1}\left(\frac{B}{\sqrt{2A}}\right)\sum_{j=0}^{2k}\frac{\Gamma(2k-j-\mu)}{\Gamma(-\mu)(2k-j)!}(-1)^j g_j(A)\ , \tag{6.11}$$

where we have used Nos. 2.5.36.5 and 2.5.36.6 from Ref. [20] and $H_k(x)$ denotes a Hermite polynomial. Since A is large, we can truncate the above result, thereby obtaining

$$S_n \cong \frac{1}{2} + \frac{1}{2}\operatorname{erf}\left(\frac{B}{\sqrt{2A}}\right) - i\,\frac{e^{-B^2/2A}}{\sqrt{2\pi A}}\left(\sum_{j=0}^{1}\frac{\Gamma(1-j-\mu)}{\Gamma(-\mu)(1-j)!}(-1)^j g_j(A)\right.$$

$$\left. - \frac{1}{2A}H_2\left(\frac{B}{2\sqrt{A}}\right)\sum_{j=0}^{3}\frac{\Gamma(3-j-\mu)}{\Gamma(-\mu)(3-j)!}(-1)^j g_j(A)+\cdots\right) - \frac{e^{-B^2/2A}}{\sqrt{\pi}}$$

$$\times\ \left(\frac{1}{4A^2}H_3\left(\frac{B}{\sqrt{2A}}\right)\sum_{j=0}^{2}\frac{\Gamma(2-j-\mu)}{\Gamma(-\mu)(2-j)!}(-1)^j g_j(A)+\cdots\right)\ . \tag{6.12}$$

Introducing specific values for the $g_j(A)$ and the Hermite polynomials, one finds that the above result for the Stokes multiplier becomes

$$S_n \cong \frac{1}{2} + \frac{1}{2}\operatorname{erf}\left(\frac{B}{\sqrt{2A}}\right) + i\,\frac{e^{-B^2/2A}}{\sqrt{2\pi A}}\left(\mu + \frac{1}{3} - \frac{B^2}{3A} + \left(\frac{1}{A} - \frac{B^2}{A^2}\right)\right.$$

$$\left.\times\ \left(\frac{\mu}{3} - \frac{\mu^2}{2} + \frac{\mu^3}{6}\right)+\cdots\right) - \frac{B}{2A}\frac{e^{-B^2/2A}}{\sqrt{2\pi A}}\left(\mu - \mu^2 +\cdots\right)\ . \tag{6.13}$$

The first line up to the $-B^2/3A$ term represents the result obtained by Olver in Ref. [11] for the Stokes multiplier, while Berry is even more restrictive by stating that only the real part dominates. Consequently, his expression for the Stokes multiplier is

$$S_n \sim \frac{1}{2} + \frac{1}{2}\operatorname{erf}\left(\frac{B}{\sqrt{2A}}\right) = \frac{1}{2} - \frac{1}{2}\operatorname{erf}\left(\Im z/\sqrt{\Re(-z)}\right)\ , \tag{6.14}$$

which means that the change in the multiplier is determined by behaviour of the error function. This result also appears as Approximation (6.1.11) in Ref. [10]. Because the error function at $(\pm\infty)$ equals ± 1 and passes smoothly through erf(0)(=0), sometimes very rapidly depending upon how quickly the magnitude of $B/\sqrt{2A}$ increases, Berry and Olver conclude that the Stokes multiplier is not subjected to discontinuities, but undergoes a smooth and rapid transition in the vicinity of a Stokes line.

It is often claimed that the above approximation for the Stokes multiplier is valid for large $|z|$, but in reality it is only valid for $|B/\sqrt{A}|$ becoming large very quickly. Since $B/\sqrt{A}$ is the ratio of the imaginary part of z over the square root of its real part, a rapid smooth transition from 0 to 1 will occur with the above approximation for the Stokes multiplier only when the imaginary part of z quickly dominates the real part outside an extremely narrow neighbourhood around $z=0$. On the other hand, large $|z|$ includes large real and small imaginary values of z and then the above approximation becomes invalid because the transition is no longer rapid.

Once it is realised that Stokes smoothing is only going to occur when the imaginary part of z dominates except for an extremely narrow region, we can compare Approximation (6.14) with Eq. (5.27). As stated previously, the latter has been determined with $S_- = -1/2$, whereas Approximation (6.14) has been determined with $S_- = 0$, which leads to the 1/2 term appearing in the latter. However, our interest is in comparing the error function term in Approximation (6.14) with that in Eq. (5.27), since they both represent the change in the Stokes multiplier. For those values of z, where $|\Im z| \gg 1$, the last two terms of Eq. (5.27) become exponentially small. Thus, the first term becomes the dominant term and we can write Eq. (5.27) alternatively as

$$S \approx \frac{1}{2i}\,\mathrm{erfi}(z) = \frac{1}{2}\,\mathrm{erf}(\pm y - ix) \ , \qquad (6.15)$$

where $z = x \pm iy$. For $y \to \infty$ and x fixed, the above result means that S experiences a smooth transition from $-1/2$ to $1/2$ as in the case of the result obtained for S_n. Therefore, if we make the typical asymptotics statement that $|\Im z| \gg 1$, while $|\Re z|$ is fixed, then the Stokes multiplier as given by Eq. (5.27) becomes approximately the same result as obtained for S_n. Hence, Eq. (5.27) possesses the smoothing behaviour as observed in S_n. Unfortunately, Approximation (6.15) represents a relatively crude approximation since we have already found that when all the terms in Eq. (5.27) are included and calculated to very high precision, there is no Stokes smoothing. That is, the Stokes multiplier is discontinuous much like the Heaviside step-function. Furthermore, this applies to all values of $|z|$ as $\arg z$ crosses the positive real axis from below, not only those where $|\Im z|$ is large.

We have seen from the preceding analysis that the smoothed result for the Stokes multiplier given by Approximation (6.14) arises from truncating Approximation (6.11) after the first few terms with the divergent remainder being neglected. Therefore, before one can say conclusively that Stokes's discontinuity is "an artefact of poor resolution" as Berry claims in Ref. [9], one needs to regularise this remainder, an issue that has not been addressed by either Berry or Olver. Furthermore, even if the rhs of Approximation (6.11) can be regularised, it would only yield the Stokes multiplier for A equal to infinity. For all other values of A, we would need to evaluate the second integral in Eq. (6.6), which has also been neglected in the above derivation of the Stokes multiplier. By including all the neglected terms, one would no longer observe a Stokes smoothing, but a Stokes discontinuity as was the case in the previous chapter. On the other hand, we have observed that to lowest order Eq. (5.27) possesses a similar smoothing term to that obtained by Berry and Olver, but including all the terms in the equation yields a completely different picture. Therefore, the purported smoothing is merely an approximation to a discontinuity caused by the truncation of an asymptotic form.

6.2   Resurgence Analysis

As mentioned in the introduction to this chapter there is another view of the Stokes phenomenon that has evolved out of the theory of resurgent functions pioneered by Ecalle [41, 43]. Although there is a complicated definition, e.g. on p. 122 of Ref. [45], a resurgent function is basically a holomorphic function that is convergent originally within a small disk, but can be extended with endless analytic continuation into the complex plane except for a countable number of singularities. From this broad definition it is not hard to see that the solutions of many singular ordinary and partial differential equations and those of general functional equations are resurgent functions. Yet the proponents of this theory are under the impression that because the power series expansions of such functions can yield subdominant exponential terms via their resummation approach, resurgence analysis is the only viable approach for understanding the Stokes phenomenon. This is despite the fact that in previous chapters we have been able to explain the origin of subdominant exponential terms in asymptotic expansions by using

Dingle's theory of terminants, although one may argue that only simple examples have been considered so far. We shall, however, consider far more intricate examples in later chapters, but before this can be accomplished, we shall be required to develop a much deeper understanding of the Stokes phenomenon. For now, we continue with the survey of contemporary developments into Stokes phenomenon by describing the view from resurgence analysis.

It should be mentioned from the outset that the view of the Stokes phenomenon evolving from resurgence analysis is not very accessible for the novice since the field is littered with jargon. Expressions like resummation direction, Stokes automorphism, Stokes graph, right and left connection isomorphism, extendible hyperfunction and ramifying singularity, to name a few, have turned the relatively simple observation made by Stokes about the behaviour of asymptotic series into a minefield of jargon in mathematical analysis. What is worse, is that none of the proponents in this field has ever been able to provide a numerical demonstration of any major problem in asymptotics at the level of accuracy of the example in the previous chapter. In fact, Delabaere et al state in the introduction of Ref. [46] that the best answer they have to deducing numerical information from purely "formal" computations of asymptotic series is the complicated and inexact hyperasymptotic approach of Berry and Howls [47], which did even not require resurgence analysis initially. Nowadays, of course, it has been incorporated in the subject because the scheme used by Berry and Howls to sum divergent series is similar to the resummation method used in resurgence analysis. Although a description of hyperasymptotics as presented by Berry and Howls in Ref. [47] is outside the scope of this book, the interested reader can pursue the various approaches emanating from their work by examining Refs. [27, 48, 49] and the vast number of references cited therein. It should be mentioned, however, that hyperasymptotics is primarily concerned with improving the accuracy of standard asymptotics by developing strategies for truncating an asymptotic series beyond the optimal point of truncation, whereas in this work we are concerned with evaluating the exact limits resulting from the regularisation of entire asymptotic series.

In view of the statements made in the preceding paragraph, perhaps the best method of introducing resurgence analysis and its deficiencies is to consider the example given in the introductory chapter of Ref. [45]. Once this is done, we can proceed to the descriptions in Refs. [46] and [50]. Therefore, we begin with what Sternin and Shatalov refer to as the Euler example, which amounts to solving the following differential equation:

$$-x^2 \frac{dy}{dx} + y = x \ , \tag{6.16}$$

By introducing $y \simeq \sum_{k=0}^{\infty} \alpha_k z^k$ into the above equation, they obtain the following formal solution:

$$y \simeq \sum_{k=1}^{\infty} (k-1)! \, x^k \ . \tag{6.17}$$

Sternin and Shatalov do not explain what they mean by the word "formal". Nor do they give a reason for introducing the $\simeq$ symbol. Presumably, the $\simeq$ symbol has been introduced to indicate an approximation has been made, while "formal" implies that the solution is divergent. From the material in Ch. 3 we know that the rhs is a specific Type II terminant. In addition, from Ch. 2 we know that the method of iteration employed by Sternin and Shatalov to obtain the above result is an asymptotic method. Hence, the solution to Euler's differential equation is equivalent to the power series expansion on the rhs of the above result, not $\simeq$. Nevertheless, Sternin and Shatalov state that the above result does not yield a solution to the equation because the series diverges. This is indeed very strange analysis. The result they obtain satisfies the original differential equation. Yet, according to them because it is divergent, it cannot be a solution! Worse still, from our study of regularisation we know that the rhs of the above result is not always divergent, namely when $\Re x < 0$.

The general solution to Eq. (6.16) can be obtained by introducing an integrating factor [22] and integrating the resulting equation. Then one finds that

$$y = Ce^{-1/x} - e^{-1/x} \int dx \, x^{-1} e^{1/x} \ . \tag{6.18}$$

By repeatedly integrating the above result by parts, Sternin and Shatalov arrive at

$$y = Ce^{-1/x} + \sum_{k=1}^{N} (k-1)! \, x^k + O\left(|x|^{N+1}\right) \ . \tag{6.19}$$

The interesting feature about this result is that now an equals sign appears due to the introduction of the vague $O()$ symbol. According to standard definition, e.g. p. 2 of Ref. [51], the $O()$ symbol means that there exist positive constants K and δ such that $|y| \leq K|x|^{N+1}$ whenever $|z| < \delta$. All this symbol does, however, is to ensure that one does not truncate beyond the optimal point of truncation, so that the remainder always can be bounded. As a consequence, it can never yield exact values for the original function or solution. Furthermore, it is vague because K and δ are continually changing as $|x| \to 0$. In fact, we know that if we attempt to include all the terms in the series, then the result can become divergent and the equals sign in Eq. (6.19) is simply invalid. On the other hand, if we stop the integration by parts at x^N, then we obtain

$$y = Ce^{-1/x} + \sum_{k=1}^{N} (k-1)!\,x^k + \Gamma(N)\,e^{-1/x} \int dx\, e^{1/x} x^{N-2} \ . \qquad (6.20)$$

In reality, the $O(|x|^{N+1})$ term in Eq. (6.19) conceals the fact that if the integration by parts is carried out continually, then an infinite series is obtained, which we have already indicated is divergent for certain values of x. That is, carrying out the integration by parts is mathematically improper because we are effectively expanding about the singularity at $x = 0$. As an aside, in terms of the key concept of regularisation in Ch. 4, if we extend N to infinity and replace the equals sign by an equivalence symbol, then a comparison of the ensuing statement with Eq. (6.20) yields the following equivalence:

$$\sum_{k=N}^{\infty} (k-1)!\,x^k \equiv \Gamma(N)\,e^{-1/x} \int dx\, e^{1/x} x^{N-2} \ . \qquad (6.21)$$

Since Sternin and Shatalov realise that the asymptotic series is divergent when N is taken to infinity, but is nevertheless a "formal" solution of Eq. (6.16), they point out that the series must be resummed. So they introduce a resummation operator σ, which they define as a homomorphism from the algebra of formal power series to the algebra of analytic functions that can be applied to convergent power series and commutes with differentiation. Then the solution to Eq. (6.18) is re-written as

$$y(x) - \sigma \left[\sum_{k=0}^{\infty} (k-1)!\,x^k \right] \simeq C_1 e^{-1/x} \ . \qquad (6.22)$$

Now this is really untenable mathematics because an operator has been introduced that only affects one part of a mathematical statement and not the other terms. In addition, there does not seem to be any need for the $\simeq$ symbol because this symbol was introduced to indicate that the asymptotic series was divergent. The resummation operator is supposed to remove the divergence in the series.

Sternin and Shatalov are, however, aware that the two terms forming the solution for $y(x)$ are dominant and recessive in different sectors of the complex plane, which they attribute to the Stokes phenomenon. From Ch. 2 we have seen that the Stokes phenomenon arises from the fact that the constants multiplying terms in the solution such as C acquire jump discontinuities at Stokes lines. According to Rule 2 in Sec. 3, at these lines the asymptotic series is at peak dominance over the exponential term and hence, the subdominant term or C acquires a jump discontinuity. We shall return to this issue when the presentation of the Stokes phenomenon via resurgence analysis has been completed. As a consequence, Sternin and Shatalov conclude that in order to study asymptotic expansions of differential equations it is crucial that a theory for resummation of a divergent series must account for the Stokes phenomenon. The basic tool they use in this resummation procedure is the Borel-Laplace transform.

The Borel transform is merely the inverse of the complex Laplace tansform. If a function $F(\zeta)$ is holomorphic in the sector $\theta_1 < \arg \zeta < \theta_2$ and is bounded by $|F(\zeta)| < C_\varepsilon \exp(c|\zeta|)$ for $\theta_1 + \varepsilon < \arg \zeta < \theta_2 - \varepsilon$, where C_ε is some positive constant and $\varepsilon > 0$, then its complex Laplace transform is given by

$$f(x) = \mathcal{L}\left[F(\zeta)\right] = \int_0^{\infty e^{i\theta}} d\zeta\, e^{-\zeta/x} F(\zeta) \ , \tag{6.23}$$

where the integration is carried out over an arbitrary ray within $\theta_1 < \arg \zeta < \theta_2$. The Borel transform of $f(x)$ is defined as

$$\mathcal{B}[f](\zeta) = \mathcal{L}^{-1}[f(x)] = -\frac{1}{2\pi i} \int_C d(1/x)\, e^{1/x} f(x) \ , \tag{6.24}$$

where C is a contour in the domain of convergence for the integral.

Now we are in a position to present the resummation method in the theory of resurgent functions to the Euler example given by Eq. (6.18). By applying the Borel transform to this equation, Sternin and Shatalov find that the equation can be written in terms of the Borel image $Y(\zeta)$ as

$$-\zeta Y(\zeta) + Y(\zeta) = 1 \ . \tag{6.25}$$

In other words, $Y(\zeta) = 1/(1 - \zeta)$. Hence, the particular solution to the Euler example becomes

$$y(x) = \int_0^{\infty e^{i\theta}} d\zeta\, \frac{e^{-\zeta/x}}{1 - \zeta} \ , \tag{6.26}$$

where the integration can be performed along any ray except along the positive real axis due to the singularity at $\zeta = 1$. For the latter situation to be included they state that one has to fix the direction of encircling the singular point in the complex plane and as a result, they denote the Laplace transform as $\mathcal{L}^{\pm}$ depending whether one decides to go in an anti-clockwise or clockwise direction around the pole. This sounds remarkably similar to the description of Dingle's terminants as presented in Ch. 3. Therefore, if an anti-clockwise direction is selected, then $y^{+}(x) = \mathcal{L}^{+}[1/(1-\zeta]$ is defined for $\Re x > 0$. To obtain the analytic continuation to $\Re x < 0$, they state that they must rotate the contour in the complex Laplace transform by an angle of π. The result will, however, be different if the contour is rotated in different directions. That is, if the rotation is carried out in the positive direction, then the analytic continuation of the function denoted by $y^{+}(x)$ is given by

$$\mathcal{A}^{+}[y^{+}(x)] = \int_{0}^{\infty e^{i\pi}} d\zeta \, \frac{e^{-1/x}}{1-\zeta} \,, \tag{6.27}$$

while if the contour is rotated in the opposite direction, then one obtains

$$\mathcal{A}^{-}[y^{+}(x)] = -2\pi i e^{-1/x} + \int_{0}^{\infty e^{i\pi}} d\zeta \, \frac{e^{-1/x}}{1-\zeta} \,. \tag{6.28}$$

The first term on the rhs of Eq. (6.28) has arisen because the contour of integration has picked up a contribution from the residue at $\zeta = 1$, just as in Ch. 3. Therefore, Sternin and Shatalov conclude that the Stokes phenomenon is directly responsible for producing the multi-valued solution obtained by using the Borel-Laplace transform. This is essentially what is meant by resummation of a divergent series in the theory of resurgent functions. In addition, we see that in this theory the direction of resummation leads to different results.

To summarise the above, there are basically three steps in the method of resummation in resurgence analysis. The first is to determine the Borel transform of the asymptotic series, which they claim is divergent everywhere except for a disk of absolute convergence in the neighbourhood of the origin. The series represents the germ or source for determining a holomorphic function. In terms of the above example it was found that

$$\mathcal{B}_{f}\left[\sum_{k=0}^{\infty} k! x^{k+1}\right] = \sum_{k=0}^{\infty} \zeta^{k} \,. \tag{6.29}$$

The second step is analytic continuation of the holomorphic function, the latter being defined everywhere except for some discrete singularities. In the above example, this means that

$$\sum_{k=0}^{\infty} \zeta^k = Y(\zeta) = \frac{1}{1-\zeta} \ . \tag{6.30}$$

Thus, there is only one singularity in this example. The final step is to evaluate the Laplace transform of $Y(\zeta)$ to obtain the exact solution to the Euler equation. Hence, the resummation operator, we have denoted above by σ, can be represented as

$$y(x) = \sigma \left[\sum_{k=0}^{\infty} k! x^{k+1} \right] = \mathcal{L} \circ \mathcal{A} \circ \mathcal{B}_f \left[\sum_{k=0}^{\infty} k! x^{k+1} \right] \ . \tag{6.31}$$

Since it is stated on p. 14 of Ref. [45] that the coefficients of the divergent series a_k must satisfy $|a_k| \le C R^k k!$ for some positive constants C and R, the resummation method will only be successful for holomorphic functions. This is because such functions belong to the family of $_pF_{p-1}$ hypergeometric functions which possess a finite radius of absolute convergence. Thus, the steps involved in converting a divergent series to a convergent integral representation in resurgence analysis are similar to Borel summation of the two types of terminant discussed in Ch. 3. As a consequence, it is no surprise that Dingle's theory of terminants and Berry's hyperasymptotic [47] method can be accommodated within resurgence analysis.

So far, other than a few points in interpretation there has not been a significant difference between the resummation method of resurgence analysis and the material in preceding chapters. It should be pointed out that the presentation has been elementary and the situation becomes far more complicated when the Laplace transform is extended outside the analytic region for the divergent series. For example, consider the Airy equation, which is

$$-h^2 \frac{d^2 u}{dx^2} - x u = 0 \ . \tag{6.32}$$

One of the (J)WKB solutions for $u(x)$ is

$$u(x, h) \equiv \exp(iS_+(x)/h) \sum_{k=0}^{\infty} (-ih)^k c_j x^{-1/4 - 3j/2} \ , \tag{6.33}$$

where $S_+(x) = 2x^{3/2}/3$, $c_k = (6k-1)(6k-5)c_{k-1}/48k$ and $c_0 = 1$. As in the Euler example, Sternin and Shatalov use the $\simeq$ symbol instead of the $\equiv$ symbol in the above result. The function $S_\pm(x)$ is known as the action and has been introduced here to show that the Airy equation is an example of the one-dimensional Schrödinger equation given by Eq. (6.1). In terms of the total energy E and potential energy $V(x)$, this equation is more familiar to physicists in the following form:

$$\frac{h^2}{2m}\frac{d^2 u}{dx^2} + \left(E - V(x)\right)u = 0 \ . \tag{6.34}$$

When solving by the (J)WKB method, one replaces u by $\exp(iS_\pm(x)/h)a(x,h)$ in the above equation, where the action is now given by

$$S_\pm = \pm \int dx \ \sqrt{2m(E - V(x))} \ , \tag{6.35}$$

and $a(x,h)$ becomes an asymptotic series in powers of h. Comparing Eq. (6.34) with Eq. (6.32), we see that $2m(E - V(x))$ has been replaced by x. Thus, we arrive at the (J)WKB solution given by Equivalence (6.33) for the positive branch of the action.

As stated in the previous section, Dingle refers to the difference between the two branches for the action as a singulant F, which is given by Eq. (6.3). In terms of resurgence analysis $S_\pm(x)$ is known as a ramifying function. By applying the resummation operator to the rhs of Equivalence (6.33), Sternin and Shatalov find that

$$u(x,h) = \sigma \left[\exp(iS_+(x)/h) \sum_{k=0}^{\infty} (-ih)^k c_k x^{-1/4-3k/2} \right]$$
$$= \frac{1}{\sqrt{h}} \int_\Gamma ds \ e^{is/h} U(x,s) \ , \tag{6.36}$$

where

$$U(x,s) = \sum_{k=0}^{\infty} \frac{(s - S(x))^{k-1/2}}{2\Gamma(k+1/2)} c_k x^{-1/4-3k/2} \ . \tag{6.37}$$

In Eq. (6.36) Γ is a special contour, which is indented around the ramifying singularities at $S_\pm(x)$. Whilst resurgence analysis avoids having to deal with divergent series by producing convergent series such as Eq. (6.37), there seems to be little benefit in doing so because evaluating the integral in Eq. (6.36) remains a formidable task. Furthermore, convergent power series are generally slowly converging. Hence, a more practical approach is to develop a theory that enables one to obtain limit values from divergent series directly without having to resort to such complicated methods for obtaining a slowly convergent power series.

Now we can discuss the view that those working with resurgence analysis have of the Stokes phenomenon. Surprisingly, this view seems to have gained acceptance by other practitioners in asymptotics, who have little or no connection with resurgence analysis, e.g. see Ch. 3 of Ref. [50]. According to Ref. [46], Stokes lines do not occur where two exponentials exchange their dominance, but where one of them is maximally dominant over the other, i.e. $\exp(iS_+(x,h))$ over $\exp(iS_-(x,h))$. There is actually nothing new here as it is consistent with Rule 2 in Ch. 3, but we shall see that it is in the implementation of the principle that differences arise.

If we replace h by i in Eq. (6.34), then we basically arrive at Eq. (6.1) used by Berry in his study of Stokes smoothing. This can be simplified to

$$y^{''} - Q(z,\lambda)y = 0 \ . \tag{6.38}$$

The zeros of $Q(z,\lambda)$, which we shall denote by $z_0(\lambda)$, are known as turning points. According to Fedoryuk on p. 90 of Ref. [50], Stokes lines are analytic curves whose maximal connected component of the level curve is given by

$$\Re S(z_0,z) = \Re \int_{z_0(\lambda)}^{z} dt \ \sqrt{Q(t,\lambda)} = 0 \ . \tag{6.39}$$

The union of all Stokes lines is called a Stokes graph, while its connected components are referred to as Stokes complexes. Furthermore, Fedoryuk states that only when there is one turning point situated at the origin in the complex plane, will the Stokes "lines" ever be rays emanating from the origin. When there are at least two turning points, the Stokes lines become complicated curves merging into one another as depicted in Ref. [46] or on p. 81 of Ref. [50]. Therefore, the regions separated by these complicated curves are no longer specific sectors in the complex plane.

This description of the Stokes phenomenon is nothing like that expounded by Stokes and elaborated by Dingle. It is clear from the presentation in Ch. 2 that Stokes did not view the lines of discontinuity as curves where one of the (J)WKB solutions was maximally dominant over the other. In Ref. [46] it is also claimed that this "natural point of view" arises from interpreting (J)WKB expansions as exact encodings of true functions through resummation, not as mere asymptotic expansions according to the Poincaré definition. In fact, the (J)WKB method was born almost 70 years after Stokes's seminal paper. As exemplified by the material in Ch. 2, Stokes's primary interest was showing that asymptotic expansions developed jump discontinuities at specific rays in the complex plane. In particular, he was never concerned with the regions of the complex plane where one solution attained peak exponential dominance over the other. Previously, we have been able to verify with spectacular numerical demonstrations that complete asymptotic expansions are alternative representations of the original functions, but without the need for the (J)WKB method or having to consider the exponential dominance of part of an asymptotic solution. Nor have we employed the resummation method of resurgence analysis.

The issue of exponential dominance of one part of an asymptotic expansion over another came well after Stokes's seminal paper emanating from the work of Heading [5, 8]. As stated previously, these investigations led to Dingle's set of rules as presented in Ch. 3. Furthermore, Stokes lines in resurgence analysis are not consistent with the first two of these rules. According to the first rule, Stokes lines are rays in the complex plane that occur when the late terms in an asymptotic series are homogeneous in phase and are all the same sign. Therefore, they are not complicated curves merging into the turning points of a (J)WKB analysis as stated in Refs. [46] and [50]. Furthermore, in the case of the (J)WKB solution given by Equivalence (6.33) Stokes lines occur whenever $\arg\left(x^{-3/2}e^{-i\pi/2}h\right) = 2n\pi$, where n is any integer. Because such lines occur whenever $\arg\left(iS_+(x)/h\right)$ is real and positive and $\arg\left(iS_-(x)/h\right)$ is real and negative, it means that the exponential factor in Equivalence (6.33) is at peak dominance compared with the exponential factor of the other solution given by $\exp(iS_-(x)/h)$. This is the actual meaning of peak exponential dominance in the second rule, not some unfounded definition given by Eq. (6.39).

In Ch. 5 we observed that when the variable is away from the limiting point such as in the case of $z = \exp(i\pi/9)/4$ in the asymptotic form for $\mathrm{erfi}(z)$, the contribution from the dominant exponential terms can yield a smaller contribution to the original function than the subdominant terms, which are represented by the Stokes discontinuity term in Eq. (5.22). Nevertheless, this does not affect the location or shape of the Stokes line as would be the case with the view of the Stokes phenomenon in resurgence analysis. As indicated previously, all the rules in Ch.

3 v indicate the conventional view of the Stokes phenomenon. That is, asymptotic expansions remain uniform in specific sectors of the complex plane, not regions, and are separated by rays emanating from the origin, not by curves merging into the turning points of (J)WKB solutions. Thus, it is ludicrous on the part of Delabaere et al to declare that their view of Stokes lines and regions in Ref. [46] is in accordance with the conventional view held by Stokes and Dingle.

In later chapters we shall demonstrate conclusively that the Stokes phenomenon is indeed a discontinuous effect, whereby an asymptotic form develops jump discontinuities across specific sectors in all branches of the complex plane. In particular, we shall derive the asymptotic forms for generalisations of both types of terminants beginning with the situation where there are several Stokes sectors situated in the principal branch of the complex plane before considering other branches. For more than one Stokes sector to reside in the branches of the complex plane the variable z in the both types of generalised terminants must be altered to z^β, where $\Re \beta > 1$. We shall derive the general results for the regularised values of the generalised terminants by investigating carefully the behaviour of the singularities in the Cauchy integrals obtained from Borel summation rather than adopting Dingle's approach of an ingenious application of the rules in Ch. 3. In order to verify these results, we shall require an alternative method of regularising a divergent series. This method, which is called Mellin-Barnes regularisation, is presented in the following chapter. As we shall see, unlike the corresponding Borel-summed results, the regularised values obtained by this method do not involve sectors bordered by lines of discontinuity, but instead possess overlapping domains of convergence.

CHAPTER 7

Mellin-Barnes Regularisation

Abstract. Ch. 7 presents the theory behind an alternative method of regularising a divergent series known as Mellin-Barnes (MB) regularisation. As a result, the regularised values for more general versions of the two types of terminants presented earlier are derived in terms of MB integrals, which are often more expedient to evaluate than the regularised values obtained via Borel summation. Furthermore, unlike the Borel-summed forms for the regularised values, the MB-regularised forms are not affected by Stokes lines and sectors, but are instead valid over domains of convergence, which extend further than Stokes sectors and overlap one another. Thus, there are two different MB-regularised forms for obtaining the regularised value in the common regions of overlapping domains of convergence, which include the Stokes lines of Borel summation. To demonstrate that MB regularisation need not only be applied to an asymptotic series, a numerical example determining the regularised value of an abbreviated version of the binomial theorem is also presented for two different values of the index ρ and for various values of the variable z outside the unit disk of absolute convergence.

As discussed in Ch. 1, Mellin-Barnes (MB) regularisation was first introduced in Ref. [13], where it was used to obtain exact values for a particular case of the generalised Euler-Jacobi series, viz. $S_3(a) = \sum_{k=0}^{\infty} \exp(-ak^3)$, from its complete asymptotic expansion. In the conclusion to that work it was mentioned that a proper theory with the regularisation technique as its lynchpin would be developed in the future. The technique was subsequently put on a more solid footing with the introduction of a definition and proof in Ref. [15]. There it was used to calculate exact values of the Bessel and Hankel functions, $J_v(z)$ and $H_v(z)$, from their corresponding large $|z|$ asymptotic expansions. In both references, however, only real values of the variable in the asymptotic expansions were studied. Nevertheless, it was stated that complex values would be considered in future work, although such an undertaking would require understanding how MB regularisation was affected by the Stokes phenomenon, if at all.

In the afore-mentioned references MB regularisation was applied to asymptotic series of the type $S(N,z) = \sum_{k=N}^{\infty} f(k)(-z)^k$. Whilst it is uncertain at this stage how or whether the MB regularisation of $S(N,z)$ is affected by the Stokes phenomenon, there is another type of asymptotic series, viz. $S_1(N,z) = \sum_{k=N}^{\infty} f(k)z^k$, which we know is very much affected by the Stokes phenomenon even when we wish to derive the regularised values by MB regularisation. This is because when an asymptotic method is used to obtain this series, it is under the condition that z is initially positive and real, which means that the regularised value of the series applies to a Stokes line from the outset. From our study of the Stokes phenomenon in Chs. 3 and 5, we have seen that as soon as z moves off a Stokes line, it automatically acquires jump discontinuities. MB regularisation has never been applied to such series in the past, but if applied in the manner as in Refs. [13] and [15], then one will simply not be able to obtain the jump discontinuities when z moves above and below this initial Stokes line. Hence, MB regularisation will need to be modified if it is to yield the regularised values of this second type of series over the complex plane. The initial Stokes line for a divergent series of the type $S_1(N,z)$ will be referred to as the primary Stokes line, while the other lines in the complex plane will be referred to as secondary Stokes lines. This is in stark contrast with $S(N,z)$, where for positive real values of z, $-z$ lies in a Stokes sector and the regularised value will remain uniform up to the Stokes lines given by $\arg(-z) = 2k\pi$, where k is any integer. In such a situation the regularised value of $S(N,z)$ obtained by MB regularisation will be valid throughout a sector initially before a jump discontinuity can occur when $-z$ encounters a Stokes line. Although the choice of this primary Stokes sector is arbitrary, we shall without loss of generality restrict it to the principal branch of the complex plane. The reader need not worry at this stage about the issues that have been discussed in these opening paragraphs since they will become clearer as we proceed in this chapter with our study into the MB regularisation of both types of series over the entire complex plane.

It should be emphasised that both types of series do not need to be divergent in order to apply the technique of MB regularisation, but our primary interest is when they are asymptotic for only then can the Stokes phenomenon occur. Therefore, we need to have a clearer idea of what we mean by an asymptotic series. As pointed out in Ref. [14], if the coefficients $f(k) \sim k^{Ak+B}$ for large k, then both types of series, $S(N,z)$ and $S_1(N,z)$, are absolutely convergent for $\Re A < 0$. For $\Re A > 0$, $S(N,z)$ is divergent when $\Re z < 0$, while $S_1(N,z)$ is divergent when $\Re z > 0$, because all the real terms are positive and increasing. For $\Re A > 0$, $S(N,z)$ is conditionally convergent when $\Re z > 0$, but it is never absolutely convergent. The same applies to $S_1(N,z)$ when $\Re A > 0$ and $\Re z < 0$. Nevertheless, when $\Re A > 0$, both types of series possess zero radius of absolute convergence. This is how we define an asymptotic series here irrespective of whether it is conditionally convergent or divergent. For $A = 0$, whether both types of series are convergent or divergent depends upon other factors, e.g. the magnitude and phase of z. Series belonging to the ${}_pF_{p-1}(z)$ family of hypergeometric functions, which include the examples discussed in Ch. 4, fall into this category. Such series are typically free of asymptotic behaviour. That is, there are no values of z where truncating the series approaches the regularised value at an optimal point and then diverges beyond it. For these series the optimal point of truncation is situated at infinity within the circle of absolute convergence. Outside the circle of absolute convergence the most accurate truncation point is the $N = 0$ value of the series, which is frequently nowhere near the regularised value of the series.

7.1 Series of the Type $S(N, z)$

Since our aim is to extend the original definition of MB regularisation in Ref. [15] before making necessary modifications to the second type of asymptotic series $S_1(N, z)$ as indicated above, we re-introduce the definition of MB regularisation by applying it to series of the type given by $S(N, z)$, but with modifications to the properties of $f(s)$. Rather than use the term definition in the proof of MB regularisation as was done in Ref. [15], we shall use the term proposition in keeping with the presentation of the results in Ch. 4.

Proposition 3. Given that (1) as $L \to \infty$, $|f(s)| = O\big(\exp(-\varepsilon_1 L)\big)$ for $s = c + iL$ and $|f(s)| = O\big(\exp(-\varepsilon_2 L)\big)$ for $s = c - iL$, where $\varepsilon_1, \varepsilon_2 > 0$, (2) $-\pi < \theta = \arg z < \pi$, (3) there exists a real number c such that the poles of $\Gamma(N - s)$ lie to the right of the line $N - 1 < c = \Re s < N$ in the complex plane and that the poles of $f(s)\Gamma(s + 1 - N)$ to the left of it and (4) $z^s f(s)\Gamma(s + 1 - N)\Gamma(N - s)$ is single-valued to the right of the line, the regularised value of the power series $S(N, z)$ as defined in the introduction to this chapter is given by

$$S(N, z) = \sum_{k=N}^{\infty} f(k)(-z)^k \equiv \int_{\substack{c-i\infty \\ N-1<c=\Re s<N}}^{c+i\infty} ds \, \frac{z^s f(s)}{e^{-i\pi s} - e^{i\pi s}} \, . \tag{7.1}$$

Remark 1. For $S(N, z)$ an asymptotic series, as the offset c increases, the MB integral in the above result will eventually increase exponentially regardless of the magnitude of z. The greatest integer lower than the value of c at which the MB integral begins to diverge rather than converge represents the optimal point of truncation.

Remark 2. The conditions on $f(s)$ have been introduced to ensure that the MB integral in Equivalence (7.1) is convergent over the entire principal branch of the complex plane, but as explained after the proof, they can be relaxed.

Proof. For a more detailed proof of this proposition the reader should consult Definition 1 in Ref. [15]. Nevertheless, a brief exposition of the proof is presented here with a greater emphasis on the contour integral along the great arc.

Consider the contour integral given by

$$I = (-1)^N \int_{c-i\infty}^{c+i\infty} ds \, z^s f(s) \Gamma(1+s-N) \Gamma(N-s) \ , \qquad (7.2)$$

where $N-1 < c = \Re s < N$. The first two conditions in the proposition are necessary for ensuring that the modulus of the integrand of this MB integral decays exponentially at the endpoints. That is,

$$\left| \frac{z^s f(s)}{e^{-i\pi s} - e^{i\pi s}} \right|^{s=c\pm iL} \underset{\sim}{} |z|^c e^{\mp L\theta} e^{-\pi L} |f(c \pm iL)| \ . \qquad (7.3)$$

The upper limit, viz. $s = c+i\infty$, decays exponentially for all values of θ in the principal branch provided $f(c+iL)| = O(\exp(-\varepsilon_1 L))$ as $L \to \infty$ and $\varepsilon_1 > 0$. The same applies to the lower limit provided $|f(c-iL)| = O(\exp(-\varepsilon_2 L))$ as $L \to \infty$ and $\varepsilon_2 > 0$. Because the MB integral is defined, we can close it to the right by introducing a contour integral along the great arc from $c+i\infty$ to $c-i\infty$. To ensure that the integrand is single-valued, we need to confine $\arg z$ to a specific branch of the complex plane such as the principal branch which we have already done by stipulating the second condition. Now we can apply Cauchy's residue theorem. The remaining conditions in the proposition ensure that when the residues within the closed contour are evaluated, they yield the series on the lhs of Equivalence (7.1). In the event that there are poles lying to the right of c, we extract all the terms in the series until all the poles lie to the left of the new value of N. Then we are left with a truncated series and a modified version of I, which we can analyse in the same manner.

Regularisation becomes an issue when we wish to evaluate the contour integral along the great arc, which can be written as

$$I_{\text{arc}} = \lim_{L\to\infty} i\pi L \int_{-\pi/2}^{\pi/2} d\gamma \, \frac{f(Le^{i\gamma}) e^{i\gamma}}{\sin(\pi Le^{i\gamma})} \, \exp(Le^{i\gamma}(\ln|z| + i\theta)) \ . \qquad (7.4)$$

In obtaining this result we have used the reflection formula for the gamma function, viz. Eq. (3.2). The magnitude of this integral is bounded by

$$|I_{arc}| \leq 2\pi \lim_{L\to\infty} L \int_{-\pi/2}^{\pi/2} d\gamma\, |f(Le^{i\gamma})| \exp\left(L\ln|z|\cos\gamma\right)$$
$$\times\ \exp\left(-L\pi|\sin\gamma| - L\theta\sin\gamma\right)\ . \tag{7.5}$$

The above integral can be split into two integrals, one between $(-\pi/2, 0)$ and and the other between $(0, \pi/2)$. Then we can apply the Jordan inequality. At the lower limits of both integrals there is no contribution from the $\sin\gamma$ terms in the exponential leaving only the terms involving $\cos\gamma$, which are bounded by unity in the integrals. Therefore, the conditions under which the contour along the great arc vanishes are $\varepsilon_1 > \ln|z|$ and $\varepsilon_2 > \ln|z|$. For $\varepsilon_1 = \ln|z|$ and $\varepsilon_2 = \ln|z|$ the integral can vanish but this will depend upon the algebraic behaviour of $f(s)$. For $\varepsilon_1 < \ln|z|$ or $\varepsilon_2 < \ln|z|$, I_{arc} will yield infinity. In this case we simply remove the infinity in accordance with the process of regularisation. Therefore, irrespective of whether the contour integral along the great arc is infinite or zero, we arrive at

$$S(N,z) = \sum_{k=N}^{\infty} f(k)(-z)^k \equiv \frac{(-1)^N}{2\pi i} \int_{\substack{c-i\infty \\ N-1<c=\Re s<N}}^{c+i\infty} ds\, z^s f(s)$$
$$\times\ \Gamma(1+s-N)\,\Gamma(N-s)\ . \tag{7.6}$$

This completes the proof.

In obtaining the regularised value of $S(N,z)$ by MB regularisation, we invoked conditions on the behaviour of $f(s)$ at the end-points $s = c \pm iL$ of the integral in Eq. (7.2). Apparently, these conditions are not necessary to ensure that the integral in Equivalence (7.1) converges. From Approximation (7.3) we see that the integral will decay exponentially even if $|f(c+iL)| = O(\exp(AL))$, where $\Re A > 0$ as $L \to \infty$, provided that $\theta < \Re A - \pi$. If $|f(c-iL)| = O(\exp(BL))$, where $\Re B > 0$ as $L \to \infty$, then it will decay exponentially provided that $\theta > \pi - \Re B$. Obviously, by placing restrictions on $\arg z$, and hence, the values of z, we may not be able to access the entire principal branch fo the complex plane, but it means that we can consider far more series than only those fulfilling the conditions in the proposition. One important example of such a series is the binomial series, which we shall use as our first example of MB regularisation soon. In addition, as our investigation into the theory behind MB regularisation progresses, we shall find that there is another method for accessing the sectors of the complex plane, where Equivalence (7.1) cannot be applied.

To clarify the above material, let us consider the following variant or abbreviation of the binomial series

$$_1\mathcal{F}_0(\rho;-z)\,|_N = \sum_{k=N}^{\infty} \frac{\Gamma(k+\rho)}{k!\,\Gamma(\rho)}\,(-z)^k\;. \tag{7.7}$$

If $N=0$, then the above series becomes the binomial series, which we found in Ch. 4 was conditionally convergent for $\Re z > -1$ and divergent elsewhere. Furthermore, for $\rho = -l$, where l is a non-negative integer less than N, the series vanishes. Since we aim to demonstrate that the MB-regularised value of the series is identical with the value obtained by Borel summation, we need to derive the latter. This can be accomplished with the aid of Proposition 1. First, we write the series as

$$_1\mathcal{F}_0(\rho;-z)\,|_N = (-z)^N \frac{\Gamma(N+\rho)}{\Gamma(\rho)} \sum_{k=0}^{\infty} \frac{\Gamma(k+N+\rho)}{\Gamma(N+\rho)\,k!}$$
$$\times\;\frac{(-z)^k}{(k+1)(k+2)\cdots(k+N)}\;. \tag{7.8}$$

An alternative representation of the above result is

$$_1\mathcal{F}_0(\rho;-z)\,|_N = (-1)^N \frac{\Gamma(N+\rho)}{\Gamma(\rho)} L_N(z) \sum_{k=0}^{\infty} \frac{\Gamma(k+N+\rho)}{\Gamma(N+\rho)\,k!}\,(-z_N)^k\;, \tag{7.9}$$

where $L_N(z)$ is the operator defined by Eq. (4.15). Now the inner series represents the entire binomial series and hence, we can introduce the regularised value given in Proposition 1. Then we obtain

$$_1\mathcal{F}_0(\rho;-z)\,|_N \equiv (-1)^N \frac{\Gamma(N+\rho)}{\Gamma(\rho)} L_N(z)(1+z_N)^{-N-\rho}\;. \tag{7.10}$$

For $N=2$, the regularised value of the abbreviated binomial series becomes

$$_1\mathcal{F}_0(\rho;-z)\,|_2 \equiv \rho z - 1 + (1+z)^{-\rho}\;, \tag{7.11}$$

which is not valid for $\rho \neq -1,0$. As indicated earlier, the series vanishes for these values of ρ.

Now we evaluate the regularised value of the series $_1\mathcal{F}_0(\rho;z)\,|_N$ via MB regularisation. For this series $f(k) = (\rho)_k/k!$, where $(\rho)_k$ is the Pochhammer notation for

$\Gamma(k+\rho)/\Gamma(\rho)$. As $k \to \infty$, we can introduce Stirling's approximation, viz. No. 8.327 in Ref. [21], which gives

$$f(k) \overset{k \to \infty}{\sim} \frac{e^{1-\rho}}{\Gamma(\rho)} k^{\rho-1} \; . \tag{7.12}$$

Thus, we see that $f(c \pm iL) \sim L^{\rho-1}$ as $L \to \infty$. For $\Re\rho < 0$, the MB integral in Equivalence (7.1) is convergent even when there is no exponential factor on the rhs of Approximation (7.3). For $\Re\rho > 0$, however, there may be a problem with convergence, but only at the endpoints of the MB integral where $\arg z = \pm\pi$. For all other values of $\arg z$ within the principal branch of the complex plane we expect to obtain a result for the MB integral in Equivalence (7.1).

The above analysis is still incomplete because there are also restrictions on the values that ρ can take. According to the third condition in Proposition 3, for $N=2$, the zeros of $\Gamma(\rho+s)/s(s-1)$ must lie to the left of the line contour through the offset c. Since $1 < c < 2$, the poles at $s=0$ and $s=1$ satisfy the third condition. However, there are also poles at $s = -\rho - k$, where k is any non-negative integer. If $\Re\rho < -2$, then there will be at least one pole to the right of the line contour, while if $-2 < \Re\rho < -1$, then c will need to be chosen so that it is to the right of $-\Re\rho$. Therefore, taking into account the third condition in Proposition 3, the regularised value of the abbreviated binomial series with $N=2$ is given by

$$_1\mathcal{F}_0(\rho;-z)\big|_2 \equiv \frac{1}{\Gamma(\rho)} \int_{\substack{c-i\infty \\ \mathrm{Max}[1,-\Re\rho]<c=\Re s<2}}^{c+i\infty} ds \, \frac{\Gamma(\rho+s)}{\Gamma(s+1)} \left(\frac{z^s}{e^{-i\pi s} - e^{i\pi s}} \right), \tag{7.13}$$

where $\Re\rho > -2$ and $|\arg z| < \pi$. For general N we find that the above equivalence becomes

$$_1\mathcal{F}_0(\rho;-z)\big|_N \equiv \frac{1}{\Gamma(\rho)} \int_{\substack{c-i\infty \\ \mathrm{Max}[N-1,-\Re\rho]<c=\Re s<N}}^{c+i\infty} ds \, \frac{\Gamma(\rho+s)}{\Gamma(s+1)} \left(\frac{z^s}{e^{-i\pi s} - e^{i\pi s}} \right), \tag{7.14}$$

where $\Re\rho > -N$ and $|\arg z| < \pi$. In addition, by putting $N=0$ and $\rho=1$ in Equivalence (7.14) we have derived the MB-regularised value of the geometric series, which is

$$\sum_{k=0}^{\infty} (-z)^k \equiv \int_{\substack{c-i\infty \\ -1<c=\Re s<0}}^{c+i\infty} ds \, \frac{z^s}{e^{-i\pi s} - e^{i\pi s}} \; . \tag{7.15}$$

In the above result $|\arg z| < \pi$. Equivalence (7.15) has undergone extensive numerical evaluation in Ref. [15].

Table 7.1: Borel-summed and MB-regularised values for $_1\mathcal{F}_0(\rho;z)|_2$

ρ	z	**Borel-summed Value**	**MB-regularised Value**
7/4	-3	$-6.039\,775\,896\,186\,571\,3$ $+0.210\,224\,103\,813\,428\,6i$	Overflow in computation
	$2i$	$-1.087\,690\,131\,338\,726\,4$ $+3.271\,692\,330\,144\,942\,9i$	$-1.087\,690\,131\,338\,726\,4$ $+3.271\,692\,330\,144\,942\,9i$
	$-4i$	$-1.057\,098\,460\,417\,246\,5$ $-6.938\,633\,341\,269\,987\,8i$	$-1.057\,098\,460\,417\,246\,5$ $-6.938\,633\,341\,269\,987\,8i$
	5	$7.793\,474\,571\,668\,702\,4$	$7.793\,474\,571\,668\,702\,4$ $+0.000\,000\,000\,000\,000\,0i$
	$8/3+7i/5$	$3.740\,048\,607\,127\,545\,4$ $+2.395\,559\,189\,489\,863\,0i$	$3.740\,048\,607\,127\,545\,4$ $+2.395\,559\,189\,489\,863\,0i$
-9/11	$-9/4$	$-0.168\,848\,000\,124\,110\,8$ $+0.648\,931\,478\,826\,048\,9i$	NIntegrate failed to converge
	$3i$	$0.338\,187\,559\,790\,484\,2$ $-0.266\,261\,583\,244\,094\,1i$	$0.338\,187\,559\,790\,484\,2$ $-0.266\,261\,583\,244\,094\,1i$
	$-5i/4$	$0.091\,982\,887\,146\,059\,9$ $+0.039\,307\,796\,770\,817\,3i$	$0.091\,982\,887\,146\,059\,9$ $+0.039\,307\,796\,770\,817\,3i$
	9/4	$-0.217\,816\,269\,084\,034\,89$	$-0.217\,816\,269\,084\,034\,89$ $+0.000\,000\,000\,000\,000\,0i$
	$7/3-9i/7$	$-0.202\,350\,311\,563\,237\,0$ $+0.211\,101\,070\,150\,001\,3i$	$-0.202\,350\,311\,563\,237\,5i$ $+0.211\,101\,070\,150\,001\,6i$

Table 7.1 presents the regularised values of the abbreviated binomial series with $N=2$ obtained from the Borel-summed form given by Equivalence (7.11) and the corresponding MB-regularised form given by Equivalence (7.13). Unlike previous tables these results have been evaluated by using Mathematica 7.0 instead of Mathematica 4.1. As a consequence, the results were able to be evaluated to 50 decimal places, which indicates that the latest version of the software package is no longer limited by machine precision, although it may be take much longer to compute such values. For the sake of brevity the results in Table 7.1 have been presented to 17 decimal places. In addition, although the results for only two values of ρ are displayed in the table, other values of ρ were considered. Similarly, far more values of z was also used in the calculations. In any case the sample presented in the table is typical of all the results obtained from this numerical

study.

The third column in Table 7.1 labelled Borel-summed values displays the results obtained by introducing the values for ρ and z into Eq. (7.11), while the next or fourth column displays the values obtained by applying the NIntegrate routine in Mathematica to the MB integral in Equivalence (7.13). Since the next chapter is devoted to the issue of implementing and calculating MB-regularised values in Mathematica, we shall only present a summary as to how the results in the fourth column were obtained. In order to apply the NIntegrate routine to Equivalence (7.14), the MB integral was expressed as two separate integrals, one evaluating the line integral in the upper half of the complex plane and the other evaluating the line integral in the lower half of the complex plane. As expected, the computational overflow and failure to converge for negative real values of z in the table occurred when Mathematica attempted to evaluate the line integral in the lower half of the complex plane, which is consistent with previous statements on this issue below Approximation (7.12). For all other values of z there was no problem with evaluating the MB-regularised values of the abbreviated binomial series.

The Borel-summed regularised values were obtained by setting the number of decimal figures in Mathematica to 50 places, while the MB-regularised values were obtained by setting the options of WorkingPrecision, AccuracyGoal, Precision-Goal, MinRecursion and MaxRecursion in the NIntegrate routine to 100, 50, 50, 3 and 10 respectively. When demanding high accuracy, WorkingPrecision must usually be set considerably higher than AccuracyGoal and PrecisionGoal or else the desired accuracy may not be obtained. Instances of this type of problem have occurred here. Consequently, the MB-regularised values were on some occasions only accurate to between 43 and 50 decimal places. Therefore, to match the accuracy of the Borel-summed values produced by Mathematica's intrinsic routines, we really needed to set WorkingPrecision much higher. Unfortunately, bolstering the working precision increases the CPU time for the evaluation of the MB-regularised integral significantly, although each set of results or each row in the table took between 15 and 20 CPU seconds to compute. The results for positive real values of z are particularly interesting because the Borel-summed values yielded a vanishing imaginary contribution, while the corresponding MB-regularised values produced a small computational error beginning at 45-th decimal place. These have been represented in the table as $0.0\ldots i$ to 17 decimal places. Nevertheless, the results in the table confirm not only that does MB regularisation give the regularised values of both divergent and convergent series, but that the result given in Proposition 1 is also valid.

Since the two forms for the regularised value of the abbreviated binomial series are equal to each other, equating them to each other yields

$$\frac{1}{2\pi i} \int_{\substack{c-i\infty \\ \text{Max}[N-1,-\Re\rho]<c=\Re s<N}}^{c+i\infty} ds \, \frac{\Gamma(\rho+s)}{\Gamma(s+1)} \left(\frac{x^{-s}}{e^{-i\pi s}-e^{i\pi s}} \right) = \frac{(-1)^N}{2\pi i} \Gamma(N+\rho)$$
$$\times \left. L_N(z)(1+z_N)^{-N-\rho} \right|_{z=1/x} , \qquad (7.16)$$

where $\Re\rho > -N$ and z has been replaced by $1/x$. The lhs of the above result is an inverse Mellin transform. Hence, by the theory of Mellin transforms [52], one finds after a little algebra that

$$\left. \int_0^\infty dx \, x^{s-1} L_N(z)(1+z_N)^{-N-\rho} \right|_{z=1/x} = \frac{(-1)^N}{(\rho)_N} B(\rho+s,-s) , \qquad (7.17)$$

where $\text{Max}[N-1,-\Re\rho] < \Re s < N$ and $\Re\rho > -N$. This unusual result demonstrates that the combination of MB regularisation and Borel summation can yield new Mellin transform pairs.

Let us consider an extension of the series $S(N,z)$ by replacing z with z^β, where β is a real number. According to Rule 1 in Ch. 3 the Stokes lines are now situated at $\arg z = 2k\pi/\beta$, where k is an arbitrary integer. Without loss of generality we take β to be positive because if it is negative, then we can substitute z by $1/z$ in our study of $S(N,z^\beta)$. The introduction of β has a minor effect on the previous proof except that we require $-\pi/\beta < \arg z < \pi/\beta$ to ensure single-valuedness of the integrand, while z^s in the regularised value, i.e. the MB integral given in Proposition 3, is replaced by $z^{\beta s}$. For $\beta > 1$, this means that the domain of convergence for the MB integral in Proposition 3 cannot encompass the entire principal branch of the complex plane. That is, we will no longer be able to obtain regularised values for all values of z lying in the principal branch unless the proof of Proposition 3 can be adapted to deal with $|\arg z| > \pi/\beta$.

In carrying out this extension of $S(N,z)$ to $S(N,z^\beta)$, let us also introduce the seemingly innocuous factor of 1^k in the form of $\exp(-2\pi i l k)$, where l is again an arbitrary integer. Then the MB-regularised value of the extended series is given by

$$S\left(N, z^\beta e^{-2\pi i l}\right) = \sum_{k=N}^\infty f(k) \left(-z^\beta e^{-2\pi i l}\right)^k$$
$$\equiv I_l\left(z^\beta\right) = \int_{\substack{c-i\infty \\ N-1<c=\Re s<N}}^{c+i\infty} ds \, \frac{z^{\beta s} e^{-2\pi i l s} f(s)}{e^{-i\pi s} - e^{i\pi s}} . \qquad (7.18)$$

Thus, the regularised value of $S(N, z^\beta \exp(-2li\pi))$ has acquired the multi-valued factor of $\exp(-2il\pi s)$, which affects the domain of convergence for the MB integral. That is, the behaviour of the modulus of the integrand at the endpoints is given by

$$\left| \frac{z^{\beta s} e^{-2i\pi l s} f(s)}{e^{-i\pi s} - e^{i\pi s}} \right|_{s=c\pm iL} \underset{\sim}{\sim} |z|^{\beta c} \left[\begin{array}{c} e^{-(\beta\theta-(2l-1)\pi)L} |f(c+iL)| \\ e^{(\beta\theta-(2l+1)\pi)L} |f(c-iL)| \end{array} \right] . \quad (7.19)$$

To ensure that the integral in Equivalence (7.18) is convergent, we require that the integrand remains exponentially decaying at the endpoints, which, in turn, means that $(2l-1)\pi - \varepsilon_1 < \beta\theta < (2l+1)\pi + \varepsilon_2$. As expected, for $l=0$ and $\beta=1$, this condition reduces to the domain of convergence for the MB integral in Proposition 3. Thus, in this extension of Proposition 3, the second condition needs to be altered so that θ lies in the sector of $((2l-1)\pi/\beta, (2l+1)\pi/\beta)$.

If we consider the $l=j$ and $l=j-1$ versions of Equivalence (7.18), then we see that the domains of convergence for both the regularised values possess a common sector, which is given by $(2j-1)\pi/\beta - \varepsilon_1/\beta < \theta < (2j-1)\pi/\beta + \varepsilon_2/\beta$. Since the MB integrals are defined over a common sector of the complex plane, we can evaluate their difference, which yields

$$\sum_{k=N}^{\infty} f(k) \left[\left(-z^\beta e^{-2\pi i j} \right)^k - \left(-z^\beta e^{-2\pi i(j-1)} \right)^k \right] \equiv \Delta I_{j,j-1} \left(z^\beta \right)$$

$$= I_j \left(z^\beta \right) - I_{j-1} \left(z^\beta \right) = \int_{\substack{c-i\infty \\ N-1 < c = \Re s < N}}^{c+i\infty} ds\, z^{\beta s} f(s)\, e^{-(2j-1)i\pi s} . \quad (7.20)$$

If $f(s)$ is the Mellin transform of another function $F(x)$, which according to Ref. [52] means that

$$F(x) = \frac{1}{2\pi i} \int_{c-i\infty}^{c+i\infty} dx\, x^{-s} f(s) ,$$

then we obtain

$$\Delta I_{j,j-1}(z^\beta) = 2\pi i F \left(z^{-\beta} \exp((2j-1)i\pi) \right) . \quad (7.21)$$

The above result is valid for all values of the offset c provided the poles of $f(s)$ lie in the left hand half of the complex plane. Its importance lies in the fact that although the original MB integral in Equivalence (7.20) is convergent for $((2j-1)\pi - \varepsilon_1)/\beta < \theta < ((2j-1)\pi + \varepsilon_2)/\beta$, it can be analytically continued to all values of θ or $\arg z$. For the situation, where $F(x) = x^\gamma g(x)$ and $g(x)$ is a function that can be expressed in terms of a convergent power series, the above result reduces to

$$\Delta I_{j,j-1}(z^\beta) = 2\pi i z^{-\beta\gamma} e^{(2j-1)i\pi\gamma} g(-1/z^\beta) \ . \tag{7.22}$$

Equivalence (7.20) is an interesting result because contrary to the expectation that subtraction of the two series should yield zero, which would be the case if we were dealing with a convergent series since the domains of convergence would not overlap, we see that there is a finite difference between them. For a series with a finite radius of absolute convergence such as the earlier example of the abbreviated binomial series, the domains of convergence just fail to abut one another because the MB integral for the regularised value is not defined for $\arg z = k\pi$, where k is an arbitrary integer. Therefore, overlapping domains of convergence will only apply to an asymptotic series. Then writing the variable z in the phasor form of $z \exp(-2i\pi l)$, where l is again an arbitrary integer, yields different regularised values of the series. Hence, we see that the phase of a variable affects an asymptotic series quite differently to both a convergent series and a series with a finite radius of absolute convergence. One upshot of this is that since the Stokes phenomenon is dependent upon the domains of convergence for MB-regularised values overlapping one another, it will only occur when a series is asymptotic.

We are now in a position to determine the regularised value of $S(N, z^\beta)$ for all arguments of z. Two issues emerge when embarking on this task. The first is that we need to specify a specific sector over the complex plane, where $I_l(z^\beta)$ represents the regularised value of $S(N, z^\beta)$ initially. According to Proposition 3, $I_l(z^\beta)$ represents the regularised value of $S(N, z^\beta)$, but it is defined only over the sector of $(2l-1)\pi/\beta < \arg z < (2l+1)\pi/\beta$. We must specify only one value of l, where it is valid because the same regularised value will apply for different values of l or sectors. For example, $I_0(z^\beta)$ and $I_1(z^\beta)$ yield the same values over their respective sectors. Specifying more than one sector means that the regularised value of $S(N, z^\beta)$ will become discontinuous, despite the fact that $S(N, z^\beta)$ could represent the asymptotic series for a continuous function over the entire complex plane. Although the choice of such a primary sector is arbitrary, we shall let $I_0(z^\beta)$ represent the regularised of $S(N, z^\beta)$ since its Stokes sector is centrally located in the principal branch of the complex plane for z.

The second issue arises once the primary sector has been decided. It is the development of a methodology for determining the regularised value of $S(N, z^\beta)$ over the other sectors of complex plane. This is necessary because when $\beta > 1$, more than one sector will be required in order to determine the values of the original function over the entire principal branch of the complex plane.

From Equivalence (7.20) we note that

$$I_0\left(z^\beta\right) = I_1\left(z^\beta\right) - \Delta I_{1,0}\left(z^\beta\right) \ , \tag{7.23}$$

which is valid over the common region of the domains of convergence for $I_0(z^\beta)$ and $I_1(z^\beta)$, viz. $(\pi - \varepsilon_1)/\beta < \arg z = \arg z < (\pi + \varepsilon_2)/\beta$. Since $I_0(z^\beta)$ represents the regularised value of $S(N, z^\beta)$ over the primary Stokes sector, the rhs of Eq. (7.23) also represents the regularised value of the series for $(\pi - \varepsilon_1)/\beta < \arg z < \pi/\beta$. In addition, $I_1(z^\beta)$ is defined for θ or $\arg z$ over $((\pi - \varepsilon_1)/\beta, (3\pi + \varepsilon_2)/\beta)$, which means that we can analytically continue the quantity on the rhs beyond the common region up to $(3\pi + \varepsilon_2)/\beta$ provided $\Delta I_{1,0}(z^\beta)$ can be expressed in the form given by Eq. (7.21). This means that over the sector of $\pi/\beta < \arg z < 3\pi/\beta$, the regularised value of $S(N, z^\beta)$ via MB regularisation is given by

$$S\left(N, z^\beta\right) \equiv I_1\left(z^\beta\right) - 2\pi i F\left(z^{-\beta} e^{i\pi}\right) \ . \tag{7.24}$$

For $\arg z$ lying between $((3\pi - \varepsilon_1)/\beta, (3\pi + \varepsilon_2)/\beta)$, we have

$$I_1\left(z^\beta\right) = I_2\left(z^\beta\right) - \Delta I_{2,1}\left(z^\beta\right) \ . \tag{7.25}$$

If we substitute $I_1(z^\beta)$ with the rhs of Eq. (7.25) in Equivalence (7.24), then we have an alternative representation for the regularised value of $S(N, z^\beta)$. However, $I_2(z^\beta)$ is defined for $\arg z$ outside the common region for $I_1(z^\beta)$ and $I_2(z^\beta)$, which means that we can analytically continue the new representation for the regularised value of $S(N, z^\beta)$ for $\arg z$ up to $(5\pi + \varepsilon_2)/\beta)$, provided that $\Delta I_{2,1}(z^\beta)$ can be expressed in the form given by Eq. (7.21). Therefore, for $\arg z$ lying in the Stokes sector of $3\pi/\beta < \arg z < 5\pi/\beta$, the regularised value of $S(N, z^\beta)$ is given by

$$S\left(N, z^\beta\right) \equiv I_2\left(z^\beta\right) - 2\pi i F\left(z^{-\beta} e^{3i\pi}\right) - 2\pi i F\left(z^{-\beta} e^{i\pi}\right) \ . \tag{7.26}$$

This process can be continued indefinitely for all values of $\arg z$. As a consequence, for $((2M-1)\pi - \varepsilon_1)/\beta < \arg z < ((2M+1)\pi + \varepsilon_2)/\beta$, we arrive at the following general result:

$$S\left(N, z^\beta\right) \equiv I_M\left(z^\beta\right) - 2\pi i \sum_{j=1}^{M} F\left(z^{-\beta} e^{(2j-1)i\pi}\right) \ . \tag{7.27}$$

Equivalence (7.27) is valid for all values of the offset c provided the poles of $f(s)$ lie in the left hand half of the complex plane. For the case, where $F(x) = x^\gamma g(x)$ and $g(x)$ can be expressed as a convergent power series, which means in turn that $g(x \exp(i(2j-1)\pi)) = g(-x)$, Equivalence (7.27) reduces to

$$S(N, z^\beta) \equiv I_M(z^\beta) - 2\pi i z^{-\beta\gamma} g(-z^{-\beta}) e^{iM\gamma\pi} \frac{\sin(M\gamma\pi)}{\sin(\gamma\pi)} \ . \tag{7.28}$$

Although all the terms in $S(N, z^\beta)$ are positive when $\arg z = (2M+1)\pi/\beta$, which according to Rule 1 in Ch. 3 indicates that a Stokes line has been encountered, we see that no discontinuities arise when $\arg z$ moves off the lines as is the case when the series is Borel-summed. In fact, the regularised values can be determined by using either the $M = M$ version or the $M = M+1$ version of Equivalence (7.27) since the Stokes line lies in the common sector of their domains of convergence. Alternatively, we can follow Dingle's prescription of averaging the two results, thereby obtaining

$$S(N, z^\beta) \equiv \frac{i}{2} \int_{\substack{c-i\infty \\ N-1 < c = \Re s < N}}^{c+i\infty} ds \, |z|^{\beta s} f(s) \cot(\pi s)$$
$$- 2\pi i \sum_{j=1}^{M} F(z^{-\beta} e^{(2j-1)i\pi}) - \pi i F(z^{-\beta} e^{(2M+1)i\pi}) \ . \tag{7.29}$$

For the case where $F(x) = x^\gamma g(x)$ and $g(x)$ can be expressed as a convergent power series, Equivalence (7.29) reduces to

$$S(N, z^\beta) \equiv \frac{i}{2} \int_{\substack{c-i\infty \\ N-1 < c = \Re s < N}}^{c+i\infty} ds \, |z|^{\beta s} f(s) \cot(\pi s)$$
$$- \frac{\pi z^{-\beta\gamma}}{\sin(\pi\gamma)} g(-z^{-\beta}) \left(1 - e^{i(2M+1)\gamma\pi} \cos(\pi\gamma)\right) \ . \tag{7.30}$$

The preceding forms for the regularised value of $S(N, z^\beta)$ have been derived by assuming that $M \geq 0$. By putting $j = -j$ in Equivalence (7.20), we find that the rhs yields

$$I_{-j}(z^\beta) - I_{-j-1}(z^\beta) = \Delta_{-j,-j-1}(z^\beta)$$
$$= \int_{\substack{c-i\infty \\ N-1 < c = \Re s < N}}^{c+i\infty} ds \, z^{\beta s} f(s) e^{(2j+1)i\pi s} \ . \tag{7.31}$$

If $f(s)$ represents the Mellin transform of $F(x)$, then we find that

$$\Delta I_{-j,-j-1}(z^\beta) = 2\pi i F\left(z^{-\beta} \exp(-(2j+1)i\pi)\right) . \qquad (7.32)$$

The common sector for the domains of convergence of $I_{-j}(z^\beta)$ and $I_{-j-1}(z^\beta)$ is $(-2j-1-\varepsilon_1)\pi/\beta < \arg z < (-2j-1+\varepsilon_2)/\beta$. Therefore, when $j=-1$, we have

$$I_0(z^\beta) = I_{-1}(z^\beta) + \Delta_{0,-1}(z^\beta) = I_{-1}(z^\beta) + 2\pi i F\left(z^{-\beta} e^{-i\pi}\right) . \qquad (7.33)$$

Although the rhs of the above equation has been derived under the condition $(-\pi - \varepsilon_1)/\beta < \arg z < (-\pi + \varepsilon_2)/\beta$, the result can be analytically continued to $(-3\pi - \varepsilon_1)/\beta < \arg z < (-\pi + \varepsilon_2)/\beta$ provided that $F(x)$ is defined over the entire complex plane. By putting $j=-1$, we find that

$$\begin{aligned}
I_0(z^\beta) &= I_{-2}(z^\beta) + \Delta_{-1,-2}(z^\beta) + \Delta_{0,-1}(z^\beta) \\
&= I_{-2}(z^\beta) + 2\pi i F(z^{-\beta} e^{-3i\pi}) + 2\pi i F(z^{-\beta} e^{-i\pi}) , \qquad (7.34)
\end{aligned}$$

where now the rhs of Eq. (7.34) is analytically continuable to the domain of convergence for $I_{-2}(z^\beta)$ or $(-5\pi - \varepsilon_1)/\beta < \arg z < (-3\pi + \varepsilon_2)/\beta$. More generally, for $((-2M-1)\pi-\varepsilon_1)/\beta < \arg z < ((-2M+1)\pi+\varepsilon_2)/\beta$, where $M \geq 0$, we have

$$S(N, z^\beta) \equiv I_{-M}(z^\beta) + 2\pi i \sum_{j=1}^{M} F\left(z^{-\beta} e^{-(2j-1)i\pi}\right) . \qquad (7.35)$$

For the case where $F(x)=x^\gamma g(x)$ and $g(x)$ can be expressed as a convergent power series, the above result reduces to

$$S(N, z^\beta) \equiv I_{-M}(z^\beta) + 2\pi i z^{-\beta\gamma} g(-z^{-\beta}) e^{-iM\gamma\pi} \frac{\sin(M\gamma\pi)}{\sin(\gamma\pi)} . \qquad (7.36)$$

On the Stokes line of $\arg z = -(2M+1)\pi/\beta$, we can also take the average of the $M=M$ and $M=(M+1)$ results of Equivalence (7.35). In fact, we can average over the entire common sector of the domains of convergence. Then we find that

$$\begin{aligned}
S(N, z^\beta) &\equiv \frac{i}{2} \int_{\substack{c-i\infty \\ N-1<c=\Re s<N}}^{c+i\infty} ds\, |z|^{\beta s} f(s) \cot(\pi s) \\
&\quad + 2\pi i \sum_{j=1}^{M} F\left(z^{-\beta} e^{-(2j-1)i\pi}\right) + \pi i F\left(z^{-\beta} e^{-(2M+1)i\pi}\right) . \qquad (7.37)
\end{aligned}$$

For the case where $F(x) = x^\gamma g(x)$ and $g(x)$ can be expressed as a convergent power series, Equivalence (7.37) simplifies to

$$
S(N, z^\beta) \equiv \frac{i}{2} \int_{\substack{c-i\infty \\ N-1 < c = \Re s < N}}^{c+i\infty} ds \, |z|^{\beta s} f(s) \, \cot(\pi s)
$$

$$
+ \frac{\pi z^{-\beta \gamma}}{\sin(\pi \gamma)} g(-z^{-\beta}) \left(1 - e^{-i(2M+1)\gamma \pi} \cos(\pi \gamma)\right) . \tag{7.38}
$$

In the MB regularisation of $S(N, z^\beta)$ we see that Rule 8a in Ch. 3 about taking the average of the two distinct asymptotic forms across adjacent Stokes sectors is redundant since the various forms for the regularised value are equal to each other on the Stokes line. That is, when an asymptotic series of the type $S(N, z^\beta)$ is MB-regularised, no jump discontinuities occur at the Stokes lines. Hence, there is no need to describe MB-regularised values in terms of Stokes sectors and lines. Instead, the regularised value of an asymptotic series takes on different forms over distinct branches in the complex plane which are in fact greater in range than the Stokes sectors obtained when Borel summation is used to obtain the regularised value of the series. These branches are determined by the domains of convergence of the resulting MB integrals that appear in the regularised value of the series. Subject to the conditions on $f(s)$ in Proposition 3, the MB-regularised value of $S(N, z^\beta)$ is given by Equivalence (7.27), but is only valid for $((2M-1)\pi - \varepsilon_1)/\beta < \arg z < ((2M+1)\pi + \varepsilon_2)/\beta$. The MB-regularised value of $S(N, z^\beta)$ for $((2M-3)\pi - \varepsilon_1)/\beta < \arg z < ((2M-1)\pi + \varepsilon_2)/\beta$ is obtained by replacing M by $M-1$ in Equivalence (7.27). This result will yield the same values as those obtained from Equivalence (7.27) over the common sector of $((2M-1)\pi - \varepsilon_1)/\beta < \arg z < ((2M-1)\pi + \varepsilon_2)/\beta$. Similarly, the MB-regularised value of $S(N, z^\beta)$ for $((2M+1)\pi - \varepsilon_1)/\beta < \arg z < ((2M+3)\pi + \varepsilon_2)/\beta$ is obtained by replacing M by $M+1$ in Equivalence (7.27). Thus, this yields the same values as those obtained from Equivalence (7.27) when $((2M+1)\pi - \varepsilon_1)/\beta < \arg z < ((2M+1)\pi + \varepsilon_2)/\beta$.

On the other hand, we have already seen that Borel summation of an asymptotic series results in specific sectors across which the regularised value acquires jump discontinuities. These Stokes sectors do not overlap and hence, there are no common sectors where different forms for the regularised value can yield the same values. Although both methods of regularisation act differently on an asymptotic series, both yield the same values. We shall see this more clearly in later chapters when we compare the values obtained via the MB regularisation of generalised terminants with those obtained by Borel summation.

The regularised value of $S(N, z^\beta \exp(-2\pi i l))$ given by Equivalence (7.18) has been derived with the specific conditions for $f(k)$ given in Proposition 3. Let us now consider extending the third condition to

$$|f(c \pm iL)| \overset{L \to \infty}{\sim} \begin{cases} e^{AL} \\ e^{BL} \end{cases} , \tag{7.39}$$

where our aim now is to determine bounds for A and B. At the upper limit the modulus of the integrand for the MB-regularised value of the Type I series behaves as

$$\left| \frac{f(c+iL) z^{\beta(c+iL)} e^{-2i\pi l(c+iL)}}{e^{-i\pi(c+iL)} - e^{i\pi(c+iL)}} \right| \overset{L \to \infty}{\sim} |z|^{\beta c} e^{(A+(2l-1)\pi - \beta\theta)L} , \tag{7.40}$$

while at the lower limit it behaves as

$$\left| \frac{f(c-iL) z^{\beta(c-iL)} e^{-2i\pi l(c-iL)}}{e^{-i\pi(c-iL)} - e^{i\pi(c-iL)}} \right| \overset{L \to \infty}{\sim} |z|^{\beta c} e^{(B-(2l+1)\pi + \beta\theta)L} . \tag{7.41}$$

To ensure that the MB integral is convergent, we require that the exponential factors in the above results are negative or decaying. From the first result we arrive at the following condition:

$$\beta\theta > (2l-1)\pi + A , \tag{7.42}$$

while the second result yields

$$\beta\theta < (2l+1)\pi - B . \tag{7.43}$$

Therefore, for the MB-regularised value given in Proposition 3 to possess a domain of convergence, we require that the rhs of the first inequality be lower than the rhs of the second inequality, which, in turn, means that

$$A < 2\pi - B . \tag{7.44}$$

If the above inequality holds, then we can write

$$S(N, z_*^\beta) = \sum_{k=N}^\infty f(k)\left(-z_*^{\beta k}\right) \equiv I_l(z^\beta) = \int_{\substack{c-i\infty \\ N-1 < c = \Re s < N}}^{c+i\infty} ds \frac{z^{\beta s} e^{-2\pi i l s} f(s)}{e^{-i\pi s} - e^{i\pi s}} , \tag{7.45}$$

where $z_* = z \exp(-2\pi i l / \beta)$.

The domain of convergence for the integral $I_{l+1}(z^{\beta})$ can be obtained by replacing l with $l+1$ in Inequalities (7.42) and (7.43) and is given by

$$(2l+1)\pi + A < \beta\theta < (2l+3)\pi - B \ . \tag{7.46}$$

For the domains of convergence of $I_l(z^{\beta})$ and $I_{l+1}(z^{\beta})$ to overlap, the lower bound of the above inequality must be lower than the upper bound of Inequality (7.43), which means that $A > -B$.

We have seen that the regularised value of $S(N, z^{\beta})$ obtained via MB regularisation begins with $I_0(z)$ whose domain of convergence according to the conditions on $f(s)$ in Proposition 3 is given by $-(\pi + \varepsilon_1)/\beta < \arg z < (\pi + \varepsilon_2)/\beta$. If we alter the conditions on $f(s)$ to those in Approximation (7.39), then the domain of convergence will be given by Inequality (7.44) or $(A - \pi)/\beta < \arg z < (\pi - B)/\beta$. In addition, if we require that the domain of convergence lies in the principal branch of the complex plane, then we must have

$$B > (1 - \beta)\pi \quad \text{and} \quad A > (1 - \beta)\pi \ . \tag{7.47}$$

From the above analysis we have observed that MB regularisation of the series $S(N, z^{\beta})$ is unaffected by the Stokes phenomenon. The MB-regularised value given by Equivalence (7.27) is valid over a domain of convergence, which with the new conditions on $f(s)$ is now given by $((2M-1)\pi + A)/\beta < \arg z < ((2M+1)\pi - B)/\beta$. For consecutive values, say M and $M+1$, the domains of convergence for the MB integrals overlap if and only if $-B < A < 2\pi - B$. In the overlapping or common sector we have two choices or forms for determining the regularised value of $S(N, z^{\beta})$. While a Stokes line often bisects the common sector, there is no jump discontinuity term that needs to be introduced to either MB-regularised form to yield the regularised value. As we shall see shortly, this will not apply to the other type of series mentioned in the introduction, viz. $S_1(N, z)$.

7.2 Series of the Type $S_1(N, z)$

We now investigate MB regularisation of the second type of series, $S_1(N, z^{\beta})$, which include the second type of terminant discussed in Ch. 3. As has already been stated, this type of series is different from $S(N, z^{\beta})$ because when it is derived via an asymptotic method, it is often done so initially for positive real values of z^β, although this is not necessary as discussed later. That is, whenever this type of asymptotic series is derived, it is usually done so for $\arg z = 2j\pi/\beta$, where j is an arbitrary integer. These, of course, represent the Stokes lines for the series. We have already seen in the case of the second type of terminant that when z moves off a Stokes line, jump discontinuities emerge, which alter the regularised value as in Equivalence (3.12). Since $S_1(N, z^\beta)$ like the second type of terminant is derived for z lying on a Stokes line from the outset, we also expect jump discontinuities to occur in its regularised value as soon as z moves off this line in either direction. As will be seen in this section, these jump discontinuities cannot be accounted for by direct application of Proposition 3 to $S_1(N, z^\beta)$. Consequently, we will find it necessary to specify the particular Stokes line or value of j for which the regularised value of $S_1(N, z^\beta)$ was derived initially. The chosen value of j will be referred to as the primary Stokes line and it will be expected to behave in accordance with the Stokes line for the second type of terminant in Ch. 3, which culminated in the regularised value being given by Equivalence (3.12). In marked contrast, the other Stokes lines, which will be referred to as secondary Stokes lines, need not necessarily possess the same properties of the primary Stokes line, especially when the regularised value is derived by MB regularisation.

Without loss of generality we shall choose the primary Stokes line to be the $j = 0$ line, while the secondary Stokes lines will be given by $\arg z = 2j\pi/\beta$, where $j \neq 0$. In making this choice for the primary Stokes line, we are basically following convention where $S_1(N, z^\beta)$ has been initially derived under the condition that $\arg z = 0$, although it should be emphasised that the results obtained in this section can be adapted to other values of j as the primary Stokes line.

For the time being, let us consider the application of Proposition 3 to the series $S_1(N, z^\beta)$. As in the previous section, we shall without loss of generality assume that β is both real and positive. Then we find that

$$S_1\left(N, z^\beta\right) = \sum_{k=N}^{\infty} f(k)(z^\beta)^k \equiv \int_{\substack{c-i\infty \\ N-1<c=\Re s<N}}^{c+i\infty} ds \, \frac{(-z^\beta)^s f(s)}{e^{-i\pi s} - e^{i\pi s}} \, . \qquad (7.48)$$

The main difference between this result and the regularised value of $S(N, z^\beta)$ is that the multi-valued phase factor of $(-1)^s$ now appears in the integrand of the MB integral. Its appearance means that the regularised value of $S_1(N, z^\beta)$ has become ambiguous because we can interpret it as either $\exp(i\pi s)$, $\exp(-i\pi s)$ or even $\exp(-i(2l+1)\pi s)$, where l is an arbitrary integer. If we adopt the first interpretation for the phase factor, then Equivalence (7.48) becomes

$$S_1(N, z^\beta) \equiv I_{-1}^*(z^\beta) = \int_{\substack{c-i\infty \\ N-1<c=\Re s<N}}^{c+i\infty} ds \, \frac{z^{\beta s} \, e^{i\pi s} f(s)}{e^{-i\pi s} - e^{i\pi s}} \,, \tag{7.49}$$

while if we adopt the second interpretation, then we obtain

$$S_1(N, z^\beta) \equiv I_0^*(z^\beta) = \int_{\substack{c-i\infty \\ N-1<c=\Re s<N}}^{c+i\infty} ds \, \frac{z^{\beta s} \, e^{-i\pi s} f(s)}{e^{-i\pi s} - e^{i\pi s}} \,. \tag{7.50}$$

Because the integrands are different, both integrals yield different results. Yet according to Proposition 3, they should yield the regularised value for the same series. In addition, since there is no reason why $(-1)^s$ cannot be interpreted as $\exp(-i(2l+1)\pi s)$, where l is any integer, we can write the MB integral more generally as

$$S_1(N, z^\beta) \equiv I_l^*(z^\beta) = \int_{\substack{c-i\infty \\ N-1<c=\Re s<N}}^{c+i\infty} ds \, z^{\beta s} f(s) \, \frac{e^{-i(2l+1)\pi s}}{e^{-i\pi s} - e^{i\pi s}} \,. \tag{7.51}$$

However, by replacing $z^{\beta k}$ by $z^{\beta k} 1^k = z^{\beta k} \exp(-2i\pi l k)$ in $S_1(N, z^\beta)$, we find also via Proposition 3 that

$$S_1\left(N, z^\beta e^{-2\pi i l}\right) = \sum_{k=N}^{\infty} f(k) z^{\beta k} \exp(-2i\pi l k) \equiv \int_{\substack{c-i\infty \\ N-1<c=\Re s<N}}^{c+i\infty} ds \, \left(-z^\beta\right)^s f(s)$$
$$\times \frac{e^{-2il\pi s}}{e^{-i\pi s} - e^{i\pi s}} \,. \tag{7.52}$$

Hence, interpreting $(-z^\beta)^s$ as $\exp(-i\pi s) z^{\beta s}$ yields the result given by Equivalence (7.51). Similarly, the regularised value of $S_1(N, z^\beta \exp(-2i(l+1)\pi))$ with $(-1)^s$ as $\exp(i\pi s)$ yields Equivalence (7.51). Therefore, Equivalence (7.51) is the regularised value for either $S_1(N, z^\beta \exp(-2il\pi))$ or $S_1(N, z^\beta \exp(-2i(l+1)\pi))$ depending upon whether $(-1)^s$ is interpreted as $\exp(-i\pi s)$ or $\exp(i\pi s)$. This means that we only need to consider the first two interpretations of $(-1)^s$, not the third option, when deriving the regularised value of $S_1(N, z^\beta)$.

Besides causing ambiguity, the multi-valued factor of $(-1)^s$ also affects the domain of convergence of the MB integral. If we consider the first interpretation of $(-1)^s$, which means that the regularised value of $S_1(N, z^\beta)$ is given by $I_{-1}^*(z)$, and assume that $f(s)$ has the modified exponential behaviour at the endpoints of the MB integral as in Approximation (7.39), then by analysing the modulus of the integrand, we find that $I_0^*(z^\beta)$ is convergent only when $A/\beta < \arg z < (2\pi - B)/\beta$ and $A < 2\pi - B$. The latter represents the condition for the existence of a domain of convergence for all $I_l^*(z^\beta)$, which applies also to all $I_l(z^\beta)$ examined in the previous section. In the case of the first condition a shift has occurred in the domain of convergence for $I_0^*(z^\beta)$ compared with the domain of convergence for $I_0(z^\beta)$ obtained by putting $l = -1$ in Inequality (7.46). For $I_{-1}^*(z^\beta)$ the domain of convergence is $(A - 2\pi)/\beta < \arg z < -B/\beta$, which overlaps that for $I_0^*(z^\beta)$ provided $A < -B$. In fact, if this condition holds, then the domains of convergence for $I_{j-1}^*(z^\beta)$ and $I_j^*(z^\beta)$ also overlap as we shall see shortly. For now, however, we see that the common sector of the domains of convergence for $I_{-1}^*(z^\beta)$ and $I_0^*(z^\beta)$ is $A/\beta < \arg z < -B/\beta$ when $S_1(N, z^\beta)$ is an asymptotic series, i.e. when the series has zero radius of absolute convergence.

The difference between consecutive values of the MB integral is given by

$$\Delta I_{j,j-1}^*(z^\beta) = I_j^*(z^\beta) - I_{j-1}^*(z^\beta) = \int_{\substack{c-i\infty \\ N-1<c=\Re s<N}}^{c+i\infty} ds \, z^{\beta s} e^{-2ij\pi s} f(s) \ . \qquad (7.53)$$

Furthermore, if $f(s)$ is the Mellin transform of a function $F(x)$, then Eq. (7.53) becomes

$$\Delta I_{j,j-1}^*(z^\beta) = 2\pi i F\left(z^{-\beta} e^{2ij\pi}\right) \ . \qquad (7.54)$$

For this result to be valid for all values of the offset c, the poles of $f(s)$ must be situated in the left hand half of the complex plane. Writing Eq. (7.54) in terms of $F(z)$ means that we can analytically continue $\Delta I_{j,j-1}^*(z^\beta)$ to all values of z. As a consequence, we can determine the regularised value of $S_1(N, z^\beta)$ for all values of z. Furthermore, for the case where $F(x) = x^\gamma g(x)$ and $g(x)$ can be written as a convergent power series, Eq. (7.54) becomes

$$\Delta I_{j,j-1}^*(z^\beta) = 2\pi i z^{-\beta\gamma} e^{2ij\pi\gamma} g\left(z^{-\beta}\right) \ . \qquad (7.55)$$

From the preceding results we see that both $I_{-1}^*(z^\beta)$ and $I_0^*(z^\beta)$ are defined over a common sector of the complex plane, yet both forms seem to represent the regularised value of $S_1(N, z^\beta)$. Clearly, this contradicts the fundamental principle of regularisation that each divergent series possesses a unique regularised value or limit. If only one of the MB-regularised forms is valid, then we need to explain why the other is invalid. At this stage there does not seem to be a logical reason for excluding either possibility. However, we can overcome this problem by defining a primary Stokes line, viz. $\theta = 0$, along which the series $S_1(N, z^\beta)$ is composed of terms that are only real and positive. According to p. 10 of Ref. [3], but which actually stems from the pioneering work of Zwaan [23], an initially real function cannot acquire an imaginary part. Previously, we have referred to this as Zwaan-Dingle principle. This means that the regularised value should always be real on the primary Stokes line, but both $I_0^*(z^\beta)$ and $I_{-1}^*(z^\beta)$ are complex. In actual fact, they are complex conjugates of one another on the primary Stokes line.

To explain this conundrum we need to introduce extra terms into the regularised value of $S_1(N, z^\beta)$ based on the fact that as soon as one moves off a Stokes line jump discontinuities emerge. Hence, we write the MB-regularised value of $S_1(N, z^\beta)$ as

$$S_1(N, z^\beta) \equiv \begin{cases} I_0^*(z^\beta) + iC(z^\beta) & , \quad 0 < \theta = \arg z < (2\pi - B)/\beta \ , \\ I_{-1}^*(z^\beta) + iD(z^\beta) & , \quad (A - 2\pi)/\beta < \theta < 0 \ . \end{cases} \tag{7.56}$$

From our study of the second type of terminant in Ch. 3 we have seen that averaging the two forms for the regularised value abutting the primary Stokes line yields the regularised value of $\bar{\Lambda}_\alpha(-z)$ along the primary Stokes line. Therefore, we average the above forms, thereby obtaining

$$S_1(N, z^\beta) \equiv \int_{\substack{c-i\infty \\ N-1<c=\Re s<N}}^{c+i\infty} ds \, \frac{z^s f(s) \cos \pi s}{e^{-i\pi s} - e^{i\pi s}} + \frac{i}{2} \left(C(z^\beta) + D(z^\beta) \right)$$

$$= \frac{i}{2} \int_{\substack{c-i\infty \\ N-1<c=\Re s<N}}^{c+i\infty} ds \, z^{\beta s} f(s) \cot \pi s + \frac{i}{2} \left(C(z^\beta) + D(z^\beta) \right) \ . \tag{7.57}$$

The MB integral in the above equivalence can be shown to be real when θ or $\arg z = 0$, which is left as an exercise for the reader. Therefore, to obtain real regularised values of $S_1(N, z^\beta)$ on the primary Stokes line that are in accordance with the forms in Equivalence (3.12), we must have $C(z^\beta) = -D(z^\beta)$. This means that we are essentially treating $(-z^\beta)^s$ by adopting the usual convention of

$$(-z^\beta)^s \equiv \begin{cases} z^{\beta s} e^{-i\pi s} & , \quad \arg z > 0 \ , \\ z^{\beta s} \cos(\pi s) & , \quad \arg z = 0 \ , \\ z^{\beta s} e^{i\pi s} & , \quad \arg z < 0 \ . \end{cases} \tag{7.58}$$

On p. 13 of Ref. [3] it is stated that the above approach for handling $(-z^\beta)^s$ in asymptotic expansions is an indication that the Stokes phenomenon has occurred. Furthermore, we have seen from Equivalence (7.18) that subtracting the results abutting the primary Stokes line as if they applied in the same Stokes sector yields the entire Stokes discontinuity, which is also imaginary. Therefore, when we subtract the regularised values in Equivalence (7.56), we obtain

$$
\begin{aligned}
C(z^\beta) &= \frac{i}{2}\,\Delta I^*_{0,-1}(z^\beta) = \frac{i}{2}\left(I^*_0(z^\beta) - I^*_{-1}(z^\beta)\right) \\
&= \frac{i}{2}\int_{\substack{c-i\infty \\ N-1<c=\Re s<N}}^{c+i\infty} ds\, z^{\beta s}\, f(s)\ .
\end{aligned}
\tag{7.59}
$$

As a consequence, Equivalence (7.56) becomes

$$
S_1(N, z^\beta) \equiv
\begin{cases}
I^*_0(z^\beta) - \frac{1}{2}\,\Delta I^*_{0,-1}(z^\beta) & , \quad 0 < \arg z < (2\pi - B)/\beta \\
I^*_{-1}(z^\beta) + \frac{1}{2}\,\Delta I^*_{0,-1}(z^\beta) & , \quad (-2\pi + A)/\beta < \arg z < 0
\end{cases}
\ ,
\tag{7.60}
$$

The upper limit on the argument of z in the first form for the regularised value of $S_1(N, z^\beta)$ represents the upper limit of the domain of convergence for $I^*_0(z^\beta)$. Because $B < 0$, this form for the regularised value is valid beyond the secondary Stokes line of $\arg z = \pi/\beta$. Conversely, the lower limit on the argument of z in the second form represents the lower limit of the domain of convergence for $I^*_{-1}(z^\beta)$. Hence, the second form for the regularised value is valid beyond the secondary Stokes line of $\arg z = -\pi/\beta$. In addition, with the aid of the $j = 0$ version of Eq. (7.53) we can re-write the above equivalence as

$$
S_1(N, z^\beta) \equiv
\begin{cases}
I^*_{-1}(z^\beta) + \frac{1}{2}\,\Delta I^*_{0,-1}(z^\beta) & , \quad 0 < \arg z < -B/\beta \\
I^*_0(z^\beta) - \frac{1}{2}\,\Delta I^*_{0,-1}(z^\beta) & , \quad A/\beta < \arg z < 0
\end{cases}
\ .
\tag{7.61}
$$

Unlike the regularised value of the first type of asymptotic series, namely $S(N, z^\beta)$, whose derivation began with a direct application of Proposition 3, we see that the regularised value of $S_1(N, z^\beta)$ as given by Equivalence (7.60) is not only composed of the corresponding MB integrals obtained by the direct application of Proposition 3 for both Stokes sectors abutting the primary Stokes line, but it also possesses an extra term representing half their difference, viz. $\Delta I^*_{0,-1}(z^\beta)$. With regard to the alternative version of the regularised value of $S_1(N, z^\beta)$ given by Equivalence (7.61), the upper limit on $\arg z$ in the first form represents the upper limit for the common region over the domains of convergence for both MB integrals, while the lower limit on $\arg z$ in the second form represents the lower limit of the common region. Consequently, the forms in Equivalence (7.61) are valid over narrower regions of $\arg z$ in the complex plane.

In summary, the extra terms in the regularised value in Equivalence (7.60) arise solely because the second type of asymptotic series denoted by $S_1(N, z^\beta)$ is generally derived for z lying on the primary Stokes line. This means that as z moves off this line, the regularised value develops jump discontinuous terms, which cannot be obtained by the direct application of Proposition 3 to $S_1(N, z^\beta)$. However, by utilising the main properties of the Stokes phenomenon presented in Ch. 1 of Ref. [3] and summarised in Ch. 3 here, we have found that the regularised value of $S_1(N, z^\beta)$ for z lying on the primary Stokes line given by $\arg = 0$ is the average of the two MB integrals in Equivalence (7.56), which is in accordance with Rule 8a presented in Ch. 3. Consequently, the regularised value of Type II series along the primary Stokes line is given by

$$S_1\left(N, z^\beta\right) \equiv \frac{i}{2} \int_{\substack{c-i\infty \\ N-1 < c = \Re s < N}}^{c+i\infty} ds\, |z|^{\beta s} f(s) \cot \pi s \ . \tag{7.62}$$

Because the above integral can be shown to be imaginary, the regularised value of $S_1(N, z^\beta)$ is real for z on the primary Stokes line in accordance with the Zwaan-Dingle principle.

To obtain the regularised value of $S_1(N, z^\beta)$ when θ or $\arg z$ lies outside the Stokes sectors abutting the primary Stokes line, we require the domain of convergence for $I_j^*(z)$, which is determined by studying the behaviour of the integrand at the endpoints of the MB integral in Equivalence (7.51). Then it is found that

$$\left| \frac{z^{\beta s} f(s) e^{-(2j+1)i\pi s}}{e^{-i\pi s} - e^{i\pi s}} \right| \overset{L \to \infty}{\sim} |z|^{\beta c} \begin{bmatrix} e^{2j\pi L - \beta \theta L + AL} \\ e^{-(2j+2)\pi L + \beta \theta L + BL} \end{bmatrix} \ . \tag{7.63}$$

From the above result we see that $I_j^*(z^\beta)$ is convergent whenever $(2j\pi + A)/\beta < \arg z < ((2j+2)\pi - B)/\beta$. Therefore, the common sector for the MB integrals, $I_j^*(z^\beta)$ and $I_{j+1}^*(z^\beta)$, is given by $((2j+2)\pi - B)/\beta < \arg z < ((2j+2)\pi + A)/\beta$, which also represents the domain of convergence for the difference between them, viz. $\Delta I_{j+1,j}^*(z^\beta)$ as given by Eq. (7.53).

If $\Delta I_{j,j-1}^*(z^\beta)$ can be expressed in the form given by Eq. (7.54), then we can extend the first form on the rhs of Equivalence (7.60) beyond $\arg z < (2\pi - B)/\beta$.

By putting $j=1$ in Eqs. (7.53) and (7.54), we obtain

$$I_0^*\left(z^\beta\right) = I_1^*\left(z^\beta\right) - 2\pi i\, F\left(z^{-\beta}e^{2i\pi}\right) \ . \tag{7.64}$$

Since the domain of convergence for $I_1^*(z^\beta)$ is $(2\pi+A)/\beta < \arg z < (4\pi-B)/\beta$, the common sector belonging to the domains of convergence for $I_1^*(z^\beta)$ and $I_0^*(z^\beta)$ is $(2\pi+A)/\beta < \arg z < (2\pi-B)/\beta$. Therefore, the regularised value of $S_1(N, z^\beta)$ over the common sector is given by

$$S_1\left(N, z^\beta\right) \equiv I_1^*\left(z^\beta\right) - 2\pi i F\left(z^{-\beta}e^{2i\pi}\right) - \pi i F\left(z^{-\beta}\right) \ . \tag{7.65}$$

However, we have seen that the domain of convergence for $I_1^*(z^\beta)$ extends to $\arg z < (4\pi-B)/\beta$. Therefore, by analytic continuation the above equivalence is valid for $(2\pi+A)/\beta < \arg z < (4\pi-B)/\beta$.

When $j=2$, Eqs. (7.53) and (7.54) yield

$$I_1^*\left(z^\beta\right) = I_2^*\left(z^\beta\right) - 2\pi i F\left(z^{-\beta}e^{4i\pi}\right) \ , \tag{7.66}$$

which is valid for $(4\pi+A)/\beta < \arg z < (4\pi-B)/\beta$. As a consequence, we can replace $I_1^*(z^\beta)$ in Equivalence (7.65) by the rhs of Eq. (7.66), thereby obtaining

$$S_1\left(N, z^\beta\right) \equiv I_2^*\left(z^\beta\right) - 2\pi i F\left(z^{-\beta}e^{4i\pi}\right) - 2\pi i F\left(z^{-\beta}e^{2i\pi}\right) - \pi i F\left(z^{-\beta}\right). \tag{7.67}$$

Since the domain of convergence for $I_2(z^\beta)$ is $(4\pi+A)/\beta < \arg z < (6\pi-B)/\beta$, Equivalence (7.67) can be continued analytically to $\arg z < (6\pi-B)/\beta$. From the $j=1$ and $j=2$ results for Eqs. (7.53) and (7.54) we see that a pattern is developing that enables us to determine the regularised value of $S_1(N, z^\beta)$ for any Stokes sector above the primary Stokes line. Therefore, for $(2l\pi+A)/\beta < \arg z < ((2l+2)\pi-B)/\beta$, the regularised value of $S_1(N, z^\beta)$ is given by

$$S_1\left(N, z^\beta\right) \equiv I_l^*\left(z^\beta\right) - 2\pi i \sum_{j=1}^{l} F\left(z^{-\beta}e^{2il\pi}\right) - \pi i F\left(z^{-\beta}\right) \ . \tag{7.68}$$

For the case where $F(x) = x^\gamma g(x)$ and $g(x)$ is a function that can be expressed in terms of a convergent power series, Equivalence (7.68) reduces to

$$S_1\left(N, z^\beta\right) \equiv I_l^*\left(z^\beta\right) - \pi i z^{-\beta\gamma} g\left(z^{-\beta}\right)\left[2e^{i(l+1)\pi\gamma}\frac{\sin(l\pi\gamma)}{\sin(\pi\gamma)} + 1\right] \ . \tag{7.69}$$

The second form on the rhs of Equivalence (7.60) is only valid for the Stokes sector immediately below the primary Stokes line, which is encompassed by $(-2\pi + A)/\beta < \arg z < 0$. To obtain the regularised value of $S_1(N, z^\beta)$ for the second Stokes sector immediately below the primary Stokes line, we first put $j = -1$ in Eqs. (7.53) and (7.54), thereby obtaining

$$I^*_{-1}(z^\beta) = I^*_{-2}(z^\beta) + 2\pi i F\left(z^\beta e^{-2i\pi}\right) \ . \tag{7.70}$$

The above result is only valid over the common sector of $(-2\pi + A)/\beta < \arg z < (-2\pi - B)/\beta$ for the MB integrals, $I_{-1}(z^\beta)$ and $I_{-2}(z^\beta)$. Next we replace $I^*_{-1}(z^\beta)$ by the rhs of the above equation in the second form of Equivalence (7.60). Since the domain of convergence for $I^*_{-2}(z^\beta)$ is $(-4\pi + A)/\beta < \arg z < (-2\pi - B)/\beta$, this means that the regularised value of $S_1(N, z^\beta)$ given by

$$S_1(N, z^\beta) \equiv I^*_{-2}(z^\beta) + 2\pi i F\left(z^{-\beta} e^{-2i\pi}\right) + \pi i F\left(z^{-\beta}\right) \ , \tag{7.71}$$

can be analytically continued to $\arg z > (-4\pi + A)/\beta$. We can continue the process indefinitely as we did for the sectors above the primary Stokes line. As a result, we find that the regularised value of $S_1(N, z^\beta)$ below the primary Stokes line can be written generally as

$$S_1(N, z^\beta) = I^*_{-l-1}(z^\beta) + 2\pi i \sum_{j=1}^{l} F\left(z^{-\beta} e^{-2il\pi}\right) + \pi i F(z^{-\beta}) \ , \tag{7.72}$$

where $(-2(l+1)\pi + A)/\beta < \arg z < (-2l\pi - B)/\beta$ and l is a non-negative integer. We also see that Equivalence (7.72) represents the complex conjugate of Equivalence (7.68). For the case where $F(x) = x^\gamma g(x)$ and $g(x)$ is a function that can be expressed in terms of a convergent power series, the regularised value of $S_1(N, z^\beta)$ reduces to

$$S_1(N, z^\beta) \equiv I^*_{-l}(z^\beta) + \pi i z^{-\beta\gamma} g(z^{-\beta})\left[2e^{-i(l+1)\pi\gamma}\frac{\sin(l\pi\gamma)}{\sin(\pi\gamma)} + 1\right] \ . \tag{7.73}$$

It should be stressed here that results such as Equivalences (7.68) and (7.72) are only valid if the real parts of A and B are negative and $S_1(N, z^\beta)$ is an asymptotic series. Otherwise, the domains of convergence for the $I^*_l(z^\beta)$ will not overlap. Stokes lines, which are given by $\theta = 2j\pi/\beta$ with j an integer, are situated in the common sectors of the overlapping domains of convergence of these integrals.

Except for the primary Stokes line, the Stokes lines in MB regularisation are no longer lines of discontinuity as they are when either $S_1(N, z^\beta)$ or $S(N, z^\beta)$ is Borel-summed. Hence, MB regularisation of the second type of series is very different to Borel summation of these series. Furthermore, there are no Stokes multipliers in the regularised values of asymptotic series obtained via MB regularisation. Nevertheless, both methods of regularisation yield the same regularised values of divergent series, which will be demonstrated numerically in Chs. 9 and 10 where Dingle's theory of terminants is generalised.

If the MB-regularised forms for the regularised value of $S_1(N, z^\beta)$ for adjacent domains of convergence are averaged across a secondary Stokes line, which amounts to averaging Equivalence (7.68) with l set equal to both l and $l+1$ and then putting $\theta = 2(l+1)\pi/\beta$, it will not necessarily produce a real regularised value for the series. According to the Zwaan-Dingle principle, mentioned earlier, whenever all the terms in the series $S_1(N, z^\beta)$ are real and positive initially, the regularised value of the series can only be real. However, when $\arg z$ lies on any Stokes line, all the terms in $S_1(N, z^\beta)$ are also real and positive. Because there seems to be no difference between the Stokes lines, this implies that whenever z lies on a Stokes line, the regularised value of $S_1(N, z^\beta)$ should be real. From Equivalences (7.68) and (7.72), we see that the only case where we can be guaranteed that a real regularised value will be produced is on the primary Stokes line or when $\arg z = 0$. For $\arg z = 2j\pi/\beta$, where $j \neq 0$, the regularised value of $S_1(N, z^\beta)$ can be complex even though all the terms in $S_1(N, z^\beta)$ are real and positive. Hence, it was necessary to introduce the concept of a primary Stokes line, which is only applicable to the initial situation, while the secondary Stokes lines simply do not obey the Zwaan-Dingle principle.

In this chapter we chose the primary Stokes line to be $\arg z = 0$ because $S_1(N, z^\beta)$ is usually determined for positive real values of z lying in the principal branch of the complex plane given by $(-\pi, \pi]$. Therefore, the choice has been dictated by convention rather than by any rule or theorem. In actual fact, we could have chosen an arbitrary secondary Stokes line, say $\theta = 2l'\pi/\beta$, as the primary Stokes line. This would result in adapting the above analysis to deal with $I_{l-l'}^*(z^\beta)$ instead of $I_l^*(z^\beta)$ and would ultimately lead to l being replaced by $l - l'$ in Equivalences (7.68) and (7.72). A different primary Stokes line merely means that the regularised value of $S_1(N, z^\beta)$ is shifted to other branches of the complex plane.

It has already been stated that when a Type II asymptotic series is derived, it need not necessarily be for positive real values of z^β. A future publication will investigate the Type II asymptotic series for the Laplace transform of $t/\sin(at)$, which is undefined along the positive real axis. In order to use the above results, the Cauchy principal value of the Laplace transform must be evaluated. Because this means neglecting semi-residue contributions, the regularised values using the above results will not give the actual values of the Laplace transform off the positive real axis. Consequently, the above results need to be adapted whenever a Type II series is only valid for a primary Stokes sector and not a primary Stokes line. Then the regularised value acquires half the Stokes discontinuity along the positive real axis, while in the next sector, it acquires the entire Stokes discontinuity. This is analogous to the situation for Type I asymptotic series.

Before concluding this chapter, let us consider two examples of the Type II series, $S_1(N, z^\beta)$: (1) where the series is not asymptotic and (2) where it is. In the first example we shall study the abbreviated binomial series given by Eq. (7.7) except that $-z$ is now replaced by z. That is, we aim to apply Proposition 3 to

$$_1\mathcal{F}_0(\rho; z)\,|_N = \sum_{k=N}^{\infty} \frac{\Gamma(k+\rho)}{\Gamma(\rho)\,k!} z^k \; . \qquad (7.74)$$

Then we find that

$$_1\mathcal{F}_0(\rho; z)\,|_N \equiv \frac{1}{\Gamma(\rho)} \int_{\substack{c-i\infty \\ \mathrm{Max}[N-1,-\Re\rho]<c=\Re s<N}}^{c+i\infty} ds\, \frac{\Gamma(\rho+s)}{\Gamma(s+1)} \left(\frac{(-z)^s}{e^{-i\pi s} - e^{i\pi s}} \right), \quad (7.75)$$

where $\Re\rho > -N$ as in Equivalence (7.14). We have two choices for $-z$; we can set it equal to either $z\exp(-i\pi)$ or $z\exp(i\pi)$. Thus, it seems that we have two different regularised values for the same series. However, we know from studying the behaviour of the integrands that these regularised values will apply over different sectors or even branches of the complex plane. That is, the asymptotic forms are different, which is in keeping within concept of regularisation. For $-z = z\exp(-i\pi)$, we find that the integrand in Equivalence (7.14) behaves as

$$\left| \frac{z^s f(s)}{e^{-i\pi s} - e^{i\pi s}} \right| \overset{L\to\infty}{\sim} |z|^c e^{-L\theta} |f(c+iL)| \; , \qquad (7.76)$$

at the endpoint $s = c+iL$, while at the lower endpoint, $s = c-iL$, it behaves as

$$\left| \frac{z^s f(s)}{e^{-i\pi s} - e^{i\pi s}} \right| \overset{L\to\infty}{\sim} |z|^c e^{L(\theta-2\pi)} |f(c-iL)| \; . \qquad (7.77)$$

When Approximation (7.12) is introduced into these results, we find that for $(-z) = z\exp(-i\pi)$, Equivalence (7.75) is valid only over $0 < \arg z < 2\pi$. If we let $-z = z \exp(i\pi)$ in Equivalence (7.75), then we find that at the upper endpoint, the integrand behaves as

$$\left| \frac{z^s f(s)}{e^{-i\pi s} - e^{i\pi s}} \right| \overset{L \to \infty}{\sim} |z|^c e^{-L(\theta + 2\pi)} |f(c + iL)| , \qquad (7.78)$$

while at the lower endpoint it behaves as

$$\left| \frac{z^s f(s)}{e^{-i\pi s} - e^{i\pi s}} \right| \overset{L \to \infty}{\sim} |z|^c e^{L\theta} |f(c - iL)| . \qquad (7.79)$$

By introducing Approximation (7.12) into the above results, we find that Equivalence (7.75) is only valid for $-2\pi < \arg z < 0$. Therefore, for the principal branch, i.e. $-\pi < \arg z \le \pi$, the MB-regularised value of the abbreviated binomial series is given by

$$_1\mathcal{F}_0(\rho; z)|_N \equiv \frac{1}{\Gamma(\rho)} \int_{\underset{\text{Max}[N-1, -\Re\rho] < c = \Re s < N}{c - i\infty}}^{c + i\infty} ds \, \frac{\Gamma(\rho + s)}{\Gamma(s+1)} \left(\frac{z^s e^{-i\pi s}}{e^{-i\pi s} - e^{i\pi s}} \right), \quad (7.80)$$

for $0 < \arg z \le \pi$, whereas for $-\pi \le \arg z \le 0$, it is given by

$$_1\mathcal{F}_0(\rho; z)|_N \equiv \frac{1}{\Gamma(\rho)} \int_{\underset{\text{Max}[N-1, -\Re\rho] < c = \Re s < N}{c - i\infty}}^{c + i\infty} ds \, \frac{\Gamma(\rho + s)}{\Gamma(s+1)} \left(\frac{z^s e^{i\pi s}}{e^{-i\pi s} - e^{i\pi s}} \right). \quad (7.81)$$

As was the case for the regularised value of $_1\mathcal{F}_0(\rho; -z)|_N$, $\Re\rho > -N$ in these results.

For $\arg z = 0$, the MB-regularised value may become undefined because the integrand displays algebraic behaviour rather than exponentially decaying behaviour at the endpoints. This is left as an exercise for the reader to investigate. Consequently, an alternative method must be sought when evaluating the regularised value of the abbreviated binomial series along the positive real axis. Let us consider expressing the abbreviated binomial series in terms of the complete binomial series. We can achieve this simply by making the appropriate modifications to Equivalence (7.9), which results in

$$_1\mathcal{F}_0(\rho; z)|_N = \frac{\Gamma(N + \rho)}{\Gamma(\rho)} L_N(z) \sum_{k=0}^{\infty} \frac{\Gamma(k + N + \rho)}{\Gamma(N + \rho) k!} z_N^k . \qquad (7.82)$$

The operator $L_N(z)$ appearing in Eq. (7.82) is given by Eq. (4.15). Therefore, according to Proposition 3 the regularised value of the above series becomes

$$_1\mathcal{F}_0(\rho; z)|_N \equiv \frac{\Gamma(N + \rho)}{\Gamma(\rho)} L_N(z)(1 - z_N)^{-N - \rho} . \qquad (7.83)$$

The above integral can be evaluated by assuming that $z, z_1, z_2, \ldots, z_{N-1}, z_N < 1$. Because the final result is independent of the z_i, the only condition that matters is $z < 1$. When $z < 1$, we can replace the equivalence symbol by an equals sign. On the other hand, if ρ is an integer, then the regularised value of the abbreviated binomial series can be extended to $z > 1$, but for all other values of ρ we encounter a branch point singularity at $z = 1$ with a cut to infinity. This is the reason why Mathematica is unable to evaluate the average of Equivalences (7.80) and (7.81) for $z > 1$ on the positive real axis and non-integral values of ρ. As stated in Ch. 4 the regularised value of a divergent series is non-analytic everywhere the original function is non-analytic, which for the abbreviated binomial series is the positive real axis from the branch point situated at $z = 1$. If we were to apply Zwaan-Dingle principle that the regularised value should only be real along the positive real axis, then the imaginary contributions must be neglected, i.e. for $z_i > 1$, $(1 - z_i)^{-N-\rho}$ should be become $(z_i - 1)^{-N-\rho} \cos((\rho + N)\pi)$. However, the integral operator $L_N(z)$ has a lower limit of zero for each integral over z_i, which means that $(1 - z_i)^{-N-\rho}$ should remain as it is. Therefore, $L_N(z)$ is discontinuous and we cannot extend the above result to $z > 1$ when ρ is no longer an integer.

In the next example we apply the analysis for the second type of asymptotic series, to the series appearing in Equivalences (4.69) and (4.70). Instead of using the cumbersome $_2\mathcal{F}_0$ notation, we shall denote this series as $T_\nu(z)$, although it can also be denoted by $S_1(N, 1/2z)$, where

$$f(s) = \frac{\Gamma(s + 1/2 + \nu)}{\Gamma(1/2 + \nu)\Gamma(s+1)} \frac{\Gamma(s + 1/2 - \nu)}{\Gamma(1/2 - \nu)} . \tag{7.84}$$

Then according to the definition of $I_l^*(z)$ on the rhs of Equivalence (7.51), we have

$$I_l^*(z) = \int_{\substack{c-i\infty \\ N-1 < c = \Re s < N}}^{c+i\infty} ds \, (1/2z)^s \frac{\Gamma(s + 1/2 + \nu)}{\Gamma(1/2 + \nu)\Gamma(s+1)}$$

$$\times \frac{\Gamma(s + 1/2 - \nu)}{\Gamma(1/2 - \nu)} \frac{e^{-i(2l+1)\pi s}}{e^{-i\pi s} - e^{i\pi s}} . \tag{7.85}$$

Since $f(c \pm iL) = O(\exp(-\pi L/2))$ as $L \to \infty$, the domain of convergence for $I_0^*(z)$ is $-5\pi/2 < \arg z < \pi/2$, while for $I_{-1}^*(z)$, it is $-\pi/2 < \arg z < 5\pi/2$. Hence, the common sector for the domains of convergence is $|\arg z| < \pi/2$.

As we have done throughout this chapter, we nominate $\arg z = 0$ as the primary Stokes line for $T_\nu(z)$. The regularised value on this line is the average of the two regularised values for the adjoining Stokes sectors. Above the line the regularised value is given by $I^*_{-1}(z)$ plus a discontinuous term. Below the line the regularised value is given by $I_0(z)$ plus another discontinuous term, which is the complex conjugate of the jump discontinuity above the primary Stokes line. Hence, the average of $I^*_{-1}(z)$ and $I^*_0(z)$ represents the regularised value of the divergent series on the primary Stokes line. Therefore, for $\arg z = 0$, we find that

$$T_v(N,z) = \sum_{k=N}^{\infty} \frac{\Gamma(k+1/2+v)}{\Gamma(1/2+v)} \frac{\Gamma(k+1/2-v)}{\Gamma(1/2-v)k!} \left(\frac{1}{2z}\right)^l \equiv \frac{i}{2}$$
$$\times \int_{\substack{c-i\infty \\ N-1<c=\Re s<N}}^{c+i\infty} ds\, (2|z|)^{-s} \frac{\Gamma(s+1/2+v)}{\Gamma(1/2+v)\Gamma(s+1)} \frac{\Gamma(s+1/2-v)}{\Gamma(1/2-v)} \cot(\pi s). \quad (7.86)$$

From the MB regularisation of $S_1(N, z^\beta)$, we know that the jump discontinuities associated with the regularised values are related to the difference between the two MB integrals, which is

$$\Delta I^*_{0,-1}(z) = I^*_0(z) - I^*_{-1}(z) = \int_{\substack{c-i\infty \\ N-1<c=\Re s<N}}^{c+i\infty} ds\, (2|z|)^{-s}$$
$$\times \frac{\Gamma(s+1/2+v)}{\Gamma(1/2+v)} \frac{\Gamma(s+1/2-v)}{\Gamma(1/2-v)\Gamma(s+1)} \,. \quad (7.87)$$

Since the variable in $T_v(z)$ is $1/2z$ rather than z, the regularised value of the series obtained from Equivalence (7.60) is given by

$$T_v(z) \equiv \begin{cases} I^*_0(1/2z) - \frac{1}{2}\Delta I^*_{0,-1}(1/2z) & , \quad -5\pi/2 < \arg z < 0 \quad , \\ I^*_{-1}(1/2z) + \frac{1}{2}\Delta I^*_{0,-1}(1/2z) & , \quad 0 < \arg z < 5\pi/2 \quad . \end{cases} \quad (7.88)$$

We can evaluate $\Delta I^*_{0,-1}(1/2z)$ in the regularised value of $T_v(z)$ by using No. 1.11.5 on p. 115 of Ref. [52], which gives

$$\int_0^\infty dx\, x^{s-1/2} e^{-ax} K_v(ax) = \sqrt{\pi}\,(2a)^{-s-1/2}$$
$$\times \frac{\Gamma(s+1/2-v)}{\Gamma(s+1)} \Gamma(s+1/2+v) \,, \quad (7.89)$$

where $\Re(s+1/2) > \pm \Re v$. From this Mellin transform we obtain the following inverse Mellin transform:

$$\frac{1}{2\pi i} \int_{\substack{c-i\infty \\ c=\Re s > |\Re v| - 1/2}}^{c+\infty} ds\, y^{-s}\, \frac{\Gamma(s+1/2-v)}{\Gamma(s+1)}\, \Gamma(s+1/2+v)$$

$$= \sqrt{\frac{\pi}{y}}\, e^{-y/2} K_v(y/2) \ . \tag{7.90}$$

By introducing an additional constraint into Eq. (7.87) so that it now becomes

$$\Delta I_{0,-1}^*(z) = I_0^*(z) - I_{-1}^*(z) = \int_{\substack{c-i\infty \\ Min[N-1,|\Re v|-1/2] < c = \Re s < N}}^{c+i\infty} ds\, (2z)^{-s}$$

$$\times\, \frac{\Gamma(s+1/2+v)}{\Gamma(1/2+v)}\, \frac{\Gamma(s+1/2-v)}{\Gamma(1/2-v)\Gamma(s+1)}\ , \tag{7.91}$$

where $|\Re v| < N + 1/2$, we can replace the MB integral by the rhs of Eq. (7.90). By noting that the half-integer Macdonald function $K_{1/2}(z) = \sqrt{\pi/2z}\exp(-z)$, we arrive at

$$\Delta I_{0,-1}^*(1/2z) = \frac{4iz\, K_{1/2}(z)K_v(z)}{\Gamma(1/2-v)\Gamma(1/2+v)}\ . \tag{7.92}$$

With the introduction of the reflection formula for the gamma function or Eq. (3.2) the regularised value of $T_v(z)$ for $|\Re v| < N + 1/2$ can be written as

$$T(N,z) \equiv I_0^*(1/2z) - \frac{2iz}{\pi}\, K_{1/2}(z)K_v(z)\cos(\pi v)\ , \tag{7.93}$$

when $-5\pi/2 < \arg z < \pi/2$, and

$$T(N,z) \equiv I_{-1}^*(1/2z) + \frac{2iz}{\pi}\, K_{1/2}(z)K_v(z)\cos(\pi v)\ , \tag{7.94}$$

when $-\pi/2 < \arg z < 5\pi/2$.

In this chapter we have shown how to derive the regularised values of the two types of asymptotic series, $S(N, z^\beta)$ and $S_1(N, z^\beta)$ via the technique known as Mellin-Barnes (MB) regularisation, which first appeared in Ref. [13]. Here, however, the technique has been developed for z over all branches of the complex plane. One of the major aims of this book is to show that the MB-regularised forms derived in this chapter are identical to the corresponding values obtained via Borel summation. Therefore, we shall be required to carry out the numerical evaluation of the MB integrals appearing in these forms for the regularised values. This issue is addressed in the following chapter.

CHAPTER 8

Numerics of MB-Regularised Forms

Abstract. This chapter is concerned with the numerics of MB-regularised forms for the regularised value of a divergent series using the Mathematica software package. To accomplish this, the remainders in the asymptotic forms for $u(a)$ given in Ch. 2 are first MB-regularised. Then an explanation of how to evaluate the resulting MB integrals follows. It is shown that when the values for the MB integrals are added to the truncated asymptotic series and the appropriate Stokes discontinuous terms, they yield exact values of $u(a)$ over the principal branch. Because the domains of convergence for the MB integrals extend beyond the Stokes sectors, the two different forms for the regularised value also give exact values of $u(a)$ over their common region of $(-\pi/4, \pi/4)$. A similar analysis is then undertaken for the error function $\mathrm{erf}(z)$, whose Stokes sectors are shifted compared with those for $u(a)$. A major problem arises when Mathematica attempts to evaluate the MB-regularised values for $|\arg z| > \pi/2$ because the factor of z^{-2s} in the MB integrals lies outside the principal branch. However, with the introduction of the seemingly innocuous factor of $\exp(2\pi i jk)$ into the asymptotic series, different MB-regularised forms are obtained with domains of convergence that encompass the previously inaccessible sectors of the principal branch. Consequently, the Stokes multiplier equals -1/2 for $j = \pm 1$, while it equals 1/2 for $j = 0$ as in Ch. 5. When a numerical analysis is undertaken for $|\arg z| > \pi/2$ with the new forms for the regularised value, exact values of the error function are obtained irrespective of the magnitude of z.

Up till now, we have presented a general theory for applying the technique of MB regularisation to an asymptotic series. Although numerical values of the MB-regularised value of the abbreviated binomial series were presented in the previous chapter, the issues concerning how the technique can be implemented numerically in a mathematical software package such as Mathematica [19] to render the exact values of the original function from its complete asymptotic expansion were only discussed very briefly. In this chapter, however, we aim to concentrate on these tricky issues, espcecially the problem of accessing the entire principal branch of the complex plane for the variable z when the series in an asymptotic expansion is in powers of z. We choose as our first example the asymptotic forms of $u(a)$, which, as stated in Ch. 2, was the first example found to exhibit the Stokes phenomenon. Since $u(a)$ is related to the error function, we shall also consider

Victor Kowalenko

MB regularisation of the asymptotic forms for the latter function thereafter in this chapter.

8.1   *Numerical Analysis for u(a)*

The asymptotic series in the asymptotic forms for $u(a)$ as given by Equivalence (2.11) is of the type $S_1(N, z^\beta)$ that was introduced in the previous chapter where $N = 0$, $f(k) = \Gamma(k+1/2)$ and $\beta = 2$. For the numerical study in this chapter, we shall, however, maintain the truncation parameter N, which will be varied at times, but when making direct comparison with the actual values of $u(a)$, we shall include the truncated or finite series summed to $N-1$. According to the previous chapter MB regularisation of the series yields

$$
\frac{1}{\sqrt{\pi} z} \sum_{k=0}^{\infty} \frac{\Gamma(k+1/2)}{z^{2k}} \equiv \sum_{k=0}^{N-1} \frac{\Gamma(k+1/2)}{z^{2k}} + \frac{1}{z} \int_{\substack{c-i\infty \\ \mathrm{Max}[-1/2,N-1]<c=\Re s<N}}^{c+i\infty} ds \left(-\frac{1}{z^2} \right)^s
$$
$$
\times \frac{\Gamma(s+1/2)}{\Gamma(1/2)} \frac{1}{(e^{-i\pi s} - e^{i\pi s})} \, . \tag{8.1}
$$

The lower bound of $\mathrm{Max}[-1/2, N-1]$ on the offset c is necessary for ensuring that the line contour remains to the right of the pole at $s = -1/2$. In the previous chapter we observed that an infinite number of versions of the above integral arise when one interprets $(-1)^s$ as $\exp(i(2j+1)\pi s)$, where j is an arbitrary integer. Because numerical packages such as Mathematica and Maple restrict the evaluation of functions to the principal branch of the complex plane, we shall consider only those interpretations that cover the principal branch, viz. the $j=0$ and $j=-1$ versions, initially. Unfortunately, we shall see that these regularised forms are unable to yield the values of $u(a)$ or the error function over the entire principal branch. As a consequence, we shall have to consider other values of j in interpreting $(-1)^s$ factor in the integrand. For the present, however, let us consider the first two versions for the regularised value of the remainder, which give

$$
R_1(N, z) = \frac{1}{z} \int_{\substack{c-i\infty \\ \mathrm{Max}[-1/2,N-1]<c=\Re s<N}}^{c+i\infty} ds \, \frac{z^{-2s} e^{i\pi s}}{(e^{-i\pi s} - e^{i\pi s})} \frac{\Gamma(s+1/2)}{\Gamma(1/2)} \, . \tag{8.2}
$$

and

$$
R_2(N, z) = \frac{1}{z} \int_{\substack{c-i\infty \\ \mathrm{Max}[-1/2,N-1]<c=\Re s<N}}^{c+i\infty} ds \, \frac{z^{-2s} e^{-i\pi s}}{(e^{-i\pi s} - e^{i\pi s})} \frac{\Gamma(s+1/2)}{\Gamma(1/2)} \, . \tag{8.3}
$$

The regions of validity for the regularised values obtained via MB regularisation are dependent upon the domains of convergence of the resulting MB integrals as opposed to Stokes sectors that arise out of Borel summation of the asymptotic series. To determine the domains of convergence for the above MB integrals, we must examine the behaviour of the integrands at the endpoints. For $s = c \pm iL$, where $L \to \infty$, the modulus of the integrand for the MB integral appearing in Equivalence (8.2) behaves as

$$\left| \frac{e^{i\pi s} \Gamma(s+1/2)}{z^{2s}(e^{-i\pi s} - e^{i\pi s})} \right| \sim \frac{\sqrt{2\pi}}{|z|^{2c}} (L^2 + (c+1/2)^2)^{c/2} e^{-c-1/2} \begin{bmatrix} e^{2L\theta - 5\pi L/2} \\ e^{-2L\theta - \pi L/2} \end{bmatrix}, \quad (8.4)$$

where $z = |z| \exp(i\theta)$ and we have introduced the leading order term of Stirling's approximation for the gamma function, which is given as either No. 8.327 in Ref. [21] or No. 6.1.37 in Ref. [30]. Specifically, this means that $\Gamma(c \pm iL + v)$ as $L \to \infty$ has been replaced by

$$|\Gamma(c \pm iL + v)| \overset{L \to \infty}{\sim} \sqrt{2\pi} e^{-(c+v)} L^{v+c-1/2} e^{-\pi L/2} . \quad (8.5)$$

Thus, the upper limit of the MB integral in Equivalence (8.2) is exponentially decaying for $\theta < 5\pi/4$, while the lower limit decays exponentially for $\theta > -\pi/4$. Hence, the integral is convergent for $-\pi/4 < \theta < 5\pi/4$, which encompasses the Stokes sector for the third asymptotic form of $u(a)$ in Equivalence (2.11). However, it is also convergent for part of the Stokes sector belonging to the first asymptotic form. Therefore, we shall investigate whether the integral is able to yield the values of $u(a)$ for $-\pi/4 < \arg z < 0$. Should this occur, we will have demonstrated that MB regularisation is very different from Borel summation.

In the case of the other MB integral appearing in Equivalence (8.3), for $s = c \pm iL$, where $L \to \infty$, the modulus of the integrand behaves as

$$\left| \frac{e^{-i\pi s} \Gamma(s+1/2)}{z^{2s}(e^{-i\pi s} - e^{i\pi s})} \right| \sim \frac{\sqrt{2\pi}}{|z|^{2c}} (L^2 + (c+1/2)^2)^{c/2} e^{-c-1/2} \begin{bmatrix} e^{2L\theta - \pi L/2} \\ e^{-2L\theta - 5\pi L/2} \end{bmatrix}. \quad (8.6)$$

Therefore, the integrand at the upper limit of the second version for the regularised remainder in Equivalence (8.1) decays exponentially when $\theta < \pi/4$, while at the lower limit it decays exponentially when $\theta > -5\pi/4$. Hence, the MB integral in Equivalence (8.3) is convergent for $-5\pi/4 < \theta < \pi/4$, which encompasses the Stokes sector for the first asymptotic form of $u(a)$ in Equivalence (2.11). As in the case of $R_1(N, z)$, the MB integral is convergent over part of another Stokes sector, but on this occasion, it is for $0 < \theta < \pi/4$, which pertains to the third asymptotic form of $u(a)$. In the following numerical study we shall investigate whether the second version yields exact values of $u(a)$ over the common sector of the complex plane in which the regularised value can be obtained by using $R_1(N, z)$.

Another point that should be made about the behaviour of the integrands given above is that as c or N increases, the integrands diverge in a similar manner to the integral in Eq. (5.5). Moreover, when there is no optimal point of truncation, i.e. $|z| < 1$, they will diverge even faster. This divergence will be cancelled by the divergence of the truncated series in the opposite direction, which means in turn that when values of $u(a)$ are being determined, cancellation of redundant decimal places occurs as was the case for the Borel-summed results in Ch. 5. Hence, it should not be surprising if problems arise in obtaining very accurate machine precision results for $u(a)$ from the MB integrals when N becomes very large.

We begin by writing the asymptotic forms for $u(a)$ given by Equivalence (2.11) in terms of the Stokes sectors that apply to the Borel-summed versions of the regularised value. Therefore, we write them as

$$
u(a) = -i\sqrt{\pi}\, e^{-a^2} + \sum_{k=0}^{N-1} \frac{\Gamma(k+1/2)}{\Gamma(1/2)}\, a^{-2k-1} + \frac{1}{a}
$$
$$
\times \int_{\substack{c-i\infty \\ \mathrm{Max}[-1/2,N-1]<c=\Re s<N}}^{c+i\infty} ds\, \frac{a^{-2s}\, e^{-i\pi s}}{(e^{-i\pi s} - e^{i\pi s})}\, \frac{\Gamma(s+1/2)}{\Gamma(1/2)}\,, \qquad (8.7)
$$

for $-\pi < \arg a < 0$, while for $0 < \arg a < \pi$,

$$
u(a) = i\sqrt{\pi}\, e^{-a^2} + \sum_{k=0}^{N-1} \frac{\Gamma(k+1/2)}{\Gamma(1/2)}\, a^{-2k-1} + \frac{1}{a}
$$
$$
\times \int_{\substack{c-i\infty \\ \mathrm{Max}[-1/2,N-1]<c=\Re s<N}}^{c+i\infty} ds\, \frac{a^{-2s}\, e^{i\pi s}}{(e^{-i\pi s} - e^{i\pi s})}\, \frac{\Gamma(s+1/2)}{\Gamma(1/2)}\,. \qquad (8.8)
$$

It should also be mentioned that the above equations can be obtained by putting $z = 1/a$, $\beta = 2$ and $f(s) = \Gamma(s+1/2)/a\Gamma(1/2)$ in the two forms of Equivalence (7.60), while the corresponding forms for $\mathrm{erfi}(a)$ are obtained simply by multiplying Eqs. (8.7) and (8.8) by $\exp(a^2)/\sqrt{\pi}$. As has been done previously, we shall refer to the first terms on the rhs of the equations as the Stokes discontinuity for the asymptotic form since, unlike Stokes smoothing, they were first evaluated by Stokes in Ref. [2].

Numerical results for Eqs. (8.7) and (8.8) can be obtained by applying the NIn-tegrate routine in Mathematica [19] to the MB integrals. In order to accomplish this, each MB integral needs to be written as the sum of two separate integrals ranging from 0 to infinity. Because NIntegrate can miss the contribution due to sudden peaks in the integrand, it is advisable to divide the range of integration into smaller intervals. Therefore, a call to the NIntegrate routine in a Mathematica module for the integrals in Eqs. (8.7) and (8.8) would typically look like:

s1=c-I r;
e4= NIntegrate[I g[z, s1],{r,0,1,5,10,100,1000,Infinity}, WorkingPrecision-> 16, PrecisionGoal->14, Accuracy Goal->14, MinRecursion->3,
MaxRecursion->10];

In the above statement g[z,s1] represents the integrand of the MB integral, while the routine evaluates the sum of the integrals over the ranges (0,1), (1,5), (5,10), etc. In this case the half of the MB integral below the real axis is being evaluated. For the half above the real axis, one would set s1 to c+I r. As in the numerical work of previous chapters, AccuracyGoal and PrecisionGoal have been set to 14, while WorkingPrecision has been set to the maximum value of 16 for a Pentium computer. All the results presented here were evaluated within a few CPU secs. As stated in the introduction we can obtain even more accurate results by setting much bigger values for WorkingPrecision, AccuracyGoal and PrecisionGoal in the more recent versions of Mathematica, namely Versions 6.0 and 7.0.

Table 8.1 presents a sample of the results obtained by introducing a relatively small value of $|a|$ into Eq. (8.7). Due to limited space, not all the decimal figures for the various contributions on the rhs have been displayed in the table, which were obtained by setting $|a|$ equal to $1/10$ and varying θ or $\arg a$ over $(-\pi, 0)$. These values of θ correspond to the Stokes sector for the second asymptotic form of $u(a)$ given by Eq. (5.22) or the form where the Stokes multiplier $S = -1/2$. The other values in the table represent the remaining contributions on the rhs of Eq. (8.7) and their sum.

It should be also noted that there is no contribution from the truncated series be-cause the truncation parameter N has been set equal to zero. Nevertheless, we observe that regardless of the value that θ takes in the table, the sums of the terms on the rhs of Eq. (8.7) and the corresponding values of $u(a)$ agree within the precision and accuracy goals specified by the NIntegrate routine. Therefore, Eq. (8.7) gives the values of $u(a)$ for the Stokes sector that applies to the first form of Equivalence (2.11) or the second form of Eq. (5.24). In other words, Eq. (8.7) is simply another representation of $u(a)$ for $-\pi < \arg a < 0$.

Table 8.1: Values of $u(\exp(i\theta)/10)$ and the various contributions on the rhs of Eq. (8.7) with N=0

θ	Term	
	Stokes Discontinuity	$0.001\,101\,880\,783\,69 - 1.754\,851\,922\,139\,01\,i$
$-\pi/100$	**MB Integral**	$0.197\,477\,268\,169\,14 + 1.748\,694\,417\,070\,99\,i$
	Sum	$0.198\,579\,148\,952\,83 - 0.006\,157\,505\,067\,01\,i$
	$u(\exp(-\pi i/100)/10)$	$0.198\,579\,148\,952\,83 - 0.006\,157\,505\,067\,02\,i$
	Stokes Discontinuity	$0.015\,426\,649\,481\,43 - 1.781\,271\,513\,036\,27\,i$
$-\pi/3$	**MB Integral**	$0.085\,909\,342\,865\,62 + 1.608\,071\,064\,278\,06\,i$
	Sum	$0.101\,335\,992\,347\,05 - 0.173\,200\,448\,758\,20\,i$
	$u(\exp(-\pi i/3)/10)$	$0.101\,335\,992\,347\,05 - 0.173\,200\,448\,758\,21\,i$
	Stokes Discontinuity	$0.0 - 1.790\,267\,308\,256\,093\,i$
$-\pi/2$	**MB Integral**	$0.0 + 1.588\,928\,626\,317\,408\,i$
	Sum	$0.0 - 0.201\,338\,681\,938\,685\,i$
	$u(\exp(-\pi i/2)/10)$	$0.0 - 0.201\,338\,681\,938\,685\,i$
	Stokes Discontinuity	$-0.017\,724\,243\,101\,55 - 1.772\,365\,228\,951\,49\,i$
$-3\pi/4$	**MB Integral**	$-0.124\,636\,140\,190\,19 + 1.631\,890\,442\,193\,06\,i$
	Sum	$-0.142\,360\,383\,291\,75 - 0.140\,474\,786\,758\,42\,i$
	$u(\exp(-3\pi i/4)/10)$	$-0.142\,360\,383\,291\,75 - 0.140\,474\,786\,758\,42\,i$
	Stokes Discontinuity	$-0.001\,101\,880\,783\,69 - 1.754\,851\,922\,138\,01\,i$
$-99\pi/100$	**MB Integral**	$-0.197\,477\,268\,169\,14 + 1.748\,694\,417\,070\,99\,i$
	Sum	$-0.198\,579\,148\,952\,83 - 0.006\,157\,505\,067\,01\,i$
	$u(\exp(-99\pi i/100)/10)$	$-0.198\,579\,148\,952\,83 - 0.006\,157\,505\,067\,02\,i$

On the other hand, Table 8.2 presents another small sample of the results obtained by introducing a larger value of $|a|$, viz. $a = 3\exp(i\theta)/20$, into Eq. (8.8). As in the previous table, not all the decimal figures for the various values on the rhs of the equation have been displayed due to limited space. Although θ varies, on this occasion the variation occurs over the sector of $(0, \pi)$. This, of course, represents the Stokes sector for the third asymptotic form of $u(a)$ in Equivalence (2.11).

Another difference between this example and the results in Table 8.1 is that the truncation parameter N has now been set equal to 3, which means that the truncated series is no longer equal to zero. Since $|a| < 1$, there is no optimal point of truncation and the truncated series diverges rapidly for small values of N. As a consequence, we see that the truncated series is of order of 10^4.

Table 8.2: Values of $u(3\exp(i\theta)/20)$ and the various contributions on the rhs of Eq. (8.8) with N=3

θ	Term	
	Stokes Discontinuity	$0.002\,448\,493\,991 - 1.733\,094\,161\,331\,i$
	Truncated Series	$9\,909.100\,612\,270 - 1\,559.183\,130\,887\,i$
$\pi/100$	**MB Integral**	$-9\,908.807\,649\,077 + 1\,557.459\,042\,745\,i$
	Sum	$0.295\,411\,687\,770 + 0.009\,006\,019\,425\,i$
	$u(3\exp(\pi i/100)/20)$	$0.295\,411\,687\,805 + 0.009\,006\,019\,475\,i$
	Stokes Discontinuity	$0.039\,876\,846\,837\,68 + 1.772\,005\,217\,45\,i$
	Truncated Series	$-7\,083.813\,193\,353\,52 + 6\,874.300\,073\,0\,i$
$\pi/4$	**MB Integral**	$7\,083.988\,601\,700\,57 - 6\,875.863\,156\,618\,i$
	Sum	$0.215\,285\,193\,884 + 0.208\,921\,601\,020\,i$
	$u(3\exp(\pi i/4)/20)$	$0.215\,285\,193\,870 + 0.208\,921\,601\,033\,i$
	Stokes Discontinuity	$0.0 + 1.812\,786\,098\,837\,i$
	Truncated Series	$0.0 - 9\,735.061\,728\,395\,061\,i$
$\pi/2$	**MB Integral**	$2.182\,787 \times 10^{-11} + 9\,733.553\,483\,057\,i$
	Sum	$2.182\,787 \times 10^{-11} + 0.304\,540\,761\,69\,i$
	$u(3\exp(\pi i/2)/20)$	$0.0 + 0.304\,540\,761\,664\,i$
	Stokes Discontinuity	$-0.034\,148\,748\,351 + 1.752\,292\,774\,279\,i$
	Truncated Series	$8\,547.563\,818\,635 - 5\,089.753\,086\,419\,i$
$5\pi/6$	**MB Integral**	$-8\,547.789\,446\,601 + 5\,088.146\,314\,024\,i$
	Sum	$-0.259\,772\,772\,599 + 0.145\,520\,378\,917\,i$
	$u(3\exp(5\pi i/6)/20)$	$-0.259\,772\,772\,577 + 0.145\,520\,378\,879\,i$
	Stokes Discontinuity	$-0.001\,224\,810\,909 + 1.733\,037\,753\,540\,i$
	Truncated Series	$-10\,000.746\,652\,686 - 781.988\,098\,015\,i$
$199\pi/200$	**MB Integral**	$10\,000.452\,369\,394\,2 + 780.259\,563\,662\,8\,i$
	Sum	$-0.295\,508\,102\,928 + 0.004\,503\,365\,319\,i$
	$u(3\exp(199\pi i/100)/20)$	$-0.295\,508\,102\,957 + 0.004\,503\,365\,371\,i$

Because the value of $|u(a)|$ is less than unity for all values of θ, the MB integral in Eq. (8.8) must cancel all the figures to the left of the decimal place in the truncated series. This represents the first four figures of the truncated series for θ equal to $\pi/100$, $\pi/4$ and $5\pi/6$ in the table, while it represents the first five figures for $\theta = 199\pi/200$. Although the cancellation does occur as expected, due to the limited machine precision of the computing system the results for the sum of the truncated series and MB integral only agree with the actual values of $u(a)$ to about 10 decimal places, not the 14 figure accuracy witnessed in the previous table. Therefore, when there is no optimal point of truncation, it is recommended to choose a truncation point as close as possible to the first term in the asymptotic series, especially if the computing system possesses limited machine precision. Even for a system with the potential for unlimited precision as in Mathematica 6.0/7.0, it is still better to keep N as close as possible to the first term in the asymptotic series, because the larger N is, the longer it will take to perform the calculations of the various quantities on the rhs of Eq. (8.8).

The selection of a truncation point well beyond the optimal point of truncation results in waning accuracy as indicated by the results in Table 8.2. Whilst we have already witnessed this behaviour in our numerical study of the Borel-summed asymptotic forms for $erfi(z)$ in Table 5.1, it is not so critical if the routines are programmed in Mathematica 7.0 with its extended precision. Nevertheless, despite this shortcoming in the above example, it is apparent that Eq. (8.8) gives the values of $u(a)$ over the Stokes sector that applies to the third asymptotic form of Equivalence (2.11) or the first form of Eq. (5.22). These, of course, represent the forms where the Stokes multiplier S is equal to 1/2. Thus, we conclude that Eq. (8.8) represents another representation of $u(a)$, whose region of validity is $0 < \arg a < \pi$.

The above results demonstrate that the Borel-summed asymptotic forms for $u(a)$ as given by Eq. (5.22) and the MB-regularised forms given by Eqs. (8.7) and (8.8) yield identical results for $u(a)$. Such demonstrations are crucial for establishing a theory of divergent series based on the concept of regularisation. However, there is a major difference between Borel-summed and MB-regularised forms. MB-regularised forms are valid over domains of convergence, which are in no way identical to the Stokes sectors that appear in the Borel-summed forms of the regularised value. Consequently, we have discovered that there is a common sector, viz. $(-\pi/4, \pi/4)$ for $u(a)$, where both the MB-regularised forms should yield the same values of $u(a)$. This remarkable feature of the latter forms, which we shall soon verify by a numerical demonstration, does not apply to Borel-summed forms because we have seen clearly from the numerical examples in Ch. 5 that Stokes sectors are bordered by lines of discontinuity and thus, there cannot be any overlap.

Tables 8.3 and 8.4 present samples of the results obtained from a Mathematica module that calculates the various contributions in Eqs. (8.7) and (8.8) for $a =$

Table 8.3: Values of $u(\exp(i\theta)/6)$ and the various contributions on the right hand sides of Eqs. (8.7) with N=1 and arg z lying in $(-\pi/4, 0)$

θ	Term	
	Truncated Series	$4.854\,101\,966\,249 + 3.526\,711\,513\,754\,i$
	Stokes Discontinuity	$-0.046\,419\,487\,713 + 1.756\,691\,368\,085\,i$
	1st MB Integral	$-4.536\,171\,050\,956 - 5.473\,461\,095\,749\,i$
$-\pi/5$	First Sum	$0.271\,511\,427\,579 - 0.190\,058\,213\,909\,i$
	2nd MB Integral	$-4.629\,010\,026\,383 - 1.960\,078\,359\,579\,i$
	Second Sum	$0.271\,511\,427\,579 - 0.190\,058\,213\,909\,i$
	$u(\exp(\text{-}i\pi/5)/6)$	$0.271\,511\,427\,579 - 0.190\,058\,213\,909\,i$
	Truncated Series	$5.543\,277\,195\,067 + 2.296\,100\,594\,190\,i$
	Stokes Discontinuity	$-0.034\,134\,941\,493 - 1.737\,644\,002\,661\,i$
	1st MB Integral	$-5.203\,570\,400\,153 - 4.155\,665\,937\,538\,i$
$-\pi/8$	First Sum	$0.305\,571\,853\,420 - 0.121\,921\,340\,686\,i$
	2nd MB Integral	$-5.271\,840\,283\,140 - 0.680\,377\,932\,215\,i$
	Second Sum	$0.305\,571\,853\,420 - 0.121\,921\,340\,686\,i$
	$u(\exp(\text{-}i\pi/8)/6)$	$0.305\,571\,853\,420 - 0.121\,921\,340\,686\,i$

$\exp(i\theta)/6$ with θ situated in the common sector of $(-\pi/4, \pi/4)$ between the domains of convergence. The first of these tables presents the results obtained for negative values of θ, while Table 8.4 presents the results for $\theta \geq 0$.

Because the Mathematica module was required to perform a greater number of calculations including two MB integrals, each set of results in the table took longer to compute than previous modules. However, the longest time taken for any set of values in the table was about 37 CPU seconds. In actual fact, the longer CPU times occurred when θ was close to the limits of the domains of convergence for the MB integrals, namely $-\pi/4$ for the first MB integral and $\pi/4$ for the second MB integral. In addition, with the exception of the $\theta = 0$ results Mathematica consistently gave error warnings. Specifically, for negative values of θ warnings of slow convergence due to a singularity or insufficient WorkingPrecision occurred during the evaluation of the lower half contour of the first MB integral, while for positive values of θ they occurred when the upper half of the second MB integral was being evaluated. In spite of these warnings, the results from the module were always within the accuracy and precision goals set in the NIntegrate routine.

The results in both tables have been obtained with the truncation parameter N

Table 8.4: Values of $u(\exp(i\theta)/6)$ and the various contributions on the right hand sides of Eqs. (8.8) with N=1 and arg z lying in $[0, \pi/4)$

θ	Term	
0	Truncated Series	6.0
	Stokes Discontinuity	$0.0 + 1.723\,896\,550\,872\,i$
	1st MB Integral	$-5.672\,771\,460\,066 - 1.723\,896\,550\,872\,i$
	First Sum	$0.327\,228\,539\,933 - 4.662\,936\,703\,425 \times 10^{-15}$
	2nd MB Integral	$-5.672\,771\,460\,066 + 1.723\,896\,550\,872\,i$
	Second Sum	$0.327\,228\,539\,933 + 4.662\,936\,703\,425 \times 10^{-15}\,i$
	$u(1/6)$	$0.327\,228\,539\,933$
$\pi/8$	Truncated Series	$5.543\,277\,195\,067 - 2.296\,100\,594\,190\,i$
	Stokes Discontinuity	$0.034\,134\,941\,493 + 1.737\,644\,002\,661\,i$
	1st MB Integral	$-5.271\,840\,283\,140 + 0.680\,377\,932\,215\,i$
	First Sum	$0.305\,571\,853\,420 + 0.121\,921\,340\,686\,i$
	2nd MB Integral	$-5.203\,570\,400\,153 + 4.155\,665\,937\,538\,i$
	Second Sum	$0.305\,571\,853\,420 + 0.121\,921\,340\,686\,i$
	$u(\exp(i\pi/8)/6)$	$0.305\,571\,853\,420 + 0.121\,921\,340\,686\,i$
$\pi/5$	Truncated Series	$4.854\,101\,966\,249 - 3.526\,711\,513\,754\,i$
	Stokes Discontinuity	$0.046\,419\,487\,713 + 1.756\,691\,368\,085\,i$
	1st MB Integral	$-4.629\,010\,026\,383 + 1.960\,078\,359\,579\,i$
	First Sum	$0.271\,511\,427\,579 + 0.190\,058\,213\,909\,i$
	2nd MB Integral	$-4.536\,171\,050\,956 + 5.473\,461\,095\,749\,i$
	Second Sum	$0.271\,511\,427\,579 + 0.190\,058\,213\,909\,i$
	$u(\exp(i\pi/5)/6)$	$0.271\,511\,427\,579 + 0.190\,058\,213\,909\,i$

set equal to unity. The contribution from the Stokes discontinuity term changes sign depending on whether the rhs of Eq. (8.7) or Eq. (8.8) is being evaluated. The rows labelled 1st MB Integral represent the results obtained from the integral on the rhs of Eq. (8.8), which in turn represents the first interpretation for the remainder as given by Eq. (8.2). The rows labelled 2nd MB Integral represent the results due to the integral on the rhs of Eq. (8.7), which represent the second interpretation for the remainder as given by Eq. (8.3). The rows labelled First Sum give the values for the sum of all the contributions on the rhs of Eq. (8.8). Thus, the Stokes discontinuity term is added to the integral on the rhs of Eq. (8.8). The rows labelled Second Sum give the values for the sum of all the contributions on the rhs of Eq. (8.7), which means that the Stokes discontinuity term is subtracted.

From the tables we see that both equations provide the correct values of $u(a)$ over the common sector of the complex plane where the domains of convergence of the MB integrals overlap. Hence, Eqs. (8.7) and (8.8) are valid over their domains of convergence rather than the narrower Stokes sectors of $(-\pi, 0)$ and $(0, \pi)$, which result from the asymptotic forms obtained via Borel summation. As a consequence, the regularised values obtained via MB regularisation for a function will not exhibit jump discontinuities at the border of adjacent Stokes sectors as the asymptotic forms obtained via Borel summation do. Furthermore, the first MB-regularised form for $u(a)$ over the sector of $(-\pi/4, 0)$ represents the complex conjugate of the second MB-regularised form for $u(a)$ over the sector of $(0, \pi/4)$ and vice-versa. Therefore, the previously mentioned error warnings that arose when obtaining the results displayed in Table 8.3 are merely the manifestation of the same problem in different parts of the common sector for the two MB-regularised forms.

Since both MB-regularised forms possess a common sector, we can average them to obtain another form for $u(a)$, which is given by

$$
u(a) = \sum_{k=0}^{N-1} \frac{\Gamma(k+1/2)}{\Gamma(1/2)} a^{-2k-1} - \frac{\sqrt{\pi}}{2\pi i} \int_{\underset{\mathrm{Max}[-1/2,N-1]<c=\Re s<N}{c-i\infty}}^{c+i\infty} ds\, a^{-2s-1}
$$
$$
\times \ \cot(\pi s)\,\Gamma(s+1/2) \ .
\tag{8.9}
$$

Eq. (8.9) is only valid for $-\pi/4 < \arg z < \pi/4$, which includes the Stokes line of $\arg a = 0$. Nevertheless, it represents another representation for $u(a)$. For $N = 0$, the above result is essentially the inverse Mellin transform given as No. 2.8.5.10 in Prudnikov et al [53], though it has been derived by using divergent mathematics. As mentioned previously, the corresponding form for $\mathrm{erfi}(a)$ is simply obtained by multiplying the above result by $\exp(a^2)/\sqrt{\pi}$.

It should also be noted that the MB-regularised forms, which yield the same regularised value over a common sector of the complex plane, can be used to determine Mellin transform pairs. For example, by equating Eqs. (8.7) with (8.8) over their common sector, we find that

$$
\frac{1}{2\pi i} \int_{\underset{\mathrm{Max}[-1/2,N-1<c=\Re s<N]}{c-i\infty}}^{c+i\infty} ds\, a^{-2s}\,\Gamma(s+1/2) = a\,e^{-a^2} \ .
\tag{8.10}
$$

The poles for this inverse Mellin transform lie to the left of the imaginary axis, which means that we can replace the condition on the offset c by $c > 0$. Then Eq. (8.9) becomes simply the inverse Mellin transform for the exponential function since according to the theory of Mellin transforms [52], the above yields

$$\int_0^\infty dy\, y^{s-1}\, \sqrt{y}\, e^{-y} = \Gamma(s+1/2) \; , \tag{8.11}$$

on making the substitution $y = a^2$.

8.2 Numerical Analysis for the Error Function

So far, we have concentrated on the MB regularisation of the asymptotic forms for $u(a)$. Now we turn our attention to the MB regularisation of the asymptotic forms for the error function given by Equivalence (5.1) and Eq. (5.2). The asymptotic series in these forms differs from those for $u(a)$ or $\mathrm{erfi}(z)$ in that a factor of $(-1)^k$ now appears in the summand. This means that the series is the first type of series studied in the previous chapter, viz. $S(N, z^\beta)$. Consequently, it can be MB-regularised according to Proposition 3. Hence, we obtain

$$\sum_{k-N}^\infty \Gamma(k+1/2)\left(-\frac{1}{z^2}\right)^k \equiv \int_{\substack{c-i\infty \\ \mathrm{Max}[-1/2,N-1]<c=\Re s<N}}^{c+i\infty} ds\, \frac{z^{-2s}\,\Gamma(s+1/2)}{(e^{-i\pi s}-e^{i\pi s})} \; . \tag{8.12}$$

If we introduce the above equivalence into Equivalence (5.1), then we find for the Stokes sector of $|\arg z| < \pi/2$ that

$$\mathrm{erf}(z) = \frac{e^{-z^2}}{\pi z}\sum_{k=0}^{N-1}(-1)^{k+1}\,\Gamma(k+1/2)\left(\frac{1}{z^2}\right)^k + 1$$
$$+ \frac{e^{-z^2}}{\pi z}\int_{\substack{c-i\infty \\ \mathrm{Max}[-1/2,N-1]<c=\Re s<N}}^{c+i\infty} ds\, \frac{z^{-2s}\,\Gamma(s+1/2)}{(e^{i\pi s}-e^{-i\pi s})} \; , \tag{8.13}$$

while for the Stokes sectors of $-3\pi/2 < \arg z < -\pi/2$ and $\pi/2 < \arg z < 3\pi/2$, we arrive at

$$\mathrm{erf}(z) = \frac{e^{-z^2}}{\pi z}\sum_{k=0}^{N-1}(-1)^{k+1}\,\Gamma(k+1/2)\left(\frac{1}{z^2}\right)^k - 1$$
$$+ \frac{e^{-z^2}}{\pi z}\int_{\substack{c-i\infty \\ \mathrm{Max}[-1/2,N-1]<c=\Re s<N}}^{c+i\infty} ds\, \frac{z^{-2s}\,\Gamma(s+1/2)}{(e^{i\pi s}-e^{-i\pi s})} \; . \tag{8.14}$$

We can simplify the above results by replacing the Stokes discontinuity with $2S$, where S is a Stokes multiplier following Berry [9]. It is given by

$$S = \begin{cases} 1/2 \ , & 0 < |\arg z| < \pi/2 \ , \\ -1/2 \ , & \pi/2 < |\arg z| < 3\pi/2 \ . \end{cases} \tag{8.15}$$

Comparing the above result with the Stokes multiplier given by Eq. (5.23), we see that the Stokes sectors have been shifted by $\pi/2$ to the left.

Before examining the asymptotic forms along the Stokes lines at $\arg z = \pm \pi/2$, let us verify that the above results are convergent over the principal branch of the complex plane. For $s = c \pm iL$, where $L \to \infty$, the modulus of the integrand for the MB integral in Eqs. (8.13) and (8.14) behaves as

$$\left| \frac{\Gamma(s+1/2)}{z^{2s}(e^{-i\pi s} - e^{i\pi s})} \right| \overset{s=c\pm iL}{\sim} \sqrt{2\pi} \, |z|^{-2c}(L^2 + (c+1/2)^2)^{c/2}$$

$$\times \ e^{-c-1/2} \begin{bmatrix} e^{2L\theta - 3\pi L/2} \\ e^{-2L\theta - 3\pi L/2} \end{bmatrix}, \tag{8.16}$$

where, once again, $z = |z| \exp(i\theta)$. Because of this behaviour, the MB integral in Eqs. (8.13) and (8.14) is convergent for $|\theta| < 3\pi/4$. Thus, at the very least we expect Eq. (8.13) to yield values of $\text{erf}(z)$ for $|\theta| < \pi/2$ and Eq. (8.14) to yield values of $\text{erf}(z)$ for $-3\pi/4 < \theta < -\pi/2$ and $\pi/2 < \theta < 3\pi/4$.

We now verify these results by conducting a numerical analysis in the vicinity of the Stokes lines, i.e. near the positive and negative imaginary axes, and at the limits of convergence of the MB integrals in Eqs. (8.13) and (8.14). Again, we use the NIntegrate routine in Mathematica [19] to evaluate the MB integrals. We shall consider small values of $|z|$, which will highlight better any difference between the actual values of the error function and those generated by Eqs. (8.13) and (8.14). Table 8.4 presents a typical sample of the results for the various terms in Eqs. (8.13) and (8.14) with $z = \exp(i\theta)/7$ and θ taking on values in all the Stokes sectors. Because this value of $|z|$ is less than unity, there is no optimal point of truncation. Hence, the truncation parameter N was set equal to unity, which represents the lowest value for which the truncated series in Eqs. (8.13) and (8.14) contributes to the error function. In addition, the arbitrary offset parameter c in the MB integrals was set equal to $N - 1/4$.

Table 8.5 displays a sample of the results for $|\theta| = \pm 7\pi/16$, which are typical of all the results obtained for the various values of $|\theta|$ that were chosen to be slightly

Table 8.5: Values of erf(exp(iθ)/7) and the various contributions on the rhs of Eqs. (8.13) and (8.14) with N=1

θ	Term	
$\pm 7\pi/16$	Stokes Discontinuity	1
	Truncated Series	$-0.754\,289\,903\,337\,632 \pm 3.953\,178\,648\,984\,191\,i$
	MB Integral	$-0.213\,647\,276\,655\,340 \mp 3.794\,163\,472\,819\,168\,i$
	Sum	$0.032\,062\,820\,007\,026 \pm 0.159\,015\,176\,165\,023\,i$
	erf(exp($\pm 7\pi i/16$)/7)	$0.032\,062\,820\,007\,029 \pm 0.159\,015\,176\,165\,021\,i$
$\pm 9\pi/16$	Stokes Discontinuity	-1
	Truncated Series	$-0.754\,289\,903\,337\,633 \pm 3.953\,178\,648\,984\,191\,i$
	MB Integral	$-0.213\,647\,276\,655\,339 \mp 3.794\,163\,472\,819\,169\,i$
	Sum	$-0.032\,062\,820\,007\,030 \pm 0.159\,015\,176\,165\,030\,i$
	erf(exp($\pm 9\pi i/16$)/7)	$-0.032\,062\,820\,007\,029 \pm 0.159\,015\,176\,165\,021\,i$
$\pm 11\pi/16$	Stokes Discontinuity	-1
	Truncated Series	$2.148\,542\,736\,116\,558 \pm 3.350\,594\,721\,601\,254\,i$
	MB Integral	$-1.239\,175\,794\,732\,259 \mp 3.216\,784\,829\,285\,402\,i$
	Sum	$-0.090\,633\,058\,615\,701 \pm 0.133\,809\,892\,315\,852\,i$
	erf(exp($\pm 11\pi i/16$)/7)	$-0.090\,633\,058\,615\,704 \pm 0.133\,809\,892\,315\,856\,i$
$\pm 15\pi/16$	Stokes Discontinuity	-1
	Truncated Series	$3.795\,072\,764\,177\,338 \pm 0.785\,747\,186\,653\,128\,i$
	MB Integral	$-2.952\,264\,384\,390\,406 \mp 0.754\,902\,883\,265\,892\,i$
	Sum	$-0.157\,191\,620\,213\,i068 \pm 0.030\,844\,303\,387\,236\,i$
	erf(exp($\pm 15\pi i/16$)/7)	$-0.157\,191\,620\,213\,064 \pm 0.030\,844\,303\,387\,243\,i$

less than $\pi/2$. As in the case of the previous tables in this chapter, not all decimal places are displayed due to limited space. From the table, however, it can be seen that all the terms on the rhs of Eq. (8.13) make a sizeable contribution to the sum. Hence, none of the terms can be neglected. Nonetheless, their sum yields the value of erf(exp($\pm 7i\pi/16$)/7) within the accuracy and precision goals, which ironically is much smaller than any of the contributions on the rhs of Eq. (8.13). In addition, it was found for all the values of $|\theta|$ less than $\pi/2$ that the Stokes discontinuity had to equal unity in order the obtain the correct values of the error function.

As yet, we have not experienced problems with using the NIntegrate routine to obtain the correct regularised value of an asymptotic series, but for $|\theta| > \pi/2$, we need to pay special attention to the factor of z^{-2s} in the MB integral. This is because Mathematica restricts the evaluation of a multi-valued integrand to the principal branch of the complex plane. To see the implications of this more clearly, we first note that the results for $|\theta| = \pm 7\pi/16$ have been obtained by writing the integrand in the NIntegrate routine as

$$\text{INTGD}[z\_, s\_] := z \wedge (-s) \, z \wedge (-s) \, \text{Gamma}[s + 1/2]$$
$$/(\text{Exp}[\text{I Pi s}] - \text{Exp}[-\text{I Pi s}]) \ .$$

For $|\arg z| < \pi/2$, however, we could have obtained similar results by writing the integrand as either

$$\text{INTGD}[z\_, s\_] := z \wedge (-2s) \, \text{Gamma}[s + 1/2]/(\text{Exp}[\text{I Pi s}] - \text{Exp}[-\text{I Pi s}]),$$

or

$$\text{INTGD}[z\_, s\_] := (z \wedge (-2)) \wedge s \, \text{Gamma}[s + 1/2]$$
$$/(\text{Exp}[\text{I Pi s}] - \text{Exp}[-\text{I Pi s}]) \ .$$

The situation becomes very different when $|\arg z| > \pi/2$. Then we find that the first expression of the integrand in Mathematica yields the values of the error function when the Stokes discontinuity term is equal to unity. Although the second expression of the integrand begins to experience convergence problems for θ equal to $\pm 9\pi/16$ and $\pm 5\pi/8$, it, nevertheless, yields the correct values of the error function to about 13 decimal places when the Stokes discontinuity is set equal to unity. Thus, both these versions for the integrand imply that Eq. (8.14) is invalid. On the other hand, if one employs the third version of the integrand, then one obtains the values of the error function when the Stokes discontinuity is equal to -1, which implies that Eq. (8.14) is now valid. Unfortunately, the third expression of the integrand in Mathematica yields the correct values of the error function when $|\arg z| > 3\pi/4$, but we have seen that the MB integral is divergent for these values of z. Thus, we can only be confident with the evaluation of the MB integral provided $|\arg z^2| < \pi$, i.e. for z lying in the Stokes sector of $|\arg z| < \pi/2$. When z lies outside this sector, Mathematica shifts the calculations back to the principal branch and then one will not be evaluating the original MB integral, but a different complex root of it as discussed on p. 564 of Ref. [19].

Our aim, however, is to develop forms of the asymptotic expansion that will yield the values of the error function not for z^2, but for z over the entire principal branch.

After all, Mathematica gives values of the error function over the entire principal branch of the complex plane, but using the above forms for the asymptotic expansion means that we are restricted to $|\arg z| < \pi/2$ since the asymptotic series in these forms is in powers of z^2 rather than z. Even if we accept the last version of the integrand in Mathematica, the asymptotic forms given by Eqs. (8.13) and (8.14) do not extend over the entire principal branch because the MB integral is divergent for $3\pi/4 < |\arg z| < \pi$. To circumvent this problem, we need to introduce the factor of $\exp(2\pi i j k)$ in the asymptotic series as we did in the previous chapter. As we have seen, although this factor is equal to unity, it results in the appearance of the exponential factor of $\exp(2\pi i j s)$ in the MB integral. Therefore, by using the material in the previous chapter, we are able to generalise Eqs. (8.13) and (8.14) to

$$
\mathrm{erf}(z) = \frac{e^{-z^2}}{\pi z} \sum_{k=0}^{N-1} (-1)^{k+1} \Gamma(k+1/2) \left(\frac{e^{2ij\pi}}{z^2} \right)^k + 2S(j)
$$

$$
+ \frac{e^{-z^2}}{\pi z} \int_{\substack{c-i\infty \\ \mathrm{Max}[-1/2,N-1]<c=\Re s<N}}^{c+i\infty} ds \, \frac{e^{2\pi i j s} \Gamma(s+1/2)}{z^{2s} \left(e^{i\pi s} - e^{-i\pi s} \right)} \, , \tag{8.17}
$$

where the Stokes multiplier is now dependent upon j. For $j=0$, it is given by

$$
S(0) = \begin{cases} 1/2 \, , & -\pi/2 < \arg z < \pi/2 \, , \\ -1/2 \, , & \pi/2 < |\arg z| < 3\pi/2 \, . \end{cases} \tag{8.18}
$$

If $j=-1$, then the modulus of the integrand in the MB integral of Eq. (8.17) behaves as

$$
\left| \frac{\Gamma(s+1/2) \, e^{-2i\pi s}}{z^{2s} (e^{-i\pi s} - e^{i\pi s})} \right| \overset{s=c\pm iL}{\sim} \sqrt{2\pi} \, |z|^{-2c} (L^2 + (c+1/2)^2)^{c/2}
$$

$$
\times \, e^{-c-1/2} \left[\begin{array}{c} e^{2L\theta + \pi L/2} \\ e^{-2L\theta - 7\pi L/2} \end{array} \right] , \tag{8.19}
$$

which means that the MB integral is convergent for $-7\pi/4 < \arg z < -\pi/4$. Thus, this result encompasses the Stokes sector of $-3\pi/2 < \arg z < -\pi/2$ for which the Stokes multiplier is equal to -1/2 from Eq. (5.2). For $-5\pi/2 < \arg z < -3\pi/2$, S equals 1/2. Hence, for $j=-1$ we find that

$$
S(-1) = \begin{cases} 1/2 \, , & -5\pi/2 < \arg z < -3\pi/2 \ \& \ -\pi/2 < \arg z < \pi/2, \\ -1/2 \, , & -3\pi/2 < \arg z < -\pi/2 \, . \end{cases} \tag{8.20}
$$

From the $j = -1$ and $j = 0$ results we can develop a general form for the Stokes multiplier for any integer j, which is given by

$$S(j) = \begin{cases} (-1)^{j+1}/2 \, , & (j-3/2)\pi < \arg z < (j-1/2)\pi \, , \\ (-1)^{j}/2 \, , & (j-1/2)\pi < \arg z < (j+1/2)\pi \, , \\ (-1)^{j+1}/2 \, , & (j+1/2)\pi < \arg z < (j+3/2)\pi \, . \end{cases} \tag{8.21}$$

We now return to the numerical results presented in Table 8.5. The results for $|\arg z| > \pi/2$ have been calculated by putting $j = 1$ and $j = -1$ in Eq. (8.17). For these cases the Stokes multiplier equals $-1/2$ and hence, the entire Stokes discontinuity is equal to -1. To avoid the problems that Mathematica experiences when evaluating multi-valued integrands as mentioned above, for $j = -1$ we write the integrand as

$$\text{INTGD}[z_-, s_-] := (z \wedge (-s) \, \text{Exp}[-\text{I Pi s}])) \wedge 2 \, \text{Gamma}[s + 1/2]$$
$$/(\text{Exp}[\text{I Pi s}] - \text{Exp}[-\text{I Pi s}]) \, .$$

For $j = 1$, we replace the factor of Exp[-I Pi s] in the numerator by Exp[I Pi s]. Other than this alteration to the integrand, the Mathematica module used to obtain the values for $|\theta| < \pi/2$ in Table 8.5 remains the same with the truncation parameter N and offset parameter c equal to 1 and 3/4, respectively. As a consequence, we are able to evaluate the regularised value of the error function when $|\theta| > \pi/2$ as displayed in Table 8.5. In an odd twist, Mathematica is unable to evaluate the MB integral due to overflow and underflow errors when θ is equal to $-\pi$, but when θ is equal to π, it is able to evaluate the MB integral and gives the value of $\text{erf}(\exp(i\pi)/7)$ to 14 decimal places, well within the accuracy and precision goals set in the module. This serves to re-inforce the fact $-\pi$ lies outside the principal branch of the complex plane and hence, will not evaluated by the software package. On the other hand, $\theta = \pi$ does belong to the principal branch and therefore, is evaluated by the software package.

From the results in Table 8.5 we see clearly that the Stokes multiplier is equal to $-1/2$ or rather that the Stokes discontinuity equals -1 for either $\pi/2 < \arg z \le \pi$ or $-\pi < \arg z < -\pi/2$. As a consequence, we are now able to determine the actual values of the error function by calculating the regularised value of the asymptotic expansion given by Equivalence (8.17) in conjunction with Eq. (8.21) for all values of z in the complex plane, not only those values of z^2 restricted to the principal branch of the complex plane. Frequently, asymptotic series in an expansion are expressed in terms of a power on the variable, say β, which means that according to the first rule in Ch. 3, the Stokes sectors for such an expansion will be determined when the powers of z^β, not z, are all of the same sign. Consequently, some Stokes sectors situated outside the principal branch of the complex plane for z^β will be projected onto the principal branch when we wish to calculate the actual values of the original function for certain values of z from its complete asymptotic expansion. E.g., in the asymptotic forms for the error function we have seen that $\beta = 2$. In order to determine the values of the error function over the entire principal branch of the complex plane for its variable z, we needed the regularised value of the asymptotic forms for the three Stokes sectors of $-3\pi/2 < \arg z < -\pi/2$, $-\pi/2 < \arg z < \pi/2$ and $\pi/2 < \arg z < 3\pi/2$. The first and third of these sectors lie outside of the principal branch for z, although not for z^2. Therefore, if we wish to determine the values of the original function for all values of z in the principal branch of the complex plane from its asymptotic expansion, then we require the asymptotic forms that apply to higher Stokes sectors. In particular, for $\beta \gg 1$, one will require numerous forms of the asymptotic expansion, not just three as in the case of $\mathrm{erf}(z)$. If the asymptotic series in these forms is MB-regularised, then the resulting MB integral will be restricted by a domain of convergence, which will not encompass all the projected Stokes sectors. To overcome this problem, a factor of $\exp(2\pi i j k)$ has been introduced into the asymptotic series for the error function, which results in an extra factor of $\exp(2\pi i j s)$ appearing in the MB integrals of its asymptotic forms. As a consequence, the domains of convergence for the MB integrals are shifted, enabling us to evaluate or access the regularised value of the asymptotic forms for each projected Stokes sector. This, in turn, has enabled us to determine the values of the original function over the entire principal branch of the complex plane for z via the MB-regularised asymptotic forms regardless of the size of β. The results in Table 8.5 vindicate this approach. In the next chapter we shall extend this idea even further so that regularised values of generalisations of Dingle's terminants in Ch. 3 may be obtained over any Stokes sector, not only those Stokes sectors encompassing the principal branch.

To complete this analysis of the MB-regularised forms for the error function, we now derive those forms that apply when z is situated on a Stokes line in the principal branch. There are two such Stokes lines, viz. the positive and negative imaginary axes. From the preceding examples we have seen that the Stokes discontinuity across a Stokes line is merely half the difference between the asymptotic forms that apply over the adjacent Stokes sectors, while the asymptotic form on a Stokes line is given by the average of the asymptotic forms for the adjacent Stokes sectors. For the Stokes line of $\arg z = (j+1/2)\pi$, this means that we must average the forms for $j = j$ and $j = j+1$ in Eq. (8.17). Because the Stokes multiplier oscillates in sign, we find that $S(j)+S(j+1)=0$. Hence, the average of both forms simplifies to

$$
\operatorname{erf}\left(|z|e^{(j+1/2)i\pi}\right) = (-1)^j i \, \frac{e^{|z|^2}}{\pi|z|} \sum_{k=0}^{N-1} \Gamma(k+1/2)|z|^{-2k} + \frac{(-1)^{j+1} e^{|z|^2}}{2\pi|z|}
$$

$$
\times \int_{\substack{c-i\infty \\ \mathrm{Max}[-1/2,N-1]<c=\Re s<N}}^{c+i\infty} ds \, z^{-2s} \Gamma(s+1/2)\cot(\pi s) \ . \quad (8.22)
$$

Eq. (8.22) has been verified for numerous values of z by using Mathematica. For example, when $z = i/10$, $j = 1$ and $N = 1$, Mathematica 4.1 yields values of $5.698\,597\,831\,295\,584\,i$ and $-5.585\,382\,657\,125\,989\,i$ respectively for the truncated series and the MB integral in the above equation. The value of the MB integral has been determined by setting the options of PrecisionGoal, AccuracyGoal and WorkingPrecision equal to 14, 14 and 16 respectively in the NIntegrate routine. Summing the value of the truncated series with that for the MB integral yields a value of $0.113\,215\,174\,169\,594\,59\,i$, which is almost identical to the value of $0.113\,215\,174\,169\,599\,84\,i$ for $\operatorname{erf}(i/10)$. However, the smaller $|z|$ is, the weaker the agreement between the actual value of the error function and that obtained from the rhs of Eq. (8.22). For $z = i/100$, Mathematica 4.1 yields a value of $2.261\,195\cdots\times 10^{-13}+0.011\,284\,167\,810\,430\,i$ for the rhs of Eq. (8.22) compared with the actual value of $0.011\,284\,167\,808\,628\,i$ for $\operatorname{erf}(i/100)$. The waning agreement between the values obtained from the rhs of Eq. (8.22) and those for the error function as $|z|$ becomes smaller, is not only due to the fact that there is a cancellation of an ever-increasing number of decimal places, but is also due to the fact that NIntegrate experiences convergence problems.

CHAPTER 9

Generalised Terminants

Abstract. Generalised terminants are produced when the coefficients of the two types of series considered in Ch. 7 are set equal to $\Gamma(pk+q)$, where p and q are both real and positive and the variable z is altered to z^β, where β can be much greater than unity. Ch. 9 is concerned with the derivation of the MB-regularised forms for the regularised values of both types of generalised terminants over the entire complex plane, which are presented in Propositions 4 and 5. These results are then simplified by considering special cases of p, the first where it is the reciprocal of a natural natural number and the second, where it equals 2. The chapter concludes by evaluating the regularised value of a Type II generalised terminant with $\beta=6$, $p=1$ and $q=1/5$ using the various MB-regularised forms that apply over the principal branch for z. Because there is no known special function equivalent for this asymptotic series, the results from this study serve as a test-bed for the results in the following chapter. Nevertheless, it is found that the regularised values obtained from the two MB-regularised forms for each of the six common regions of the overlapping domains of convergence equal one another for both small and large values of $|z|$.

In Ch. 7 we developed general forms for the regularised values of both types of asymptotic series, $S(N, z^\beta)$ and $S_1(N, z^\beta)$ via **MB** regularisation, but we have yet to demonstrate that these forms yield the same values as the asymptotic forms obtained via Borel summation. Moreover, from Ch. 3 we have seen that in order to carry out Borel summation of an asymptotic series, we need to introduce the integral representation for the gamma function. This means that the coefficients $f(k)$ in both types of asymptotic series will have to be limited to the gamma function if the proposed comparison between Borel-summed and MB-regularised values is to proceed.

It was also mentioned in Ch. 3 that a major drawback with Dingle's theory of terminants is that it is virtually limited to the principal branch of the complex plane. For example, a result such as Equivalence (3.12) is only applicable to the Stokes sectors of $0 < \theta = \arg z < 2\pi$ and $-2\pi < \arg z < 0$, but cannot be applied to the other Stokes sectors given by $2l\pi < \arg z < 2(l+1)\pi$, where l is an arbitrary integer not equal to -1 or 0. We have seen from the last chapter that when an asymptotic expansion is derived, it may no longer be dependent upon the variable

z, but on z to some real power, say β. When this occurs, the Stokes sectors are given by $2l\pi/|\beta| < \arg z < 2(l+1)\pi/|\beta|$. For $|\beta| > 1$, it means that higher Stokes sectors are projected onto the principal branch of $-\pi < \arg z \leq \pi$. Traversing these higher Stokes sectors results in more Stokes discontinuities that need to be evaluated in order to obtain the regularised value of the asymptotic series over the entire principal branch. Unfortunately, because Dingle's theory of terminants is limited to $\beta = 1$, we do not have a general formula that enables us to evaluate the Stokes discontinuities for higher Stokes sectors.

It might be argued that Dingle's rules given in Ch. 3 can access the higher Stokes sectors, but they are awkward to implement and require some ingenuity as can be seen by the example of the asymptotics for the parabolic cylinder function in Ch. 1 of Ref. [3]. Furthermore, it remains to be seen whether the rules are valid for all asymptotic series. What we require is the ability to access the higher Stokes sectors by using general asymptotic forms like those obtained in Ch. 7 via MB regularisation. Moreover, Dingle's theory of terminants is limited to asymptotic series whose coefficients a_k are of the form $\Gamma(k+\alpha)$, but there are many special functions with more complicated coefficients in their asymptotic series such as the family of Bessel functions [15]. Whilst we are restricted to considering series with gamma function coefficients in order to be able to make a comparison with Borel-summed regularised values from the discussion in the opening paragraph of this chapter, we can, however, consider more general forms of Dingle's terminants by putting $f(k) = \Gamma(pk+q)$, where p and q are both real and positive. Therefore, in this chapter we shall study the following types of asymptotic series

$$S_{p,q}^{I}\left(N, z^\beta\right) = \sum_{k=N}^{\infty} \Gamma(pk+q)\left(-z^\beta\right)^k , \qquad (9.1)$$

and

$$S_{p,q}^{II}\left(N, z^\beta\right) = \sum_{k=N}^{\infty} \Gamma(pk+q)\left(z^\beta\right)^k . \qquad (9.2)$$

We shall refer to the first series as a Type I generalised terminant and the second series as a Type II generalised terminant. In addition, in these representations we shall regard p and q as parameters since it will be seen that the regularised values of both types of the series do not vary, once they are fixed. We could also regard β as a parameter, but it has an important role in determining the range of Stokes sectors and appears whenever z appears, whereas p and q do not. Therefore, β has been retained in the notation for both types of series together with z and N, the latter two being definitely variables that affect the regularised values of the series. Finally, without loss of generality we shall assume that β is real and positive as was the case in Ch. 7.

Our aim in this chapter is to derive the regularised values of both types of generalised terminants via MB regularisation over all Stokes sectors in the complex plane. We shall conclude by presenting an interesting numerical example, whose primary purpose is to give the reader an understanding of how to obtain the regularised values of both types of generalised terminants via numerical integration. The example will also serve as the basis for enabling a spectacular comparison between MB-regularised and Borel-summed values of a Type II generalised terminant in the following chapter.

It should be noted that when an asymptotic method is used to derive a generalised terminant, p and q invariably turn out to be rational numbers despite the fact that there are more irrational numbers than rational numbers. If, however, p is a rational number, then a generalised terminant can be expressed in terms of other generalised terminants or even the standard terminants of Dingle given in Ch. 3. Therefore, it can be argued that we do not really require generalised terminants when p is rational, but as we shall see, it is much simpler to deal with one generalised terminant rather than replacing it by a finite sum of terminants.

To observe how generalised terminants can be written as a finite sum of terminants when p is rational, we assume that $p = m/n$, where m and n are non-negative integers with no common factors. Then the generalised terminants as defined above can be written as

$$S_{m/n,q}^{\{I,II\}}(N,z^\beta) = \Theta(\mathrm{mod}(N,n)) \sum_{k=N}^{nN^*-1} \Gamma(mk/n+q) \left(\mp z^\beta\right)^k$$
$$+ \sum_{i=0}^{n-1} \left(\mp z^\beta\right)^i \sum_{k_1=N^*}^{\infty} \Gamma(mk_1+mi/n+q) \left(\mp z^\beta\right)^{nk_1} , \qquad (9.3)$$

where $\mathrm{mod}(N,n) = N-n[N/n]$, the Heaviside step-function $\Theta(z)$ is defined as

$$\Theta(z) = \begin{cases} 1, & z>1 , \\ 0, & z\leq 0 , \end{cases} \qquad (9.4)$$

and N^* is given by

$$
N^* = \begin{cases} [N/n] + 1, & \mod(N,n) \neq 0 , \\ N/n, & \mod(N,n) = 0 . \end{cases} \tag{9.5}
$$

In Eq. (9.3) $\{I,II\}$ means that the upper and lower signs on the rhs refer respectively to $S^I_{m/n,q}$ and $S^{II}_{m/n,q}$. In addition, k has been replaced by $nk_1 + i$ so that $\sum_{k=N}^{\infty}$ becomes $\sum_{i=0}^{n-1} \sum_{k_1=N^*}^{\infty}$ plus a finite sum, which vanishes when $N = 0$. The latter appears as the first sum in Eq. (9.3). The inner sum of the second double sum in Eq. (9.3) represents another generalised terminant. Hence, Eq. (9.3) can be written as

$$
S^{\{I,II\}}_{m/n,q}(N,z^\beta) = \Theta\left(\mod(N,n)\right) \sum_{k=N}^{nN^*-1} \Gamma(mk/n + q) \left(\mp z^\beta\right)^k
$$

$$
+ \sum_{i=0}^{n-1} \left(\mp z^\beta\right)^i S^{\{I,II\}}_{m,mi/n+q}\left(N^*, z^{n\beta}\right) . \tag{9.6}
$$

Therefore, it has been shown that $S^{\{I,II\}}_{m/n,q}(N,z^\beta)$ can be expressed in terms of a finite sum of other generalised terminants. By determining the regularised values of all these terminants we obtain the regularised value of $S^{\{I,II\}}_{m/n,q}(N,z^\beta)$. However, this approach is unwieldy because all the terminants in Eq. (9.6) possess $z^{n\beta}$ as their variable, which means that there will be more Stokes sectors projected on to the principal branch for z or more domains of convergence to be considered for the MB-regularised values than there would be in evaluating the regularised value of $S^{\{I,II\}}_{m/n,q}(N,z^\beta)$ directly.

It has been mentioned that $S^{\{I,II\}}_{m/n,q}(N,z^\beta)$ can be expressed in terms of standard terminants. All that is required here is to show that each generalised terminant appearing in Eq. (9.6) can be expressed as a finite sum of them. We can accomplish this simply by putting $k' = mk$ in $S^{\{I,II\}}_{m,\alpha}(N,z^\beta)$, which yields

$$
S^{\{I,II\}}_{m,\alpha}(N,z^\beta) = \sum_{k'=mN,m(N+1),m(N+2),\ldots}^{\infty} \Gamma(k' + \alpha) \left(\mp z^{\beta/m}\right)^{k'} . \tag{9.7}
$$

The sum in the above equation increases incrementally in m. For it to increase with successive integers, we require the following identity:

$$
\sum_{k=mN,m(N+1),m(N+2),\ldots}^{\infty} = \frac{1}{m} \sum_{k=mN}^{\infty} \sum_{l=0}^{m-1} \exp(2i\pi kl/m) . \tag{9.8}
$$

By interchanging the sums over k and l, we find that Eq. (9.7) becomes

$$S_{m,\alpha}^{\{I,II\}}(N,z^\beta) = \frac{1}{m}\sum_{l=0}^{m-1}S_{1,\alpha}^{\{I,II\}}\left(mN,z^{\beta/m}\exp(2i\pi l/m)\right) . \qquad (9.9)$$

Now we have obtained a finite sum of terminants, which means that we can express the results on the rhs of Eq. (9.9) in terms of $\Lambda_{mN+\alpha-1}(z^{\beta/m}\exp(2\pi il/m))$ and $\bar{\Lambda}_{mN+\alpha-1}(z^{\beta/m}\exp(2\pi il/m))$ as defined by Eqs. (3.3) and (3.5). On this occasion, however, an extra factor of $\exp(2\pi il/m)$ appears with $z^{\beta/m}$ and consequently, we are no longer restricting z to the principal branch. This means that the Stokes sectors for each terminant will be different since for the first type of terminant in Eq. (9.9) the Stokes lines occur at $\arg z = ((2n-1)m-2l)\pi/\beta$, while for the second type of terminant they occur at $\arg z = (2nm-2l)\pi/\beta$. Because l ranges from 0 to $m-1$, the Stokes sectors for each terminant will be shifted and the regularised value of each terminant will develop jump discontinuities at different locations in the complex plane. This becomes unwieldy if both β and m are large because not only do we need to contend with a large number of projected sectors in the principal branch for z due to β, but each terminant in the finite sum will have its Stokes sectors shifted for each value of l.

A similar situation applies to the MB-regularised value of the series on the lhs of Eq. (9.9) if we wish to determine it by using the rhs of the equation. This is because the domains of convergence of the MB integrals for the terminants on the rhs of Eq. (9.9) are also affected by the factor of $\exp(2\pi il/m)$ appearing in the series. However, the situation will not be as unwieldy as the Borel-summed forms because there are no Stokes lines involved in MB regularisation. Hence, we do not require three distinct asymptotic forms using Borel-summed regularised values at each Stokes line, but at most two with a common sector when considering the MB-regularised values. Nevertheless, deriving regularised values either by MB-regularisation or Borel summation of Eqs. (9.3) or (9.9) is unwieldy as can be seen by the following example.

Suppose in solving a problem the asymptotic series for the solution is found to be given by the following Type I generalised terminant:

$$S_{3/2,1}^I(0,z) = \sum_{k=0}^\infty \Gamma(3k/2+1)(-z)^k . \qquad (9.10)$$

To apply Eq. (9.6) to the above series, we note that $m=3$, $n=2$, $q=1$, $N=0$ and $\beta=1$. Hence, both $\mathrm{mod}(N,n)$ and N^* are equal to zero, which means that Eq.

(9.6) yields

$$S_{3/2,1}^I(0,z) = \sum_{i=0}^{1} (-z)^i S_{3,3i/2+1}^I\left(0,z^2\right) = S_{3,1}^I(0,z^2) - zS_{3,5/2}^I(0,z^2) . \quad (9.11)$$

By introducing Eq. (9.9) into the above result, we find that

$$\begin{aligned}
S_{3/2,1}^I(0,z) = \frac{1}{3} &\left(S_{1,1}^I\left(0,z^{2/3}\right) - zS_{1,5/2}^I\left(0,z^{2/3}\right) \right.\\
&+ S_{1,1}^I\left(0,z^{2/3}e^{2i\pi/3}\right) - zS_{1,5/2}^I\left(0,z^{2/3}e^{2i\pi/3}\right)\\
&+ \left. S_{1,1}^I\left(0,z^{2/3}e^{-2i\pi/3}\right) - zS_{1,5/2}^I\left(0,z^{2/3}e^{-2i\pi/3}\right) \right) .
\end{aligned} \quad (9.12)$$

The various series on the rhs of Eq. (9.12) can be expressed in terms of Dingle's Λ-notation via Eq. (3.3). Furthermore, the regularised value of each terminant can be obtained from Equivalence (3.13). Hence, we arrive at

$$\begin{aligned}
S_{3/2,1}^I(0,z) \equiv \frac{1}{3} \int_0^\infty dt\, e^{-t} &\left(\frac{1-zt^{3/2}}{1+z^{2/3}t} + \frac{1-zt^{3/2}}{1+z^{2/3}e^{2i\pi/3}t} \right.\\
&+ \left. \frac{1-zt^{3/2}}{1+z^{2/3}e^{-2i\pi/3}t} \right) ,
\end{aligned} \quad (9.13)$$

which is only valid for $|\arg z| < \pi/2$. This last condition represents the common sector of the complex plane over which all the terminants are valid. By introducing Eq. (5.5) into Eq. (9.13), one eventually obtains

$$\begin{aligned}
S_{3/2,1}^I(0,z) \equiv \frac{z^{-2/3}}{3} &\left(e^{z^{-2/3}}\left(\Gamma(0,z^{-2/3}) - \Gamma(5/2)\Gamma(-3/2,z^{-2/3}) \right) \right.\\
&+ e^{z^{-2/3}e^{2i\pi/3}}\left(e^{2i\pi/3}\Gamma(0,z^{-2/3}e^{2i\pi/3}) - e^{-i\pi/3}\Gamma(5/2) \right.\\
&\times \left. \Gamma(-3/2,z^{-2/3}e^{2i\pi/3}) \right) + e^{z^{-2/3}e^{-2i\pi/3}}\left(e^{-2i\pi/3}\Gamma(0,z^{-2/3}e^{-2i\pi/3}) \right.\\
&- \left. \left. e^{i\pi/3}\Gamma(5/2)\Gamma(-3/2,z^{-2/3}e^{-2i\pi/3}) \right) \right) .
\end{aligned} \quad (9.14)$$

Whilst we have been able to express the regularised value of $S_{3/2,1}^I(0,z)$ in terms of special functions or integral identities via Dingle's theory of terminants, some issues have arisen with Equivalences (9.13) and (9.14). The first and perhaps, more critical issue, is that these equivalences are limited to $|\arg z| < \pi/2$, which is due to the fact that Dingle's theory of terminants is simply incapable of providing regularised value of the generalised terminant for all values of z in the principal branch of the complex plane. The second issue is that the result is very unwieldy. A better option is to use a general asymptotic form that applies to each Stokes sector or domain of convergence covering the entire complex plane rather than having to manipulate the regularised values of numerous terminants. In response to these issues we now turn our attention to the derivation of the asymptotic forms of the two types of generalised terminants via MB regularisation in the remaining sections of this chapter, while in the next chapter we shall derive the regularised values of the generalised terminants via Borel summation.

9.1 MB Regularisation of Type I Series

In Ch. 7 we found that the regularised value of power series with alternating coefficients could be obtained by applying the main result in Proposition 3, whereas the regularised value of power series with single-sign coefficients presented an ambiguity due to an extra factor of $(-1)^s$ appearing in the integrand of the MB integral. Since Type I generalised terminants belong to the former type of series, we expect no problems by directly applying Proposition 3 to them. Hence, we arrive at the following proposition.

Proposition 4. The MB-regularised value of the Type I generalised terminant as defined by Eq. (9.1) is given by

$$
S_{p,q}^{I}(N,z^{\beta}) \equiv \int_{\substack{c-i\infty \\ \mathrm{Max}[N-1,-q/p]<c=\Re s<N}}^{c+i\infty} ds\, \frac{z^{\beta s}\, e^{\mp 2\pi i M s}}{e^{-i\pi s}-e^{i\pi s}}\, \Gamma(ps+q)
$$

$$
\mp \frac{2\pi i}{p} z^{-\beta q/p}\, e^{\mp i q\pi/p} \sum_{j=1}^{M} e^{\pm 2ijq\pi/p} \exp\left(-z^{-\beta/p} e^{\pm(2j-1)i\pi/p}\right)\,, \qquad (9.15)
$$

provided that $(\pm 2M-1-p/2)\pi/\beta < \arg z < (\pm 2M+1+p/2)\pi/\beta$, $M\geq 0$ and $N > -q/p$.

Remark. If $N < -q/p$, then we can separate the terms in $S_{p,q}^{I}(N,z^{\beta})$ up to the first value that is greater than $-q/p$ and set this value to N. Then the result in Proposition 4 applies with the new value of N, while the regularised value of the original series is obtained by adding the separated terms to the regularised value for the modified series.

Proof. According to Equivalence (7.27) we have

$$S^I_{p,q}(N, z^\beta) \equiv \int_{\substack{c-i\infty \\ N-1<c=\Re s<N}}^{c+i\infty} ds \, \frac{z^{\beta s} e^{-2\pi i M s}}{e^{-i\pi s} - e^{i\pi s}} \Gamma(ps+q)$$

$$- 2\pi i \sum_{j=1}^{M} F\left(z^{-\beta} e^{i(2j-1)\pi}\right) . \tag{9.16}$$

To determine the domain of convergence for the MB integral, we need to analyse the behaviour of the integrand at its endpoints. From Stirling's approximation for the gamma function, i.e. No. 8.327 in Ref. [21], or introducing Approximation (8.5), we have

$$|\Gamma(p(c \pm iL) + q)| \overset{L \to \infty}{\sim} \sqrt{2\pi} \left((pc+q)^2 + p^2 L^2\right)^{(pc+q+1/2)/2} e^{-\pi p L/2} . \tag{9.17}$$

This means that the exponential behaviour of the coefficients in $S^I_{p,q}(N, z^\beta)$ at the endpoints of the MB integral is consistent with Approximation (7.39), provided $A = B = -\pi p/2$. Hence, the domain of convergence for the MB integral in Equivalence (9.16) is found to be

$$(2M - 1 - p/2)\pi/\beta < \theta = \arg z < (2M + 1 + p/2)\pi/\beta . \tag{9.18}$$

As explained in Ch. 7, the function $F(x)$ in Equivalence (9.16) is determined from the MB integral $\Delta I_{j,j-1}(z^\beta)$, which for $j \geq 0$, is given by

$$\Delta I_{j,j-1}(z^\beta) = \int_{\substack{c-i\infty \\ N-1<c=\Re s<N}}^{c+i\infty} ds \left(e^{i(2j-1)\pi s}/z^\beta\right)^{-s} \Gamma(ps+q) . \tag{9.19}$$

The domain of convergence for this integral is $(2j-1-p/2)\pi < \beta\theta < (2j-1+p/2)\pi$, which, as expected, represents the common region of the domains of convergence for $I_{j-1,j-2}(z^\beta)$ and $I_{j,j-1}(z^\beta)$. To derive the functional form for $\Delta I_{j,j-1}(z^\beta)$, we note that the integral representation for the gamma function can be written via a change of variable as

$$\int_0^\infty dy \, y^{s+q/p-1} \exp\left(-y^{1/p}\right) = p\Gamma(ps+q) . \tag{9.20}$$

The integral on the lhs of Eq. (9.20) is a Mellin transform, which yields the gamma function appearing in Eq. (9.19) provided $\Re s > -q/p$. Hence, the function $F(x)$ is given by

$$F(x) = p^{-1} x^{q/p} \exp\left(-x^{1/p}\right) . \tag{9.21}$$

The condition on the real part of s means that $N > -q/p$. If $-q/p > N-1$, then the lower bound in the MB integral becomes $-q/p$. Hence, the lower bound in the MB integral in Equivalence (9.15) has been altered to allow for this possibility. Then by using Eq. (9.21), we arrive at

$$\Delta I_{j,j-1}\left(z^{\beta}\right) = 2\pi i F\left(z^{-\beta} e^{i(2j-1)\pi}\right) = 2\pi i p^{-1} z^{-\beta q/p} e^{i(2j-1)q\pi/p}$$
$$\times \ \exp\left(-z^{-\beta/p} e^{i(2j-1)\pi/p}\right) \ . \tag{9.22}$$

The rhs of Eq. (9.22) represents the analytic continuation of the MB integral representation for $\Delta_{j,j-1}(z^{\beta})$. As a consequence, the first form for the regularised value in Proposition 4 is restricted to the domain of convergence for $I_M(z^{\beta})$ which is given immediately below Eq. (9.19). When Eq. (9.22) is introduced into the rhs of Equivalence (9.16) and j is replaced by M, we finally obtain the first form for the regularised value of the series given in the proposition.

For j a negative integer, we must use Equivalence (7.35) with $F(x)$ again given by Eq. (9.21). This form for the regularised value is only valid for $(-2|j|-1-p/2)\pi/\beta < \arg z < (-2|j|+1+p/2)\pi/\beta$. As before, by replacing $|j|$ with M, where $M \geq 0$, and carrying out a little algebra, one arrives at the second form or the lower sign version for the regularised value in the proposition. This completes the proof of Proposition 4.

If p is the reciprocal of an odd number including unity, then the exponential factor involving $z^{-\beta/p}$ in the second term of the regularised value reduces to $\exp(z^{-\beta/p})$, whereas if p is the reciprocal of an even integer, it reduces to $\exp(-z^{-\beta/p})$. In both cases the sum over j yields

$$Q_{I,\pm}^{MB}(M) = \sum_{j=1}^{M} e^{\pm i(2j-1)q\pi/p} = e^{\pm iMq\pi/p} \frac{\sin(Mq\pi/p)}{\sin(q\pi/p)} \ . \tag{9.23}$$

As a consequence, we can express the second term in the regularised value of Proposition 4 in terms of $Q_{I,\pm}^{MB}(M)$. Then by setting $p = 1/l$, we find that the upper sign version of Equivalence (9.15) yields

$$S_{p,q}^{l}(N, z^{\beta}) \equiv \int_{\underset{\text{Max}[N-1,-q/p]<c=\Re s<N}{c-i\infty}}^{c+i\infty} ds \, \frac{z^{\beta s} e^{-2\pi iMs}}{e^{-i\pi s} - e^{i\pi s}} \Gamma(s/l+q)$$
$$- 2\pi i l \, e^{-ilq\pi} Q_{I,+}^{MB}(M) z^{-\beta lq} \exp\left((-1)^{l+1} z^{-\beta l}\right) \ , \tag{9.24}$$

which is valid for $(2M-1-l/2)\pi/\beta < \arg z < (2M+1+l/2)\pi/\beta$. For $(-2M-1-l/2)\pi/\beta < \arg z < (-2M+1+l/2)\pi/\beta$, we introduce $p=1/l$ into the lower sign version of Equivalence (9.15), thereby obtaining

$$
S_{p,q}^I(N,z^\beta) \equiv \int_{\substack{c-i\infty \\ \text{Max}[N-1,-q/p]<c=\Re s<N}}^{c+i\infty} ds\, \frac{z^{\beta s}\, e^{2\pi i M s}}{e^{-i\pi s}-e^{i\pi s}}\, \Gamma(s/l+q)
$$
$$
+ 2\pi i l\, e^{ilq\pi}\, Q_{l,-}^{MB}(M)\, z^{-\beta lq}\, \exp\Big((-1)^{l+1} z^{-\beta l}\Big) \ . \tag{9.25}
$$

In both of the above equivalences if l is an even integer, then the extra terms are subdominant because of the decaying exponential, but if l is an odd integer, then they become the dominant terms in the regularised value.

In order to determine the regularised value of $S_{p,q}^I(N,z^\beta)$ for $p=2$, we need to consider odd and even values of M separately. For $M=2J+1$, we find that the sum over the $\Delta I_{j,j-1}(z^\beta)$, which ultimately yields the sum over j in Equivalence (9.15), can be expressed as

$$
\sum_{j=1}^M F\Big(z^{-\beta} e^{\pm i(2j-1)\pi}\Big) = z^{-\beta q/2} \cos\Big(q\pi/2 - z^{-\beta/2}\Big) e^{\pm(J+1)iq\pi}
$$
$$
\times\ \frac{\sin(Jq\pi)}{\sin(q\pi)} + \frac{1}{2}\, z^{-\beta q/2} \exp\Big(\pm i\big(q\pi/2 - z^{-\beta/2}\big)\Big) \ . \tag{9.26}
$$

Then the upper sign version of Equivalence (9.15) reduces to

$$
S_{p,q}^I(N,z^\beta) \equiv \int_{\substack{c-i\infty \\ \text{Max}[N-1,-q/2]<c=\Re s<N}}^{c+i\infty} ds\, \frac{z^{\beta s}\, e^{-(4J+2)\pi i s}}{e^{-i\pi s}-e^{i\pi s}}\, \Gamma(2s+q)
$$
$$
-\ 2\pi i\, z^{-\beta q/2} \cos\Big(q\pi/2 - z^{-\beta/2}\Big) e^{(J+1)iq\pi}\, \frac{\sin(Jq\pi)}{\sin(q\pi)}
$$
$$
-\ \pi i\, z^{-\beta q/2} \exp\Big(i\big(q\pi/2 - z^{-\beta/2}\big)\Big) \ , \tag{9.27}
$$

where $4J\pi/\beta < \arg z < (4J+4)\pi/\beta$. On the other hand, the lower sign version of Equivalence (9.15) yields

$$
S_{p,q}^I(N,z^\beta) \equiv \int_{\substack{c-i\infty \\ \text{Max}[N-1,-q/2]<c=\Re s<N}}^{c+i\infty} ds\, \frac{z^{\beta s}\, e^{(4J+2)\pi i s}}{e^{-i\pi s}-e^{i\pi s}}\, \Gamma(2s+q)
$$
$$
+\ 2\pi i\, z^{-\beta q/2} \cos\Big(q\pi/2 - z^{-\beta/2}\Big) e^{-(J+1)iq\pi}\, \frac{\sin(Jq\pi)}{\sin(q\pi)}
$$
$$
+\ \pi i\, z^{-\beta q/2} \exp\Big(-i\big(q\pi/2 - z^{-\beta/2}\big)\Big) \ , \tag{9.28}
$$

where $(-4J-4)\pi/\beta < \arg z < -4J\pi/\beta$. Equivalence (9.28) is basically the complex conjugate of Equivalence (9.27).

For $M=2J$, the sum over j in Equivalence (9.15) simplifies to

$$\mp 2\pi i \sum_{j=1}^{M} F\left(z^{-\beta} e^{\pm i(2j-1)\pi}\right) = \mp 2\pi i z^{-\beta q/2} \cos\left(q\pi/2 + z^{-\beta/2}\right)$$
$$\times\ e^{\pm Jiq\pi}\ \frac{\sin(Jq\pi)}{\sin(q\pi)}\ . \tag{9.29}$$

Therefore, Equivalence (9.15) becomes

$$S_{p,q}^{I}(N,z^{\beta}) \equiv \int_{\substack{c-i\infty \\ \mathrm{Max}[N-1,-q/2]<c=\Re s<N}}^{c+i\infty} ds\ \frac{z^{\beta s} e^{\mp 4J\pi i s}}{e^{-i\pi s} - e^{i\pi s}}\ \Gamma(2s+q)$$
$$\mp\ 2\pi i z^{-\beta q/2} \cos\left(q\pi/2 + z^{-\beta/2}\right) e^{\pm Jiq\pi}\ \frac{\sin(Jq\pi)}{\sin(q\pi)}\ , \tag{9.30}$$

where the upper sign is valid for $(4J-2)\pi/\beta < \arg z < (4J+2)\pi/\beta$, while the lower sign is valid for $(-4J-2)\pi/\beta < \arg z < (-4J+2)\pi/\beta$.

In the previous section we studied the generalised terminant $S_{3/2,1}^{I}(0,z)$, which resulted in the presentation of two different versions of the regularised value . These were given by Equivalences (9.13) and (9.14). Both results, however, were restricted to $|\arg z| < \pi/2$. According to Equivalence (9.15), the MB-regularised value of $S_{3/2,1}^{I}(0,z)$ is

$$S_{3/2,1}^{I}(0,z) \equiv \int_{\substack{c-i\infty \\ -2/3<c=\Re s<0}}^{c+i\infty} ds\ \frac{z^{s} e^{\mp 2\pi i M s}}{e^{-i\pi s} - e^{i\pi s}}\ \Gamma(3s/2+1)$$
$$\mp \frac{4\pi i}{3} z^{-2/3} e^{\mp 2i\pi/3} \sum_{j=1}^{M} e^{\pm 4ij\pi/3} \exp\left(-z^{-2/3} e^{\pm 2(2j-1)i\pi/3}\right)\ . \tag{9.31}$$

This result is only valid for $(\pm 2M - 7/4)\pi < \arg z < (\pm 2M + 7/4)\pi$, where $M \geq 0$. Not only is this a more compact result than either Equivalence (9.13) or Equivalence (9.14), it is also a more general result since it covers all branches of the complex plane for z. Hence, Equivalence (9.31) incorporates the values of $\arg z$ for Equivalences (9.13) and (9.14) in addition to all the remaining values of z in the entire principal branch. Specifically, for $M=0$ Equivalence (9.31) reduces to

$$S_{3/2,1}^{I}(0,z) \equiv \int_{\substack{c-i\infty \\ -2/3<c=\Re s<0}}^{c+i\infty} ds\ \frac{z^{s}}{e^{-i\pi s} - e^{i\pi s}}\ \Gamma(3s/2+1)\ , \tag{9.32}$$

where $-7\pi/4 < \arg z < 7\pi/4$.

Before turning our attention to the MB regularisation of Type II generalised terminants, we conclude this section by considering a more sophisticated example of the regularised value given by Equivalence (9.15) in deriving the asymptotic forms of a special function. From No. 3.389(7) in Ref. [21] we have

$$I = \int_0^\infty dx \, \frac{x^{\nu-1} e^{-zx}}{1+x^2} = \pi \csc(\nu\pi) V_\nu(2z,0) \; , \qquad (9.33)$$

where $\Re \nu, z > 0$ and $V_\nu(z,y)$ is a Lommel function of two variables. According to No. 8.578(9) of Ref. [21], which has actually been taken from No. 16.52(10) in Ref. [54], we can express the Lommel function of two variables in Eq. (9.33) in terms of the Lommel function $S_{\mu,\nu}(z)$ as

$$V_\nu(2z,0) = \frac{\sqrt{z}}{\Gamma(1-\nu)} \, S_{1/2-\nu,1/2}(z) \; . \qquad (9.34)$$

By noting that the denominator of the integrand in Eq. (9.33) can be regarded as the regularised value of the geometric series, we can introduce the lhs of Equivalence (4.7), thereby obtaining

$$S_{1/2-\nu,1/2}(z) \equiv \sum_{k=0}^\infty (-1)^k \frac{\Gamma(2k+\nu)}{\Gamma(\nu)} \, z^{-2k-\nu-1/2} \; . \qquad (9.35)$$

The series in the above equivalence is of the form given by Eq. (9.1) with $p=2$, $q=\nu$, $\beta=2$ and z replaced by $1/z$. Hence, we can apply the $p=2$ results for Proposition 4 given by Equivalences (9.27), (9.28) and (9.30). If we truncate the asymptotic series at N, then Equivalence (9.35) becomes

$$S_{1/2-\nu,1/2}(z) = \sum_{k=0}^{N-1} (-1)^k \frac{\Gamma(2k+\nu)}{\Gamma(\nu)} \, z^{-2k-\nu-1/2} + R_N(M) \; , \qquad (9.36)$$

where

$$R_N(2J+1) = \frac{z^{-\nu-1/2}}{\Gamma(\nu)} \int_{\mathrm{Max}[N-1,-\nu/2] < c = \Re s < N}^{c+i\infty}{}_{c-i\infty} ds \, \frac{z^{-2s} e^{\pm(4J+2)\pi is}}{e^{-i\pi s} - e^{i\pi s}} \Gamma(2s+\nu)$$

$$\pm \frac{\pi i}{\sqrt{z}\Gamma(\nu)} \left(2\cos(\nu\pi/2 - z) e^{\mp(J+1)i\nu\pi} \frac{\sin(J\nu\pi)}{\sin(\nu\pi)} + e^{\mp i(\nu\pi/2-z)} \right) \; , \qquad (9.37)$$

and

$$R_N(2J) = \frac{z^{-\nu-1/2}}{\Gamma(\nu)} \int_{\underset{\text{Max}[N-1,-\nu/2]<c=\Re s<N}{c-i\infty}}^{c+i\infty} ds \, \frac{z^{-2s} e^{\pm 4J\pi is}}{e^{-i\pi s} - e^{i\pi s}} \Gamma(2s+\nu)$$

$$\pm \frac{2\pi i}{\sqrt{z}\Gamma(\nu)} \cos(\nu\pi/2+z) e^{\mp Ji\nu\pi} \frac{\sin(J\nu\pi)}{\sin(\nu\pi)} \, . \tag{9.38}$$

Note that the above mathematical statement for the Lommel function has become an equation since the remainder is now finite. The upper sign in Eq. (9.37) is valid for $2J\pi < \arg z < (2J+2)\pi$, while the lower sign is valid for $-(2J+2)\pi < \arg z < -2J\pi$. In the case of Eq. (9.38) the upper sign is valid for $(2J-1)\pi < \arg z < (2J+1)\pi$, while the lower sign is valid for $-(2J+1)\pi < \arg z < -(2J-1)\pi$.

9.2   MB Regularisation of Type II Series

Now we turn our attention to the second type of generalised terminant $S_{p,q}^{II}(N,z^\beta)$ as given by Eq. (9.2). As discussed in Ch. 7, this type of asymptotic series is different from the Type I generalised terminant studied in the previous section because when it is derived by an asymptotic method, it is generally done so initially along a line of discontinuity. In Ch. 7 this line was referred to as the primary Stokes linc. As soon as the argument of the variable in the asymptotic series, namely z^β, moves off this line, the regularised value acquires jump discontinuities, which are not rendered by the direct application of Proposition 3. Therefore, a proposition that aims to provide the regularised value of Type II generalised terminants must be able to account for this anomalous behaviour, which is achieved in the following proposition.

Proposition 5. Given that the regularised value of the Type II generalised terminant, $S_{p,q}^{II}(N,z^\beta)$, as defined by Eq. (9.2), is real along the primary Stokes line, which is taken to be $\arg z = 0$, the regularised value of the series via MB regularisation is

$$S_{p,q}^{II}(N,z^\beta) \equiv \int_{\underset{\text{Max}[N-1,-q/p]<c=\Re s<N}{c-i\infty}}^{c+i\infty} ds \, \frac{z^{\beta s} e^{-i(2M+1)\pi s}}{e^{-i\pi s} - e^{i\pi s}} \Gamma(ps+q) - \frac{2\pi i}{p} z^{-\beta q/p}$$

$$\times \sum_{j=1}^{M} e^{2ijq\pi/p} \exp\left(-z^{-\beta/p} e^{2ij\pi/p}\right) - \frac{\pi i}{p} z^{-\beta q/p} \exp\left(-z^{-\beta/p}\right) \, , \tag{9.39}$$

provided that $N > -q/p$ and $(2M - p/2)\pi/\beta < \arg z < (2M+2+p/2)\pi/\beta$, where $M \geq 0$. For $N > -q/p$ and $(-2(M+1) - p/2)\pi/\beta < \arg z < (-2M+p/2)\pi/\beta$, the regularised value is the complex conjugate of the previous result, which is

$$S_{p,q}^{II}(N, z^\beta) \equiv \int_{\substack{c-i\infty \\ \text{Max}[N-1,-q/p]<c=\Re s<N}}^{c+i\infty} ds \, \frac{z^{\beta s} e^{i(2M+1)\pi s}}{e^{-i\pi s} - e^{i\pi s}} \, \Gamma(ps+q) + \frac{2\pi i}{p} z^{-\beta q/p}$$

$$\times \sum_{j=1}^{M} e^{-2ijq\pi/p} \exp\left(-z^{-\beta/p} e^{-2ij\pi/p}\right) + \frac{\pi i}{p} z^{-\beta q/p} \exp\left(-z^{-\beta/p}\right) \,. \quad (9.40)$$

Remark. As was the case for the previous proposition, if $N \leq -q/p$, then we separate all the terms up to the first value that is greater than $-q/p$ and apply the proposition to the modified series with N set to the new value. The regularised value of the original series is obtained by adding the separated terms to the regularised value of the modified series.

Proof. Without loss of generality, once more, we assume that β is real and positive since if it is negative, we can take its modulus and replace z by $1/z$ in order to calculate the regularised value of the series. From Approximation (9.17) we see that A and B in Approximation (7.39) are both equal to $-p\pi/2$ again. Furthermore, since the primary Stokes line is assumed to be $\arg z$ or $\theta = 0$, we can apply Equivalences (7.68) and (7.72) to derive the regularised value of a Type II generalised terminant over the entire complex plane. Hence, by applying Equivalence (7.68) to the Type II generalised terminant, we find that

$$S_{p,q}^{II}(N, z^\beta) \equiv I_M^*(z^\beta) - 2\pi i \sum_{j=1}^{M} F\left(z^{-\beta} e^{2ij\pi}\right) - \pi i F\left(z^{-\beta}\right) \,, \quad (9.41)$$

where the MB integral $I_M^*(z^\beta)$ is given by

$$I_M^*(z^\beta) = \int_{\substack{c-i\infty \\ N-1<c=\Re s<N}}^{c+i\infty} ds \, \frac{z^{\beta s} e^{-i(2M+1)\pi s}}{e^{-i\pi s} - e^{i\pi s}} \, \Gamma(ps+q) \,, \quad (9.42)$$

and $(2M - p/2)\pi/\beta < \arg z < (2M+2+p/2)\pi/\beta$. In Equivalence (9.41) $F(x)$ is the same function appearing in Proposition 4, whose Mellin transform yields $p\Gamma(ps+q)$. Note, however, that the phase factor multiplying the variable of $z^{-\beta}$ in the function is now $\exp(2ij\pi)$ rather than $\exp((2j-1)i\pi)$. That is,

$$2\pi i F\left(z^{-\beta} e^{2ij\pi}\right) = \frac{2\pi i}{p} z^{-\beta q/p} e^{2ijq\pi/p} \exp\left(-z^{-\beta/p} e^{2ij\pi/p}\right) \,, \quad (9.43)$$

while

$$\pi i F\left(z^{-\beta}\right) = \frac{\pi i}{p} \, z^{-\beta q/p} \exp\left(-z^{-\beta/p}\right) \ . \tag{9.44}$$

Substituting the above results into Equivalence (9.41) yields the extra terms to the MB integral on the rhs of the proposition. In obtaining these terms, we require that $N > -q/p$, while the remark addresses how to adjust the regularised value in the case of $N \leq -q/p$. If, however, $-q/p > N-1$, then we have to modify $I_M^*(z^\beta)$ so that the offset c is greater than $-q/p$ rather than $N-1$. Hence, the offset has to be shifted to the right of the maximum of $N-1$ and $-q/p$ as indicated in the proposition. Furthermore, by examining the behaviour of the integrand of the MB integral at the endpoints in the same fashion as in the previous chapter, we see that its domain of convergence is $(2M-p/2)\pi/\beta < \arg z < (2M+2+p/2)\pi/\beta$. As expected, this is the condition for which Equivalence (7.68) is valid.

We can also apply Equivalence (7.72) to obtain the regularised value of the Type II generalised terminant. This result, which is only valid for $(-2M-2-p/2)\pi/\beta < \arg z < (-2M\pi+p/2)\pi/\beta$, is given by

$$S_{p,q}^{II}\left(N, z^\beta\right) \equiv I_{-M-1}^*\left(z^\beta\right) + 2\pi i \sum_{j=1}^{M} F\left(z^{-\beta} e^{2ij\pi}\right) + \pi i F\left(z^{-\beta}\right) \ , \tag{9.45}$$

where the MB integral $I_{-M-1}^*(z^\beta)$ is given by

$$I_{-M-1}^*\left(z^\beta\right) = \int_{\substack{c-i\infty \\ N-1<c=\Re s<N}}^{c+i\infty} ds \, \frac{z^{\beta s} \, e^{i(2M+1)\pi s}}{e^{-i\pi s} - e^{i\pi s}} \, \Gamma(ps+q) \ . \tag{9.46}$$

To obtain the extra terms to the MB integral, we take the complex conjugate of Eqs. (9.43) and (9.44) and introduce the resulting terms into Equivalence (9.45). Again, we require that $N > -q/p$ but if $-q/p > N-1$, then we need to ensure the offset in $I_{-M-1}^*(z^\beta)$ is to the right of $-q/p$ instead of $N-1$ as was the case when we applied Equivalence (7.68). Analogously, the offset c needs to be adjusted so that it is greater than the maximum of $-q/p$ and $N-1$. When this is done, we arrive at the second result in the proposition given by Equivalence (9.40) or the complex conjugate of Equivalence (9.39). This completes the proof of Proposition 5.

As was found for the previous proposition, the extra terms appearing in both forms for the regularised value can be simplified greatly for specific values of p. If p is the reciprocal of a natural number including unity, then the exponential factor of $\exp(\pm 2ij\pi/p)$ becomes unity and we can evaluate the finite sum in the extra terms. Then we obtain

$$2\sum_{j=1}^{M} F\left(z^{-\beta}e^{\pm 2ij\pi}\right) + F\left(z^{-\beta}\right) = p^{-1}z^{-\beta q/p}\exp(-z^{-\beta/p})\,Q_{II,\pm}^{MB}(M)\ ,\quad (9.47)$$

where

$$Q_{II,\pm}^{MB}(M) = 2e^{\pm i(M+1)q\pi/p}\,\frac{\sin\left(Mq\pi/p\right)}{\sin(q\pi/p)} + 1\ . \tag{9.48}$$

Hence, the sums in the two forms for the regularised value of $S_{p,q}^{II}(N,z^{\beta})$ given in Proposition 5 can be expressed in terms of a multiplier and a subdominant exponential term, which demonstrates that for these values of p the discontinuities in the MB-regularised value of $S_{p,q}^{II}(N,z^{\beta})$ behave in a similar manner to the conventional view of the Stokes phenomenon. However, the multiplier is not affected by Stokes lines, but by domains of convergence. Nor does the multiplier experience jumps of unity for consecutive values of M.

For $p=2$, however, odd- and even-numbered values of M must be considered separately in the derivation of the regularised value of the Type II generalised terminant. For M odd, i.e. $M=2J+1$, the MB-regularised value according to Equivalences (9.39) and (9.40) becomes

$$S_{p,q}^{II}(N,z^{\beta}) \equiv \int_{\mathrm{Max}[N-1,-q/2]<c=\Re s<N}^{c+i\infty}\!\!\!\!\!\!\!\!\!\! ds\ \frac{z^{\beta s}\,e^{\mp i(4J+3)\pi s}}{e^{-i\pi s}-e^{i\pi s}}\,\Gamma(2s+q) \mp \pi i z^{-\beta q/2}$$

$$\times\left(e^{\pm iq\pi}\exp\left(z^{-\beta/2}\right) + 2\cosh\left(z^{-\beta/2}\pm iq\pi/2\right)e^{\pm i(J+3/2)q\pi}\right.$$

$$\times\left.\frac{\sin(Jq\pi)}{\sin(q\pi)} + \frac{1}{2}\exp\left(-z^{-\beta/2}\right)\right)\ . \tag{9.49}$$

The upper sign result is valid for $(4J+1)\pi/\beta < \arg z < (4J+5)\pi/\beta$, while the lower sign result is valid for $(-4J-5)\pi/\beta < \arg z < (-4J-1)\pi/\beta$. On the other hand, for $M=2J$ the regularised value of the asymptotic series is found to be

$$S_{p,q}^{II}(N,z^{\beta}) \equiv \int_{\mathrm{Max}[N-1,-q/2]<c=\Re s<N}^{c+i\infty}\!\!\!\!\!\!\!\!\!\! ds\ \frac{z^{\beta s}\,e^{\mp i(4J+1)\pi s}}{e^{-i\pi s}-e^{i\pi s}}\,\Gamma(2s+q) \mp \pi i z^{-\beta q/2}$$

$$\times\left(\frac{1}{2}\exp\left(z^{-\beta/2}\right) + 2\cosh\left(z^{-\beta/2}\mp iq\pi/2\right)e^{\pm i(J+1/2)q\pi}\frac{\sin(Jq\pi)}{\sin(q\pi)}\right)\ . \tag{9.50}$$

Here, the upper sign result is valid for $(4J-1)\pi/\beta < \arg z < (4J+3)\pi/\beta$, while the lower sign result is valid for $(-4J-3)\pi/\beta < \arg z < (-4J+1)\pi/\beta$. All the $p=2$ results given above demonstrate that it is no longer possible to express the finite sum in the MB-regularised value of the Type II generalised terminant as the product of a multiplier and a subdominant exponential as we observed when p was equal to the reciprocal of a natural number.

Before we turn our attention to the regularised values of the two types of generalised terminants via Borel summation, let us make the preceding material more concrete by considering a numerical example. We shall consider a Type II generalised terminant where the parameters p and q are equal to 1 and 1/5 respectively. According to Equivalence (9.39) in Proposition 5, the regularised value of this series is given by

$$S^{II}_{1,1/5}(N,z^6) \equiv \int_{\substack{c-i\infty \\ \mathrm{Max}[N-1,-1/5]<c=\Re s<N}}^{c+i\infty} ds \, \frac{z^{6s} \, e^{-i(2M+1)\pi s}}{e^{-i\pi s} - e^{i\pi s}} \, \Gamma(s+1/5) - 2\pi i z^{-6/5}$$

$$\times \; \exp(-z^{-6}) \left(\frac{1}{2} + e^{(M+1)i\pi/5} \, \frac{\sin(M\pi/5)}{\sin(\pi/5)} \right) , \qquad (9.51)$$

which is valid for $(2M-1/2)\pi/6 < \arg z < (2M+5/2)\pi/6$ and $M \geq 0$. For $(-2M-5/2)\pi/6 < \arg z < (-2M+1/2)\pi/6$, we require Equivalence (9.40), which yields

$$S^{II}_{1,1/5}(N,z^6) \equiv \int_{\substack{c-i\infty \\ \mathrm{Max}[N-1,-1/5]<c=\Re s<N}}^{c+i\infty} ds \, \frac{z^{6s} \, e^{i(2M+1)\pi s}}{e^{-i\pi s} - e^{i\pi s}} \, \Gamma(s+1/5) + 2\pi i z^{-6/5}$$

$$\times \; \exp(-z^{-6}) \left(\frac{1}{2} + e^{-(M+1)i\pi/5} \, \frac{\sin(M\pi/5)}{\sin(\pi/5)} \right) . \qquad (9.52)$$

Equivalences (9.51) and (9.52) can be combined into one result by replacing M by j, where j can be a negative integer. As a consequence, we introduce another multiplier, which equals the factors in the large parentheses of the above equivalences. Then we can express the preceding results as

$$S^{II}_{1,1/5}(N,z^6) \equiv \int_{\substack{c-i\infty \\ \mathrm{Max}[N-1,-1/5]<c=\Re s<N}}^{c+i\infty} ds \, \frac{z^{6s} \, e^{-i(2j+1)\pi s}}{e^{-i\pi s} - e^{i\pi s}} \, \Gamma(s+1/5)$$

$$+ \; 2\pi i z^{-6/5} \exp(-z^{-6}) \, T(j) , \qquad (9.53)$$

where $T(j)$ is given by

$$T(j) = \begin{cases} -\frac{1}{2} - e^{(j+1)i\pi/5} \, \sin(j\pi/5)/\sin(\pi/5) , & j \geq 0 , \\ \frac{1}{2} + e^{-(j+1)i\pi/5} \, \sin(j\pi/5)/\sin(\pi/5) , & j < 0 . \end{cases} \qquad (9.54)$$

The forms for $j \geq 0$ are only valid if $(2j-1/2)\pi/6 < \arg z < (2j+5/2)\pi/6$, while those for $j < 0$ are only valid if $(2j-1/2)\pi/6 < \arg z < (2j+5/2)\pi/6$.

Equivalence (9.53) is the MB-regularised form which we shall programme into a Mathematica module to calculate the regularised value of this particular example of a Type II generalised terminant. From the the analysis in Ch. 7 we know that the domains of convergence for the MB integral in Equivalence (9.53) for a specific value of j will overlap the domains of convergence for the MB integrals where $j = j - 1$ and $j = j + 1$. For example, the domain of convergence for the $j = 0$ MB integral is $(-\pi/12, 5\pi/12)$, while those for the $j = -1$ and $j = 1$ MB integrals are respectively $(-5\pi/12, \pi/12)$ and $(\pi/4, 3\pi/4)$. When $\arg z$ lies in the common sector of two domains of convergence, we shall ensure that the Mathematica module evaluates both forms or versions of the regularised value. This serves as a necessary check because if both forms do not give the same values in the common sectors, then we know immediately that the regularised value given in Proposition 5 cannot possibly be correct.

The problem with the domains of convergence for the $j = 0$ and $j = \pm 1$ MB integrals is that they do not cover all values of $\arg z$ in the principal branch. Thus, the module must be constructed so that it will enquire whether there is another form for the regularised value, should it exist. To accomplish this, we restrict the evaluation of each MB integral in Equivalence (9.53) to the Stokes sector that it encompasses rather than its entire domain of convergence. The Stokes sectors for this Type II generalised terminant are bordered by the lines given by $\arg z = j\pi/3$, where j is an integer. Therefore, the evaluation of the $j = 0$ MB integral is restricted to $\arg z$ lying in $[0, \pi/3)$, while the $j = 1$ MB integral is restricted to $[\pi/3, 2\pi/3)$ and so on. Technically, these are not Stokes sectors since the Stokes lines are included at the lower ends. We shall refer to the regularised values obtained when the $\arg z$ values lie directly in these pseudo Stokes sectors as the first form for the regularised value. For example, if $\pi/3 < \arg z < 5\pi/12$, then the first form of the regularised value is given by Equivalence (9.53) with $j = 1$. On the other hand, the regularised value can also be evaluated by the $j = 0$ result of Equivalence (9.53), whose pseudo Stokes sector is $[0, \pi/3)$. Those cases where $\arg z$ does not lie directly in a pseudo Stokes sector, but still yields the regularised value will be referred to as the second form for the regularised value. Therefore, the Mathematica module will be required to determine the value of j corresponding to the second form, when it exists. Basically, this means that if $\arg z$ lies between $[\pi j/3, \pi(j + 1/4)/3)$, then the $j = j - 1$ result in Equivalence (9.53) is the second form for the regularised value, while if it lies between

$[\pi(j+1/4)/3, \pi(j+1)/3)$, then the second form is given by the $j=j+1$ value of Equivalence (9.53).

Table 9.1: MB-regularised forms for the regularised value of $S_{1,1/5}^{II}(0,z^6)$ in Proposition 5 with $|z|=7/3$ and arg z assigned various values over the principal branch of the complex plane

arg z	Form	Regularised Value
$\pm 15\pi/17$	1	$-1.563\,354\,963\,772\,398\,2 \pm 0.324\,634\,838\,578\,441\,7i$
$\pm 13\pi/17$	1	$-1.522\,003\,005\,250\,221 \mp 0.444\,176\,963\,902\,584\,54i$
$\pm 11\pi/17$	1	$-1.156\,473\,556\,261\,043\,8 \mp 0.989\,173\,042\,444\,333\,6i$
$\pm 11\pi/17$	2	$-1.156\,473\,556\,261\,044 \mp 0.989\,173\,042\,444\,337i$
$\pm 9\pi/17$	1	$-0.683\,407\,869\,429\,495\,8 \mp 1.411\,677\,602\,015\,271\,3i$
$\pm 7\pi/17$	1	$0.041\,934\,793\,840\,417\,095 \mp 1.598\,302\,795\,235\,478\,2i$
$\pm 7\pi/17$	2	$0.041\,934\,793\,840\,411\,99 \mp 1.598\,302\,795\,235\,511\,8i$
$\pm 5\pi/17$	1	$0.714\,519\,183\,875\,366\,8 \mp 1.372\,946\,978\,975\,308\,6i$
$\pm 5\pi/17$	2	$0.714\,519\,183\,875\,367\,8 \mp 1.372\,946\,978\,975\,312\,8i$
$\pm 3\pi/17$	1	$1.202\,040\,278\,119\,570\,7 \mp 0.966\,331\,167\,394\,292\,9i$
$\pm \pi/17$	1	$1.539\,575\,647\,279\,512\,1 \mp 0.366\,397\,084\,777\,980\,3i$
$\pm \pi/17$	2	$1.539\,575\,647\,279\,513\,5 \mp 0.366\,397\,084\,777\,977\,6i$

Table 9.1 presents a sample of the regularised values obtained from Equivalence (9.53) with $|z| = 7/3$. Since there is no optimal point of truncation for this value of $|z|$, N has been set equal to zero. These results cover most of the principal branch of the complex plane with arg z initially set equal to $-15\pi/17$ and then incremented continuously by $2\pi/17$ till it reaches the final value of $15\pi/17$. As a consequence, the regularised values of this Type II generalised terminant have been evaluated by employing all the Stokes sectors for z^6 that are projected onto the principal branch of the complex plane for z. That is, the regularised values have been evaluated by using the six asymptotic forms ranging from $j = -3$ to $j = 2$ in Equivalence (9.53). From the table we see that these results give the complex conjugate values when z is complex-conjugated, which is also necessary for establishing the validity of the MB-regularised forms in Proposition 5. More importantly, however, is that on the occasions, where there is a second form for the regularised value, the value the latter gives is identical to the regularised value obtained from the first form within the precision and accuracy goals set in the Mathematica module. That is, even though the MB integrals evaluated in both

forms are different from each other, when the subdominant exponential terms are added to them, the regularised values are identical. This can only occur because the multipliers given by Eq. (9.54) are also different for the various values of j. Although not sufficient, the agreement that we have just observed between the two forms for the regularised value of this generalised Type II terminant validates the concept of regularisation.

Table 9.2: MB-regularised values of $S_{1,1/5}^{II}(0,z^6)$ with $|z|=7/3$ and arg z lying on the Stokes lines in the principal branch of the complex plane

arg z	Form	MB-Regularised Value
0	1	$1.590\,061\,161\,675\,929\,7 - 4.662\,936\,703\,425\,657\,5 \times 10^{-15}\,i$
0	2	$-1.590\,061\,161\,675\,929\,7 - 4.662\,936\,703\,425\,657\,5 \times 10^{-15}\,i$
$\pm\pi/3$	1	$0.515\,841\,753\,698\,810\,2 \mp 1.478\,536\,171\,758\,948\,7\,i$
$\pm\pi/3$	2	$0.515\,841\,753\,698\,810\,2 \mp 1.478\,536\,171\,758\,959\,i$
$\pm 2\pi/3$	1	$-1.222\,281\,759\,782\,957\,8 \mp 0.913\,785\,607\,743\,181\,5\,i$
$\pm 2\pi/3$	2	$-1.222\,281\,759\,782\,957\,6 \mp 0.913\,785\,607\,743\,190\,5\,i$
π	1	$-1.222\,281\,759\,782\,958\,5 + 0.913\,785\,607\,743\,191\,6\,i$
π	2	$-1.222\,281\,759\,782\,958 + 0.913\,785\,607\,743\,182\,4\,i$

So far, we have studied the regularised values when arg z is situated within the Stokes sectors of the principal branch of the complex plane. Now we present the regularised values of $S_{1,1/5}^{II}(0,z^6)$, again for $|z|=7/3$, but with arg z lying on the Stokes lines within the principal branch of the complex plane. Specifically, this means that arg $z = j\pi/3$, where $j=0, \pm 1, \pm 2$ and 3. Except for the primary Stokes line, the MB-regularised values of $S_{1,1/5}^{II}(N,z^6)$ are not affected by the Stokes phenomenon. As a consequence, we can use the same Mathematica module which was used to obtain the results in Table 9.1.

Table 9.2 presents the regularised values obtained from Equivalence (9.53) for arg z lying on the various Stokes lines in the principal branch of the complex plane. As stated previously, all the Stokes lines lie in the common region of the domains of convergence between consecutive values of j in Equivalence (9.53). There-fore, all the results in Table 9.2 have been evaluated via the two different forms of Equivalence (9.53). As we found for the values of arg z, where this applied in Table 9.1, both forms agree with each other for each Stokes line within the accuracy and precision goals set in Mathematica. It should also be noted that only on the primary Stokes line, viz. arg $z=0$, is the regularised value real within the accuracy and precision goals. We see that for the secondary Stokes lines the regularised values are complex with imaginary parts that cannot be dismissed as round-off numerical errors like those values obtained for the primary Stokes line. Therefore, the Zwaan-Dingle principle can only be applied to the primary Stokes line despite the fact that the terms in a generalised Type II terminant are homogeneous in phase and all the same sign when $\arg z$ is situated on a secondary Stokes line.

CHAPTER 10

Borel Summation of Generalised Terminants

Abstract.   In this chapter general Borel-summed forms for the regularised values of the two types of generalised terminants introduced in the previous chapter are derived for the entire complex plane. This is done by expressing both asymptotic series in terms of Cauchy integrals and analysing the singular behaviour as the variable z moves across Stokes sectors. For both types of generalised terminants the Stokes lines represent the complex branches of the singularities in the Cauchy integrals, the difference being that the singularity in the Type I case occurs at $-z^{-\beta}$, while for the Type II case it occurs at $z^{-\beta}$. Consequently, for a Type I generalised terminant the Cauchy integral represents the regularised value over a primary Stokes sector, whereas for the Type II case, it is the regularised value for a primary Stokes line provided the Cauchy principal value is evaluated. For the other Stokes sectors and lines, the regularised values acquire extra contributions due to the residues of the Cauchy integrals, which emerge each time $z^{-\beta}$ undergoes a complete revolution. In the case of the Type II generalised terminant, it also acquires an equal and opposite semi-residue contribution once $z^{-\beta}$ moves off the primary Stokes line in either direction. Hence, the results for the regularised values of both types of generalised terminants are treated separately depending upon whether the singularity undergoes clockwise or anti-clockwise rotations continuously. By referring to the special cases of p studied in the previous chapter, we find that the Borel-summed forms for the regularised values seldom conform to the conventional view of the Stokes phenomenon. The chapter concludes with the numerical evaluation of the Borel-summed forms for the regularised value of the same Type II generalised terminant at the end of Ch. 9. Though there are more Borel-summed forms to evaluate, in all cases the regularised value obtained from the Borel-summed forms agrees with that obtained from the corresponding MB-regularised forms.

At this stage we do not know whether the impressive numerical results presented in Tables 9.1 and 9.2 are indeed the actual or true regularised values of the particular generalised terminant denoted by $S^{II}_{1,1/5}(0, z^6)$. The agreement between the two forms of the MB-regularised values over the common sectors within the domains of convergence is extremely promising, but it is not conclusive since the common sectors do not cover the entire principal branch of the complex plane. In fact, the

Victor Kowalenko

agreement between the two forms is necessary, but not sufficient to establish the validity of the MB-regularised values of the generalised terminants in the previous section. In particular, we cannot say with any certainty, where only one form for the MB-regularised value has been used such as for $\arg z = \pm 13\pi/17$, that the value in Table 9.1 is indeed the actual regularised value of this Type II generalised terminant. Previously, we have been able to compare the MB-regularised values of asymptotic series with the values obtained by using Mathematica's intrinsic routines for evaluating the corresponding functional representations of the series. In the numerical example of the previous chapter, however, we do not know the function or integral which the particular Type II generalised terminant represents. Hence, we require a completely different method of determining the regularised values of generalised terminants. If both methods produce the same results, then we can be more assured that the MB-regularised values in Tables 9.1 and 9.2 are indeed the actual regularised values of $S^{II}_{1,1/5}(0,z^6)$. Naturally, the alternative method to which we are referring is Borel summation.

In Ch. 3 we found that Dingle's theory of terminants [3] was also based on Borel summation. However, this theory cannot be employed in determining the regularised values of both types of generalised terminants over the entire complex plane because it is limited to $\beta \leq 2$ and $p = 1$ in Eqs. (9.1) and (9.2). For the example studied at the end of the previous chapter, namely $S^{II}_{1,1/5}(0,z^6)$, we have $\beta = 6$ and $p = 1$. Although this Type II terminant is a series expansion in powers of z^6, we are assuming that it represents the asymptotic series for a function whose variable is z. Then according to the first of Dingle's rules given in Ch. 3, there are six Stokes sectors for z^6 which are projected onto the principal branch of the complex plane for z. Unfortunately, Dingle's theory of terminants is only capable of providing the regularised value for two of these Stokes sectors, $-\pi/3 < \arg z < 0$ and $0 < \arg z < \pi/3$. So, before we can establish that the values presented in Tables 9.1 and 9.2 are indeed correct over the entire principal branch for z, we need to derive the Borel-summed forms for the regularised value of the Type II generalised terminant over the four remaining Stokes sectors. Furthermore, the situation would have been even worse had we chosen to study $S^{I}_{1,1/5}(0,z^6)$ instead, for then Dingle's theory of terminants would only be able to provide the regularised value over one Stokes sector, viz. $|\arg z| < \pi/6$. Rather than put β to a specific value such as 6 in our example, we shall let β be an arbitrary positive number in the analysis in this chapter. In addition, we shall not consider only those asymptotic forms that are projected onto the principal branch for z. That is, we aim to derive the various Borel-summed forms for the regularised values of both types of generalised terminants over the entire complex plane. Of course, when it comes to numerical evaluation, we shall only be able to study those Stokes sectors situated in the principal branch of the complex plane for z.

10.1 Type I Generalised Terminants

As explained in Ch. 3, the first step in Borel summation is to replace the gamma function in a generalised terminant by its integral representation. For a Type I generalised terminant this results in

$$
S_{p,q}^{I}\left(N,z^{\beta}\right) = \sum_{k=N}^{\infty} \int_{0}^{\infty} dt\, e^{-t} t^{pk+q-1} \left(-z^{\beta}\right)^{k} = \int_{0}^{\infty} dt\, e^{-t}
$$
$$
\times\; t^{q-1} \sum_{k=N}^{\infty} \left(-z^{\beta} t^{p}\right)^{k}\;. \tag{10.1}
$$

The final form of Eq. (10.1) is only absolutely convergent when $|z^{\beta} t^{p}| < 1$, a condition which cannot be met since t ranges from zero to infinity. Based on the material presented in Ch. 4, we know that the series is conditionally convergent for $\Re z^{\beta} > 0$ and divergent for $\Re z^{\beta} < 0$. For the latter case, we need to regularise the series in order to obtain finite values for the series. We observed in Sec. 4.1 that the regularised value of the geometric series is the same limit value when the series is either conditionally or absolutely convergent. Thus, Eq. (10.1) becomes

$$
S_{p,q}^{I}\left(N,z^{\beta}\right) \equiv (-1)^{N} z^{\beta N} \int_{0}^{\infty} dt\, \frac{e^{-t} t^{pN+q-1}}{1+z^{\beta} t^{p}}\;. \tag{10.2}
$$

Alternatively, we can express the rhs of the above equivalence as

$$
S_{p,q}^{I}\left(N,z^{\beta}\right) \equiv (-1)^{N} p^{-1} z^{\beta(N-1)} \int_{C} ds\, \frac{s^{N+q/p-1} e^{-s^{1/p}}}{s-\left(-z^{-\beta}\right)}\;, \tag{10.3}
$$

where C is the line contour along the positive real axis. If we let

$$
f(s) = p^{-1} z^{\beta(N-1)} s^{N+q/p-1} e^{-s^{1/p}}\;, \tag{10.4}
$$

then the integral on the rhs of Equivalence (10.3) represents a Cauchy integral with the singularity at $s = -z^{-\beta}$. Because $z^{-\beta}$ appears in the Cauchy integral rather than

z^β as in the case of the MB-regularised values presented in Propositions 4 and 5, we see immediately that Borel summation operates or acts in a different manner compared with MB regularisation.

In the previous chapter we carried out the MB-regularisation of the generalised terminant defined by Equivalence (9.1) with z^β replaced by $z^\beta \exp(-2li\pi)$. Therefore, let us consider the same modification. Then Borel summation of the series yields

$$S^I_{p,q}\left(N, z^\beta e^{-2li\pi}\right) \equiv (-1)^N p^{-1} z^{\beta(N-1)} \int_C ds \, \frac{s^{N+q/p-1} e^{-s^{1/p}}}{s - \left(-z^{-\beta} e^{2li\pi}\right)} \, . \qquad (10.5)$$

It appears that since $\exp(2li\pi)$ is equal to unity for all values l, we should simply write the above result as Equivalence (10.3). However, we know from Ch. 7 that the regularised value of $S^I_{p,q}\left(N, z^\beta \exp(-2li\pi)\right)$ is different for each value of l. Therefore, the above equivalence can only be valid for one value of l or one Stokes sector, which we shall refer to as the primary Stokes sector. For the other values of l we can use Equivalence (7.20), which in turn has been derived from Proposition 3, to determine the difference between the regularised values of $S^I_{p,q}\left(N, z^\beta \exp(-2li\pi)\right)$ and $S^I_{p,q}\left(N, z^\beta \exp(-2(l-1)i\pi)\right)$. Therefore, we find that

$$\Delta S^I_l(N, z^\beta) = S^I_{p,q}\left(N, z^\beta e^{-2li\pi}\right) - S^I_{p,q}\left(N, z^\beta e^{-2(l-1)i\pi}\right)$$

$$\equiv \int_{\substack{c-i\infty \\ \text{Max}[N-1,-q/p]<c=\Re s<N}}^{c+i\infty} ds \, z^{\beta s} \, \Gamma(ps+q) e^{-(2l-1)i\pi s} \, , \qquad (10.6)$$

where the lower bound on the offset c has been adjusted to ensure that the poles for the MB integral remain to the right of $-q/p$, if it should be greater than $N-1$. This adjustment has been made in accordance with the conditions stipulated in Proposition 3. Furthermore, since

$$|\Gamma(p(c \pm iL) + q)| \stackrel{L \to \infty}{\sim} \begin{cases} e^{-p\pi L/2} \\ e^{-p\pi L/2} \end{cases} \, , \qquad (10.7)$$

we find that ε_1 and ε_2 as defined in Proposition 3 are both equal to $p\pi/2$ and the MB integral in Equivalence (10.6) exists only when $(2l-1-p/2)\pi/\beta < \arg z < (2l-1+p/2)\pi/\beta$. By making the change of variable, $y = ps+q$, we can express Equivalence (10.6) as

$$\Delta S^I_l(N, z^\beta) \equiv (z^{-\beta} e^{(2l-1)i\pi})^{q/p} p^{-1} \int_{\substack{c-i\infty \\ c=\Re s>-q/p}}^{c+i\infty} dy \left(z^{-\beta} e^{(2l-1)i\pi}\right)^{-y/p} \Gamma(y) \, . \qquad (10.8)$$

Since the MB integral in Equivalence (10.8) can be regarded as the inverse Mellin transform of $\exp(-x)$, where $x = z^{-\beta/p}\exp((2l-1)i\pi/p)$, we arrive at

$$\Delta S_l^I(N, z^\beta) \equiv 2\pi i\, p^{-1}\, z^{-\beta q/p}\, e^{(2l-1)iq\pi/p} \exp\left(-z^{-\beta/p}e^{(2l-1)i\pi/p}\right) \ . \quad (10.9)$$

Since the above result can be analytically continued over the entire complex plane, we can drop the condition on $\arg z$.

From p. 412 of Ref. [24] we know that the Cauchy integral on the rhs of Equivalence (10.5) develops jump discontinuities as $-z^{-\beta}\exp(2il\pi)$ moves across the line contour. This means that while $z^{-\beta}\exp(2li\pi)$ is located in a branch of the complex plane, say $(2j-1)\pi < \arg\left(z^{-\beta}\exp(2li\pi)\right) < (2j+1)\pi$, the regularised value is given by the Cauchy integral, but once $z^{-\beta}\exp(2li\pi)$ moves outside of this branch, it acquires extra terms or else the regularised value would be the same for all Stokes sectors. Therefore, the regularised value of $S_{p,q}^I(N, z^\beta\exp(-2li\pi))$ cannot be represented solely by a Cauchy integral. Consequently, another problem has emerged, which is that we need to determine the specific Stokes sector over which the Cauchy integral is the sole contribution to the regularised value. This appears to be arbitrary, much like selecting a principal branch in the complex plane. A similar problem arose earlier in when we analysed the MB regularisation of the Type II generalised terminant. There we solved the problem by nominating a primary Stokes line. So, we do likewise here except now we nominate a primary Stokes sector. Hence, we nominate the $j = l = 0$ branch of the complex plane as the primary Stokes sector. Then the Cauchy integral on the rhs of Equivalence (10.5) represents the regularised value only when $-\pi/\beta < \arg z < \pi/\beta$. Unlike the situation for the MB regularisation of a Type II generalised terminant, which was completely arbitrary, the choice of the primary Stokes sector for the Borel-summed regularised value of a Type I generalised terminant has been motivated by the fact that the $j = l = 0$ branch reduces to the principal branch of the complex plane when $\beta = 1$ and also yields the regularised value of the first type of terminant given on p. 406 of Ref. [3]. In other words, we obtain Equivalence (3.13) when β and p are set equal to unity in Eq. (9.1). Therefore, for $l = 0$ we arrive at

$$S_{p,q}^I\left(N, z^\beta\right) \equiv (-1)^N p^{-1} z^{\beta(N-1)} \int_C ds\, \frac{s^{N+q/p-1}\, e^{-s^{1/p}}}{s - \left(-z^{-\beta}\right)} \ , \quad (10.10)$$

which is only valid for $-\pi/\beta < \arg z < \pi/\beta$.

We are now in a position to determine the jump discontinuities that apply over the secondary Stokes sectors in the complex plane. First, we note that the Cauchy integral in Equivalence (10.5) is singular whenever $\arg z^{-\beta} = (2j+1)\pi$, where j is any integer. As $z^{-\beta} \exp((2l-1)i\pi)$ crosses from the primary sector Stokes sector to the adjacent sector or $j=l+1$ branch of the complex plane, $-z^{-\beta} \exp(2li\pi)$ moves from below the line contour or positive real axis to above the axis. During this transition the Cauchy integral becomes undefined when $-z^{-\beta} \exp(2li\pi)$ is situated on the positive real axis. Let us evaluate the residue by introducing an infinitesimal circle around the pole at $s = -z^{-\beta} \exp(-2li\pi)$. At this stage, we shall not be concerned with whether we are considering a complete rotation or a semi-circular rotation around the pole. Nor will we be concerned with the direction of the indentation for the moment. Therefore, we shall assume that the infinitesimal indentation begins at an angle, γ_1 in the complex plane, and ends at another angle, γ_2. Then we find that the contribution from the pole in the Cauchy integral in Equivalence (10.5) is given by

$$
I^I = ip^{-1}(-1)^N z^{\beta(N-1)} \lim_{\varepsilon \to 0} \int_{\gamma_1}^{\gamma_2} d\gamma \left(z^{-\beta} e^{i(2l-1)\pi} \right)^{N+q/p-1}
$$
$$
\times \; \exp\left(-z^{-\beta/p} e^{i(2l-1)\pi/p} \right) = i(-1)^N \Delta\gamma f(s)\Big|_{s=z^{-\beta} \exp((2l-1)i\pi)} , \qquad (10.11)
$$

where $\Delta\gamma = \gamma_2 - \gamma_1$ and $f(s)$ is given by Eq. (10.4). Thus, aside from the phase factor of $(-1)^N$, $f(s)$ with $s = -z^{-\beta} \exp((2l-1)i\pi)$ basically represents the residue of the Cauchy integral. Moreover, if $\Delta\gamma = 2\pi$, which corresponds to a complete rotation in a clockwise direction, then we see that the residue of the Cauchy integral in Equivalence (10.5) is identical to the difference of the regularised values for $S_{p,q}^I(N, z^\beta \exp(-2li\pi))$ and $S_{p,q}^I(N, z^\beta \exp(-2(l-1)i\pi))$ given by Equivalence (10.9). The above result also confirms the remarkable insight made by Dingle on p. 412 of Ref. [3] that the jump discontinuity due to crossing Stokes sectors is dependent upon the singular behaviour of the Cauchy integral that emerges from the introduction of the regularised value of the geometric series during Borel summation. This is truly a remarkable result because it has been derived by using the theory of Mellin transforms instead of complex analysis and Cauchy's residue theorem.

If we put $l=1$ in Equivalence (10.6), then we obtain

$$
S_{p,q}^I\left(N, z^\beta e^{-2i\pi} \right) - S_{p,q}^I\left(N, z^\beta \right) \equiv \frac{2\pi i}{p} z^{-\beta q/p}
$$
$$
\times \; e^{iq\pi/p} \exp\left(-z^{-\beta/p} e^{i\pi/p} \right) . \qquad (10.12)
$$

Since the above equivalence is valid only for $-\pi/\beta < \arg z < \pi/\beta$, we can replace $S_{p,q}^I(N, z^\beta)$ by its regularised value since the Cauchy integral is also valid over this sector. Thus, Equivalence (10.12) becomes

$$
S_{p,q}^I\left(N, z^\beta e^{-2i\pi}\right) \equiv (-1)^N p^{-1} z^{\beta(N-1)} \int_C ds \, \frac{s^{N+q/p-1} \, e^{-s^{1/p}}}{s - (-z^{-\beta})}
$$
$$
+ \frac{2\pi i}{p} z^{-\beta q/p} e^{iq\pi/p} \exp\left(-z^{-\beta/p} e^{i\pi/p}\right) . \tag{10.13}
$$

Now we replace $z \exp(-2i\pi/\beta)$ by z_*, which gives

$$
S_{p,q}^I\left(N, z_*^\beta\right) \equiv (-1)^N p^{-1} z^{\beta(N-1)} \int_C ds \, \frac{s^{N+q/p-1} \, e^{-s^{1/p}}}{s + z^{-\beta}}
$$
$$
+ \frac{2\pi i}{p} z^{-\beta q/p} e^{iq\pi/p} \exp\left(-z^{-\beta/p} e^{i\pi/p}\right) , \tag{10.14}
$$

where $-3\pi/\beta < \arg z_* < -\pi/\beta$ and $-\pi/\beta < \arg z < \pi/\beta$. The terms on the rhs of Equivalence (10.14) have been left in terms of z to emphasise the fact that when they are evaluated, they are done so in the primary Stokes sector. That is, the regularised values of the series on the lhs of Equivalence (10.14) for values of z_* lying in the sector of $(-3\pi/\beta, -\pi/\beta)$ are determined by evaluating the terms on the rhs for the corresponding values of z_* lying in the primary Stokes sector. This anomaly arises from the fact that if software packages such as Mathematica are used to carry out calculations of the rhs of the above equivalence in determining regularised value of the Type I generalised terminant outside the primary sector, then they will only do so for values of the complex variable lying in the principal branch of the complex plane. That is, in order to obtain regularised value outside the primary sector we need to evaluate forms where the complex variable lies inside it.

To make Equivalence (10.14) appear less awkward, we replace z_* and z, respectively by z and z_1, where the latter is defined as $z_1 = z \exp(2i\pi/\beta)$. Then Equivalence (10.14) can be expressed as

$$
S_{p,q}^I\left(N, z^\beta\right) \equiv (-1)^N p^{-1} z_1^{\beta(N-1)} \int_C ds \, \frac{s^{N+q/p-1} \, e^{s^{1/p}}}{s - (-z_1^{-\beta})}
$$
$$
+ \frac{2\pi i}{p} z_1^{-\beta q/p} e^{iq\pi/p} \exp\left(-z_1^{-\beta/p} e^{i\pi/p}\right) , \tag{10.15}
$$

where $-3\pi/\beta < \arg z < -\pi/\beta$. We shall use this result to derive the regularised value of $S^I_{p,q}(N, z^\beta)$ for any Stokes sector as z^β is rotated continuously in the complex plane.

To obtain the Borel-summed regularised value of $S^I_{p,q}(N, z^\beta)$ on the Stokes line, where $\arg z = -\pi/\beta$, all we need to do is invoke Rule 8a in Ch. 3. This means that the regularised values for the two Stokes sectors abutting the line are averaged after z_1 in Equivalence (10.15) has been replaced by $z \exp(2i\pi/\beta)$, while the resultant contour integral is modified so that only the Cauchy principal value is evaluated. Therefore, by carrying out this averaging process the regularised value of the Type I generalised terminant for $\arg z = -\pi/\beta$ is given by

$$
S^I_{p,q}(N, z) \equiv -p^{-1} |z|^{\beta(N-1)} P \int_0^\infty dt\, \frac{t^{N+q/p-1}\, e^{-t^{1/p}}}{t - |z|^{-\beta}}
$$
$$
+ \frac{\pi i}{p} |z|^{-\beta q/p} \exp\left(-|z|^{-\beta/p}\right) . \tag{10.16}
$$

From this result we see that the regularised value is composed of the regularised value in the primary Stokes sector except that now the only principal value of the Cauchy integral is evaluated and the semi-residue contribution taken in a clockwise direction ($\Delta \gamma = \pi$) of the $l = 1$ version of Equivalence (10.15).

If we put $l = 2$ in Equivalence (10.6), then with the aid of Equivalence (10.9) we find that

$$
S^I_{p,q}\left(N, z^\beta e^{-4i\pi}\right) - S^I_{p,q}\left(N, z^\beta e^{-2i\pi}\right) \equiv \frac{2\pi i}{p} z^{-\beta q/p}
$$
$$
\times\, e^{3iq\pi/p} \exp\left(-z^{-\beta/p} e^{3iq\pi/p}\right) . \tag{10.17}
$$

We can replace or remove $S^I_{p,q}(N, z^\beta \exp(-2i\pi))$ by introducing Equivalence (10.13) into the above result. This yields

$$
S^I_{p,q}\left(N, z^\beta e^{-4i\pi}\right) \equiv (-1)^N p^{-1} z^{\beta(N-1)} \int_C ds\, \frac{s^{N+q/p-1}\, e^{-s^{1/p}}}{s - (-z^{-\beta})} + \frac{2\pi i}{p} z^{-\beta q/p}
$$
$$
\times\, e^{3iq\pi/p} \exp\left(-z^{-\beta/p} e^{3iq\pi/p}\right) + \frac{2\pi i}{p} z^{-\beta q/p} e^{iq\pi/p} \exp\left(-z^{-\beta/p} e^{i\pi/p}\right). \tag{10.18}
$$

Next we replace $z \exp(-4i\pi/\beta)$ by z on the lhs, while on the rhs z is replaced by

z_2, the latter now being equal to $z \exp(4i\pi/\beta)$. Hence, we arrive at

$$
S_{p,q}^I(N, z^\beta) \equiv (-1)^N p^{-1} z_2^{\beta(N-1)} \int_C ds \, \frac{s^{N+q/p-1} \, e^{-s^{1/p}}}{s + z_2^{-\beta}} + \frac{2\pi i}{p} z_2^{-\beta q/p}
$$

$$
\times \sum_{j=1}^2 e^{i(2j-1)q\pi/p} \exp\left(-z_2^{-\beta/p} e^{i(2j-1)\pi/p}\right) , \tag{10.19}
$$

where $-(2(2)+1)\pi/\beta < \arg z < -(2(2)-1)\pi/\beta$ or $-5\pi/\beta < \arg z < -3\pi/\beta$. When z lies on the Stokes line that borders the $l=1$ and $l=2$ sectors, i.e. where $\arg z = -3\pi/\beta$, all we need to do is average Equivalences (10.15) and (10.19) and take the Cauchy principal value of the resulting contour integral. Before we can average the two equivalences we must replace z_1 in Equivalence (10.15) by $z \exp(2i\pi/\beta)$ and z_2 in Equivalence (10.19) by $z \exp(4i\pi/\beta)$ so that z is the same variable in both equivalences. Then taking the average of the two modified equivalences and setting $\arg z$ equal to $-3\pi/\beta$, we find that the regularised value of the Type I generalised terminant is given by

$$
S_{p,q}^I(N, z^\beta) \equiv -p^{-1} |z|^{\beta(N-1)} P \int_0^\infty dt \, \frac{t^{N+q/p-1} \, e^{-t^{1/p}}}{t - |z|^{-\beta}} + \frac{2\pi i}{p} |z|^{-\beta q/p} e^{2iq\pi/p}
$$

$$
\times \, \exp\left(-|z|^{-\beta/p} e^{2i\pi/p}\right) + \frac{\pi i}{p} |z|^{-\beta q/p} \exp\left(-|z|^{-\beta/p}\right) . \tag{10.20}
$$

A pattern is now emerging that will allow us to determine the regularised value of a Type I generalised terminant for the Stokes sectors resulting from clockwise rotations of z. We simply replace the upper limit 2 by M in the sum on the rhs of Equivalence (10.19). As a consequence, we find that the regularised value of $S_{p,q}^I(N, z^\beta)$ for $-(2M+1)\pi/\beta < \arg z < -(2M-1)\pi/\beta$ is given by

$$
S_{p,q}^I(N, z^\beta) \equiv (-1)^N z_M^{\beta(N-1)} p^{-1} \int_0^\infty dt \, \frac{t^{N+q/p-1} \, e^{-t^{1/p}}}{t + z_M^{-\beta}} + \frac{2\pi i}{p} z_M^{-\beta q/p}
$$

$$
\times \sum_{j=1}^M e^{i(2j-1)q\pi/p} \exp\left(-z_M^{-\beta/p} e^{i(2j-1)\pi/p}\right) , \tag{10.21}
$$

where $z_M = z \exp(2Mi\pi/\beta)$. For the Stokes line given by $\arg z = -(2M+1)\pi/\beta$, one takes the average of the regularised values for the two adjacent Stokes sectors bordered by the line and replaces the Cauchy integral by its principal value.

After a little algebra one finds that the regularised value of the Type I generalised terminant reduces to

$$S_{p,q}^I(N,z^\beta) \equiv -|z|^{\beta(N-1)} p^{-1} P \int_0^\infty dt \, \frac{t^{N+q/p-1} e^{-t^{1/p}}}{t - |z|^{-\beta}} + \frac{2\pi i}{p} |z|^{-\beta q/p}$$

$$\times \sum_{j=1}^M e^{2ijq\pi/p} \exp\left(-|z|^{-\beta/p} e^{2ij\pi/p}\right) + \frac{\pi i}{p} |z|^{-\beta q/p} \exp\left(-|z|^{-\beta/p}\right) . \quad (10.22)$$

In the next section we consider special values of p and q in the general results presented above.

10.2   Type I Special Values with Arg z < 0

If p is the reciprocal of a positive integer including unity, then with the aid of Eq. (9.23) we can write Equivalence (10.21) as

$$S_{p,q}^I(N,z^\beta) \equiv (-1)^N z_M^{\beta(N-1)} p^{-1} \int_C ds \, \frac{s^{N+q/p-1} e^{-s^{1/p}}}{s + z_M^{-\beta}} + \frac{2\pi i}{p} z_M^{-\beta q/p}$$

$$\times \; Q_{I,+}^{BS}(M) \exp\left((-1)^{(p+1)/p} z_M^{-\beta/p}\right) , \quad (10.23)$$

where C is the contour along the positive real axis and the multiplier $Q_{I,\pm}^{BS}(M)$ is defined as

$$Q_{I,\pm}^{BS}(M) = e^{\pm iMq\pi/p} \frac{\sin(Mq\pi/p)}{\sin(q\pi/p)} . \quad (10.24)$$

We see that the multiplier in the jump discontinuous terms for the Borel-summed form of the regularised value of $S_{p,q}^I(N,z^\beta)$ is the complex conjugate of the multiplier in the corresponding terms of the MB-regularised value given by Equivalence (9.25). However, if we replace z_M by z in these terms, then the multiplier becomes the same as that in Equivalence (9.25). It can also be seen from Equivalence (10.23) that the jump discontinuous terms can be expressed in terms of a multiplier and an exponential term. However, if $1/p$ is an odd number, then the exponential term is no longer subdominant. In addition, the jump discontinuous term is not necessarily $\pi/2$ out of phase with the Cauchy integral as stipulated by Dingle's sixth rule in Ch. 3.

To obtain the regularised value of the Type I generalised terminant when $\arg z = -(2M+1)\pi/\beta$, we need to consider Equivalence (10.22). After some algebra this result becomes

$$
\begin{aligned}
S_{p,q}^{I}\left(N, |z|^{\beta} e^{-(2M+1)i\pi}\right) &= \sum_{k=N}^{\infty} \Gamma(pk+q)|z|^{\beta k} e^{-(2M+1)ik\pi} \equiv -|z|^{\beta(N-1)} \\
&\times P \int_{0}^{\infty} dt\, \frac{t^{N+q/p-1} e^{-t^{1/p}}}{p(t-|z|^{-\beta})} + \frac{\pi}{p} |z|^{-\beta q/p} \exp\left(-|z|^{-\beta/p}\right) \\
&\times \left(\frac{e^{(2M+1)iq\pi/p} - \cos(q\pi/p)}{\sin(q\pi/p)}\right) .
\end{aligned}
\tag{10.25}
$$

Hence, we find that the jump discontinuous term can be expressed in terms of a multiplier and a subdominant exponential factor, although the multiplier is far more complicated than in the conventional view of the Stokes phenomenon. If q is an integer, then the regularised value will be real along the Stokes lines because the factor in the large parenthesis vanishes. When q is a non-integer, however, the regularised value will be complex. Thus, the regularised value of a Type I generalised terminant is not always real when all of the terms in the series are real and positive. This might appear to be a violation of the Zwaan-Dingle principle, but it should be noted that the principle refers to the terms in an asymptotic series being real and positive initially, which in turn means that it only applies to Type II generalised terminants.

We can also develop the general Borel-summed forms for the regularised value of $S_{p,q}^{I}(N, z^{\beta})$ when $p=2$ as we did with the corresponding MB-regularised forms in the previous chapter. There, we found that two different forms emerged depending upon whether M was even or odd. For $M=2J$, where J is an integer, Equivalence (10.21) yields

$$
\begin{aligned}
S_{2,q}^{I}\left(N, z^{\beta}\right) &\equiv (-1)^{N} z_{2J}^{\beta(N-1)} \int_{C} ds\, \frac{s^{N+q/2-1} e^{-\sqrt{s}}}{2(s+z_{2J}^{-\beta})} + 2\pi i z_{2J}^{-\beta q/2} \\
&\times \cos\left(q\pi/2 + z_{2J}^{-\beta/2}\right) e^{Jiq\pi} \frac{\sin(Jq\pi)}{\sin(q\pi)} ,
\end{aligned}
\tag{10.26}
$$

while for $M = 2J+1$, we obtain

$$S_{2,q}^I(N, z^\beta) \equiv (-1)^N z_{2J+1}^{\beta(N-1)} \int_C ds \, \frac{s^{N+q/2-1} e^{-\sqrt{s}}}{2(s + z_{2J+1}^{-\beta})} + \pi i z_{2J+1}^{-\beta q/2}$$

$$\times \left(2 \cos \left(\frac{q\pi}{2} - z_{2J+1}^{-\beta/2} \right) e^{(J+1)iq\pi} \frac{\sin(Jq\pi)}{\sin(q\pi)} + \exp \left(\frac{iq\pi}{2} - i z_{2J+1}^{-\beta/2} \right) \right). \quad (10.27)$$

In these results z_M is defined as below Equivalence (10.21).

For the case of the Stokes lines, where $\arg z^{-\beta} = (2M+1)\pi$, we put $p=2$ in Equivalence (10.22). Then after some algebra we find for $M = 2J+1$ that

$$S_{2,q}^I(N, z^\beta) \equiv -|z|^{\beta(N-1)} P \int_0^\infty dt \, \frac{t^{N+q/2-1} e^{-\sqrt{t}}}{2(t - |z|^{-\beta})} + \pi i \, |z|^{-\beta q/2} e^{iq\pi/2} \left(\left(1 + 2 \right. \right.$$

$$\times \, e^{(J+1)iq\pi} \frac{\sin(Jq\pi)}{\sin(q\pi)} \right) \cosh \left(|z|^{-\beta/2} + iq\pi/2 \right) + \frac{1}{2} e^{|z|^{-\beta/2 + iq\pi/2}} \right), \quad (10.28)$$

while for $M = 2J$, we obtain

$$S_{2,q}^I(N, z^\beta) \equiv -|z|^{\beta(N-1)} P \int_0^\infty dt \, \frac{t^{N+q/2-1} e^{-\sqrt{t}}}{2(t - |z|^{-\beta})} + \pi i \, |z|^{-\beta q/2} e^{-iq\pi/2}$$

$$\times \left(2 \cosh \left(|z|^{-\beta/2} - iq\pi/2 \right) e^{(J+1)iq\pi} \frac{\sin(Jq\pi)}{\sin(q\pi)} + e^{-|z|^{-\beta/2 + iq\pi/2}} \right). \quad (10.29)$$

As was the case for the $p = 2$ MB-regularised forms, we cannot express the jump discontinuous terms in the Borel-summed forms for the regularised value of $S_{2,q}^I(N, z^\beta)$ in terms of a multiplier and a subdominant exponential function. Instead, a hyperbolic cosine has usurped the role of the "subdominant" exponential term, but its argument is no longer uniform for successive values of M.

10.3 Type I Generalised Terminants for Arg z > 0

Now we consider negative values of $\arg z^{-\beta}$ in the complex plane or clockwise rotations of $z^{-\beta}$. We begin by putting $l = 0$ in Equivalences (10.6) and (10.9), which yields

$$S_{p,q}^I \left(N, z^\beta e^{2i\pi} \right) - S_{p,q}^I \left(N, z^\beta \right) \equiv -\frac{2\pi i}{p} z^{-\beta q/p}$$

$$\times \, e^{-iq\pi/p} \exp \left(-z^{-\beta/p} e^{-i\pi/p} \right). \quad (10.30)$$

From this result we see that the regularised value for the lower Stokes sector given by $S_{p,q}^I(N, z^\beta)$ is related to the regularised value for the Stokes sector immediately above, viz. $S_{p,q}^I(N, z^\beta \exp(2i\pi))$, plus the residue contribution of the Cauchy integral taken in a clockwise direction. In fact, the regularised value for any of the lower Stokes sectors is related to the regularised value for the Stokes sector immediately above it plus the residue contribution of the Cauchy integral taken in a clockwise direction. Moreover, the above result represents the complex conjugate of Equivalence (10.12). Since Equivalence (10.30) is only valid over the primary Stokes sector, i.e. for $-\pi/\beta < \arg z < \pi/\beta$, we can introduce the regularised value of $S_{p,q}^I(N, z^\beta)$ as given by Equivalence (10.3) into it. Then by replacing $z\exp(2i\pi/\beta)$ in the resulting equivalence with z, one obtains

$$S_{p,q}^I\left(N, z^\beta\right) \equiv (-1)^N p^{-1} z_{-1}^{\beta(N-1)} \int_C ds \, \frac{s^{N+q/p-1}}{s - \left(-z_{-1}^{-\beta}\right)} \, e^{-s^{1/p}}$$
$$- \frac{2\pi i}{p} z_{-1}^{-\beta q/p} e^{-iq\pi/p} \exp\left(-z_{-1}^{-\beta/p} e^{-i\pi/p}\right) \, , \qquad (10.31)$$

where $\pi/\beta < \arg z < 3\pi/\beta$ and $z_{-1} = z\exp(-2i\pi/\beta)$.

For $\arg z = \pi/\beta$ we first re-write the rhs of Equivalence (10.31) in terms of z. Then we average the resulting equivalence with Equivalence (10.3). Next we take the Cauchy principal value of the resulting integral in accordance with Rule 8a in Ch. 3. In the final step we replace z by $|z|\exp(i\pi/\beta)$, which yields the Borel-summed regularised value of a Type I generalised terminant along this Stokes line. Thus, we arrive at

$$S_{p,q}^I\left(N, z^\beta\right) \equiv -p^{-1} |z|^{\beta(N-1)} P \int_0^\infty dt \, \frac{t^{N+q/p-1} e^{-t^{1/p}}}{t - |z|^{-\beta}}$$
$$- \frac{\pi i}{p} |z|^{-\beta q/p} \exp\left(-|z|^{-\beta/p}\right) \, . \qquad (10.32)$$

As expected, Equivalence (10.32) represents the complex conjugate of Equivalence (10.16).

For $l = -1$, the combination of Equivalences (10.6) and (10.9) yields

$$S_{p,q}^I\left(N, z^\beta e^{4i\pi}\right) - S_{p,q}^I\left(N, z^\beta e^{2i\pi}\right) \equiv -\frac{2\pi i}{p} z^{-\beta q/p}$$
$$\times \, e^{-3iq\pi/p} \exp\left(-z^{-\beta/p} e^{-3i\pi/p}\right) \, . \qquad (10.33)$$

We can express $S^I_{p,q}(N, z^\beta \exp(2i\pi))$ in terms of $S^I_{p,q}(N, z^\beta)$ by putting $l = 0$ in Equivalences (10.6) and (10.9). Then we find that

$$S^I_{p,q}\left(N, z^\beta e^{4i\pi}\right) \equiv S^I_{p,q}\left(N, z^\beta\right) - \frac{2\pi i}{p} z^{-\beta q/p}$$

$$\times \sum_{j=1}^2 e^{-i(2j-1)q\pi/p} \exp\left(-z^{-\beta/p} e^{-i(2j-1)\pi/p}\right) , \qquad (10.34)$$

where $(2(2) - 1)\pi/\beta < \arg z < (2(2) + 1)\pi/\beta$. Setting z equal to $z\exp(-4i\pi/\beta)$ and introducing the appropriately modified version of Equivalence (10.10), one eventually obtains

$$S^I_{p,q}\left(N, z^\beta\right) \equiv (-1)^N p^{-1} z_{-2}^{\beta(N-1)} \int_C ds \, \frac{s^{N+q/p-1} \, e^{-s^{1/p}}}{s + z_{-2}^{-\beta}} - \frac{2\pi i}{p} z_{-2}^{-\beta q/p}$$

$$\times \sum_{j=1}^2 e^{-i(2j-1)q\pi/p} \exp\left(-z_{-2}^{-\beta/p} e^{-i(2j-1)q\pi/p}\right) , \qquad (10.35)$$

which, as expected, is the complex conjugate of Equivalence (10.19). For $\arg z = 3\pi/\beta$, we first express the rhs's of Equivalences (10.31) and (10.35) in terms of z rather than z_{-1} and z_{-2}. Then we average the resulting equivalences and evaluate the Cauchy principal value of the resulting contour integral. Alternatively, we can take the complex conjugate of Equivalence (10.20). Hence, we find that

$$S^I_{p,q}\left(N, z^\beta\right) \equiv -p^{-1} |z|^{\beta(N-1)} P \int_0^\infty dt \, \frac{t^{N+q/p-1} \, e^{-t^{1/p}}}{t - |z|^{-\beta}} - \frac{2\pi i}{p} |z|^{-\beta q/p}$$

$$\times \, e^{-2iq\pi/p} \exp\left(-|z|^{-\beta/p} e^{-2i\pi/p}\right) - \frac{\pi i}{p} |z|^{-\beta q/p} \exp\left(-|z|^{-\beta/p}\right) . \qquad (10.36)$$

From the above results we see a similar pattern emerging as in the case of the anti-clockwise rotations of $z^{-\beta}$ in Sec. 10.1. Therefore, by extending Equivalences (10.35) and (10.36) we are able to derive general forms for the Borel-summed regularised value of a Type I generalised terminant for any Stokes sector or line, where $\arg z > 0$. In particular, the generalisation of Equivalence (10.35) to $(2M - 1)\pi/\beta < \arg z < (2M+1)\pi/\beta$, where $M > 0$, can be carried out simply by replacing

2 with M. This gives

$$S^I_{p,q}(N, z^\beta) \equiv (-1)^N z^{\beta(N-1)}_{-M} \int_0^\infty dt\, \frac{t^{N+q/p-1}\, e^{-t^{1/p}}}{p(t + z^{-\beta}_{-M})} - \frac{2\pi i}{p} z^{-\beta q/p}_{-M}$$

$$\times\, \sum_{j=1}^{M} e^{-i(2j-1)q\pi/p} \exp\left(-z^{-\beta/p}_{-M} e^{-i(2j-1)\pi/p}\right) . \quad (10.37)$$

In the above result $z_{-M} = z \exp(-2Mi\pi/\beta)$. Hence, we see that Equivalence (10.37) represents the complex conjugate of Equivalence (10.21). For the Stokes line where $\arg z = (2M+1)\pi/\beta$, generalising Equivalence (10.36) yields

$$S^I_{p,q}(N, z^\beta) \equiv -|z|^{\beta(N-1)} P \int_0^\infty dt\, \frac{t^{N+q/p-1}\, e^{-t^{1/p}}}{p(t - |z|^{-\beta})} - \frac{2\pi i}{p} |z|^{-\beta q/p} \sum_{j=1}^{M} e^{-2ijq\pi/p}$$

$$\times\, \exp\left(-|z|^{-\beta/p} e^{-2ij\pi/p}\right) - \frac{\pi i}{p} |z|^{-\beta q/p} \exp\left(-|z|^{-\beta/p}\right) , \quad (10.38)$$

which is the complex conjugate of Equivalence (10.22).

10.4 Type I Special Values with Arg z > 0

In this section we consider special values of p and q for the results obtained in the previous section. For p equal to a reciprocal of a positive integer including unity, the sums in Equivalences (10.37) and (10.38) can be evaluated in a similar manner to the case of the regularised value of a Type I generalised terminant undergoing anti-clockwise rotations of $z^{-\beta}$ in the complex plane. Therefore, we find that Equivalence (10.37) simplifies to

$$S^I_{p,q}(N, z^\beta) \equiv (-1)^N z^{\beta(N-1)}_{-M} \int_C ds\, \frac{s^{N+q/p-1}\, e^{-s^{1/p}}}{p(s + z^{-\beta}_{-M})} - \frac{2\pi i}{p} z^{-\beta q/p}_{-M}$$

$$\times\, Q^{BS}_{I,-}(M) \exp\left((-1)^{(p+1)/p} z^{-\beta/p}_{-M}\right) , \quad (10.39)$$

where the multiplier $Q^{BS}_{I,-}(M)$ is given by Eq. (10.24). Moreover, the remarks appearing below Eq. (10.24) also apply here. In the case of the Stokes lines, for which $\arg z = (2M+1)\pi/\beta$, Equivalence (10.38) yields

$$S^I_{p,q}\left(N, |z|^\beta e^{(2M+1)i\pi}\right) \equiv -|z|^{\beta(N-1)} \, P \int_0^\infty dt \, \frac{t^{N+q/p-1} e^{-t^{1/p}}}{p(t - |z|^{-\beta})}$$

$$- \frac{\pi}{p} |z|^{-\beta q/p} \exp\left(-|z|^{-\beta/p}\right) \left(\frac{\cos(q\pi/p) - e^{-(2M+1)iq\pi/p}}{\sin(q\pi/p)}\right) , \qquad (10.40)$$

which represents the complex conjugate of Equivalence (10.25).

Once again, odd and even values of M require separate consideration when $p=2$. For z lying in the Stokes sector of $(2M-1)\pi/\beta < \arg z < (2M+1)\pi/\beta$, where $M = 2J+1$, Equivalence (10.37) reduces to

$$S^I_{2,q}(N, z^\beta) \equiv (-1)^N z^{\beta(N-1)}_{-(2J+1)} \int_C ds \, \frac{s^{N+q/2-1} e^{-\sqrt{s}}}{2(s + z^{-\beta}_{-(2J+1)})} - \pi i \, z^{-\beta q/2}_{-(2J+1)} \left(2 e^{(J+1)iq\pi}\right.$$

$$\times \left. \cos\left(\frac{q\pi}{2} - z^{-\beta/2}_{-(2J+1)}\right) \frac{\sin(Jq\pi)}{\sin(q\pi)} + \exp\left(-\frac{iq\pi}{2} + iz^{-\beta/2}_{-(2J+1)}\right)\right), \qquad (10.41)$$

while for $M = 2J$, one obtains

$$S^I_{2,q}(N, z^\beta) \equiv (-1)^N z^{\beta(N-1)}_{-2J} \int_C ds \, \frac{s^{N+q/2-1} e^{-\sqrt{s}}}{2(s + z^{-\beta}_{-2J})} - 2\pi i \, z^{-\beta q/2}_{-2J} e^{-Jiq\pi}$$

$$\times \cos\left(\frac{q\pi}{2} + z^{-\beta/2}_{2J}\right) \frac{\sin(Jq\pi)}{\sin(q\pi)} . \qquad (10.42)$$

To conclude this section, we evaluate the $p=2$ regularised value of a Type I generalised terminant along the Stokes lines given by $\arg z = (2M+1)\pi/\beta$. For $M = 2J+1$ we find that Equivalence (10.34) reduces to

$$S^I_{2,q}(N, z^\beta) \equiv -|z|^{\beta(N-1)} \, P \int_0^\infty dt \, \frac{t^{N+q/2-1} e^{-\sqrt{t}}}{2(t - |z|^{-\beta})} - \pi i \, |z|^{-\beta q/2} e^{-iq\pi/2} \left(\left(1 + 2\right.\right.$$

$$\times e^{-(J+1)iq\pi} \left.\frac{\sin(Jq\pi)}{\sin(q\pi)}\right) \cosh\left(|z|^{-\beta/2} - iq\pi/2\right) + \frac{1}{2} e^{|z|^{-\beta/2 - iq\pi/2}}\right), \qquad (10.43)$$

whereas for $M = 2J$, we arrive at

$$S^I_{2,q}(N, z^\beta) \equiv -|z|^{\beta(N-1)} \, P \int_0^\infty dt \, \frac{t^{N+q/2-1} e^{-\sqrt{t}}}{2(t - |z|^{-\beta})} - \pi i \, |z|^{-\beta q/2} e^{iq\pi/2}$$

$$\times \left(2 \cosh\left(|z|^{-\beta/2} + iq\pi/2\right) e^{-(J+1)iq\pi} \frac{\sin(Jq\pi)}{\sin(q\pi)} + e^{-|z|^{-\beta/2 - iq\pi/2}}\right). \qquad (10.44)$$

All the $p=2$ results presented so far confirm that it is not always possible to express the extra terms to the Cauchy integral in the regularised value of a Type I generalised terminant as a product of a simple multiplier and a subdominant exponential term that retains the same form irrespective of the Stokes sector or line pertaining to $\arg z$. Thus, the concept of a conventional view of the Stokes phenomenon where constant multipliers associated with subdominant terms in a complete asymptotic expansion develop discontinuous jumps, often of unity, across a Stokes line is generally not valid. In fact, the concept seems to be only valid for $p=1$, which represents standard terminants.

10.5 Type I Examples

Before investigating Borel summation of Type II generalised terminants, let us consider some examples of Type I generalised terminants. At the beginning of Ch. 5 the asymptotic forms for the error function, which had been derived in Ch. 2 by following the approach in Stokes's seminal paper [2], were presented in terms of the Stokes multiplier S. Then the asymptotic series appearing in these forms was Borel-summed according to the description of the technique in Ch. 3, which resulted in Equivalence (5.6).

Since the asymptotic series in Equivalence (5.1) is an example of a Type I generalised terminant, we re-write the result as

$$S^I_{1,1/2}\left(0,z^2\right) \equiv -\frac{\pi}{z}\, e^{1/z^2} \mathrm{erf}\left(1/z\right) + \frac{2\pi}{z}\, S e^{1/z^2} \ , \qquad (10.45)$$

where the Stokes multiplier S is still given by Eq. (5.2). For $|\arg z| < \pi/2$, $S=1/2$, while for $\pi/2 < |\arg z| < 3\pi/2$, $S=-1/2$. Now our aim is to see whether any of the above forms can be obtained from the theory presented so far, namely from Equivalences (10.21), (10.22), (10.37) and (10.38). Therefore, let us introduce $p=1$, $q=1/2$, $\beta=2$ and $N=0$ into Equivalence (10.37). After some algebra we arrive at

$$S^I_{1,1/2}\left(0,z^2\right) \equiv \int_0^\infty dt\, \frac{t^{-1/2} e^{-t}}{1+z^2_{-M}\, t} + \frac{2\pi}{z_{-M}}\, e^{1/z^2_{-M}} \sum_{j=1}^M e^{-ij\pi} \ , \qquad (10.46)$$

where $(M-1/2)\pi < \arg z < (M+1/2)\pi$ and $z_{-M} = (-1)^M z$. Hence, we see that $z^2_{-M} = z^2$, which means that we need not be concerned with forcing z to remain in the principal branch through z_{-M}. The integral in Equivalence (10.46) can be evaluated with the aid of Eq. (5.5). According to Ref. [21], Eq. (5.5) is only valid for $|\arg z| < \pi/2$. The reason for this restriction is that there are poles along the positive and imaginary axes. However, the integral is defined everywhere else in the complex plane. Thus, we can introduce Eq. (5.5) into Equivalence (10.46), which yields

$$S^I_{1,1/2}(0,z^2) \equiv \frac{\sqrt{\pi}}{z_{-M}} e^{1/z^2_{-M}} \Gamma(1/2, 1/z^2_{-M}) + \frac{2\pi}{z_{-M}} e^{1/z^2_{-M}} \sum_{j=1}^{M} e^{-ij\pi}. \quad (10.47)$$

According to No. 8.359(3) in Ref. [21], $\Gamma(1/2, x^2) = \sqrt{\pi} - \sqrt{\pi}\,\mathrm{erf}(x)$. We have already seen from Eq. (5.8) that this result is valid for $|\arg z| < \pi/2$, i.e. $M = 0$, but not for $\pi/2 < |\arg z| < 3\pi/2$ or $M = \pm 1$. For the latter case $\Gamma(1/2, x^2) = -\sqrt{\pi} - \sqrt{\pi}\,\mathrm{erf}(x)$. However, because z_{-M} appears in the integral, it means that we are always evaluating the integral as if $|\arg z| < \pi/2$ regardless of the value for M. Hence, we introduce the first result for $\Gamma(1/2, x^2)$ into Equivalence (10.47), thereby obtaining

$$S^I_{1,1/2}(0,z^2) \equiv \frac{\pi e^{1/z^2_{-M}}}{z_{-M}} \left(1 + 2\sum_{j=1}^{M} e^{-ij\pi} - \mathrm{erf}(1/z_{-M}) \right) . \quad (10.48)$$

For $M = 0$ the sum on the rhs of Equivalence (10.48) vanishes and we obtain

$$S^I_{1,1/2}(0,z^2) \equiv -\frac{\pi}{z} e^{1/z^2} \mathrm{erf}(1/z) + \frac{\pi}{z} e^{1/z^2} , \quad (10.49)$$

which is only valid for $|\arg z| < \pi/2$. This is, of course, the $l = 0$ asymptotic form of Eq. (5.2) introduced into Equivalence (5.1). In actual fact, Equivalence (10.49) is valid whenever $M = 2l$ and l is a non-negative integer, which corresponds to $(2l - 1/2) < \arg z < (2l + 1/2)\pi$. Therefore, for these values of z Equivalence (10.49) can be written as

$$\mathrm{erf}(1/z) - 1 \equiv -\frac{z}{\pi} e^{-1/z^2} S^I_{1,1/2}(0,z^2) . \quad (10.50)$$

On the other hand, if M is an odd number equal to $2l + 1$, where l is again a non-negative integer, then the sum over j on the rhs of Equivalence (10.48) yields -1. Hence, we find that for $(2l + 1/2)\pi < \arg z < (2l + 3/2)\pi$, which includes $\pi/2 < \arg z < 3\pi/2$, Equivalence (10.48) yields

$$S^I_{1,1/2}(0,z^2) \equiv \frac{\pi}{z} e^{1/z^2} \mathrm{erf}(-1/z) + \frac{\pi}{z} e^{1/z^2} . \quad (10.51)$$

Noting that $\text{erf}(-z) = -\text{erf}(z)$ and multiplying both sides of Equivalence (10.51) by $-z\exp(-1/z^2)/z$, we again arrive at Equivalence (10.50). Hence, we see that the regularised value of $S^I_{1,1/2}(0, z^2)$ equals the function $\text{erf}(1/z) - 1$ provided the series is obtained in the primary Stokes sector of $|\arg z| < \pi/2$.

If we introduce $p = 1$, $q = 1/2$, $\beta = 2$ and $N = 0$ into Equivalence (10.21), then after some algebra we arrive at

$$S^I_{1,1/2}\left(0, z^2\right) \equiv \int_0^\infty dt\, \frac{t^{-1/2}\, e^{-t}}{1 + z_M^2 t} + \frac{2\pi}{z_M}\, e^{1/z_M^2} \sum_{j=1}^M e^{ij\pi}\,, \tag{10.52}$$

where $-(M+1/2)\pi < \arg z < -(M-1/2)\pi$. The only difference between the above result and Equivalence (10.46) occurs in the summation over j, whose summand is now the complex-conjugate of that in the latter equivalence. However, this has no effect on the regularised value as the sum is real. Consequently, whenever M is an even or odd number, Equivalence (10.52) reduces to Equivalence (10.50). Hence, we have verified the first form of Eq. (5.2) when it is introduced into Equivalence (5.1). All that remains is to examine the situation on the Stokes lines.

To derive the regularised value along the Stokes lines, we introduce $p = 1$, $q = 1/2$, $\beta = 2$ and $N = 0$ into Equivalences (10.22) and (10.38), thereby obtaining

$$S^I_{1,1/2}\left(0, z^2\right) \equiv P \int_0^\infty dt\, \frac{t^{-1/2}\, e^{-t}}{1 - |z|^2 t} \mp \frac{\pi i}{|z|}\, e^{-1/|z|^2} \left(1 + 2\sum_{j=1}^M e^{\mp ij\pi}\right), \tag{10.53}$$

for $\arg z = \pm(M+1/2)\pi$. Equivalence (10.53) can be further simplified to

$$S^I_{1,1/2}\left(0, z^2\right) \equiv P \int_0^\infty dt\, \frac{t^{-1/2}\, e^{-t}}{1 - |z|^2 t} \mp \frac{\pi i}{|z|}\, e^{-1/|z|^2}\,, \tag{10.54}$$

where M is an even integer for the upper sign version and an odd integer for the lower sign version. Unfortunately, we do not know the Cauchy principal value of the integral in the above result. Since we have enjoyed success by averaging the regularised value across adjacent Stokes sectors to obtain the regularised value on the Stokes line, we now average the $M = 2l$ and $M = (2l + 1)$ versions of Equivalence (10.48). Then by setting $z = |z| \exp(i(2l \pm 1/2)\pi)$, we obtain

$$S^I_{1,1/2}\left(0, -|z|^2\right) \equiv -\frac{i\pi}{|z|}\, e^{-1/|z|^2}\, \text{erf}(i/|z|) \mp \frac{i\pi}{|z|}\, e^{-1/|z|^2}\,. \tag{10.55}$$

Thus, the regularised value of $S^I_{1,1/2}(0, -z^2)$ along the positive imaginary axis is identical to the corresponding value along the negative imaginary axis, while the principal value of the Cauchy integral is given by

$$P \int_0^\infty dt \, \frac{t^{-1/2} e^{-t}}{1 - |z|^2 t} = \frac{\pi}{|z|} e^{-1/|z|^2} \operatorname{erfi}(1/|z|) \ . \tag{10.56}$$

As a result, we have discovered a method for evaluating the Cauchy principal value of a complex integral.

With regard to the second asymptotic form given in Equivalence (10.45), i.e. the form that applies when $S = -1/2$, we cannot apply the general forms for the regularised value of a Type I generalised terminant derived in the previous sections. Those results have been derived by assuming that the primary sector is given by $-\pi/\beta < \arg z < \pi/\beta$, whereas the series $S^I_{1,1/2}(0, z^2)$ in the second asymptotic form given in Equivalence (10.45) has been derived for a different sector, either for $-3\pi/2 < \arg z < \pi/2$ or $\pi/2 < \arg z < 3\pi/2$. It also represents the regularised value of a different function, namely $-\pi \exp(1/z^2)(1 + \operatorname{erf}(1/z))/z$ rather than $\pi \exp(1/z^2)(1 - \operatorname{erf}(1/z))/z$ as in Equivalence (10.48). Consequently, the primary Stokes sector for the Borel-summed regularised value of $S^I_{1,1/2}(0, z^2)$ is no longer $-\pi/\beta < \arg z < \pi/\beta$ as in Equivalence (10.10) upon which all the results in Secs. 10.1 to 10.4 are based. Instead, we need to consider shifting the regularised value given by the Cauchy integral to $(2l-1)\pi/\beta < \arg z < (2l+1)\pi/\beta$, where l equals ± 1. Although we would obtain different results for both values of l, they would become equal to each other when $p = 1$, $q = 1/2$, $N = 0$ and $\beta = 2$. This is left as an exercise for the reader. It also highlights the importance of specifying the particular values of $\arg z$ for which each form for the regularised value of an asymptotic series is valid.

In the previous chapter we derived the regularised value of $S^I_{3/2,1}(0, z)$ by using Dingle's theory of terminants. Two forms given by Equivalences (9.13) and (9.14) were obtained. We found that both results were unwieldy and limited to the right hand half of the principal branch. On the other hand, via MB regularisation we were able to obtain one form for the regularised value of the generalised terminant that covered the entire principal branch. This result was given as Equivalence (9.31). To complete the comparison, let us now determine the equivalent Borel-summed values for the generalised terminant by using the preceding analysis. According to Equivalences (10.21) and (10.37), the Borel-summed value of

$S_{3/2,1}^I(0,z)$ is given by

$$
S_{3/2,1}^I(0,z) \equiv \frac{2}{3} \int_0^\infty dt \, \frac{t^{-1/3} e^{-t^{2/3}}}{1+z_{\pm M} t} \pm \frac{4\pi i}{3} z_{\pm M}^{-2/3}
$$
$$
\times \sum_{j=1}^M e^{\pm i(4j-2)\pi/3} \exp\left(-z_{\pm M}^{-2/3} e^{\pm i(4j-2)\pi/3}\right) , \qquad (10.57)
$$

where $(\mp 2M-1)\pi < \arg z < (\mp 2M+1)\pi$ and $z_{\pm M} = z\exp(\pm 2iM\pi)$. For $M=0$, which means that $|\arg z| < \pi$, the above result reduces to

$$
S_{3/2,1}^I(0,z) \equiv \int_0^\infty dy \, \frac{e^{-y}}{1+zy^{3/2}} , \qquad (10.58)
$$

after making the substitution $y = t^{2/3}$. For the Stokes lines, where $\arg z = \pm(2M+1)\pi$, the regularised value obtained from Equivalences (10.22) and (10.38) is found to be

$$
S_{3/2,1}^I(0,z) \equiv P \int_0^\infty \frac{t^{-1/3} e^{-t}}{1-|z|t} \mp \frac{4\pi i}{3} |z|^{-2/3} \sum_{j=1}^M e^{\mp 4ij\pi/3}
$$
$$
\times \exp\left(-|z|^{-2/3} e^{\mp 4ij\pi/3}\right) \mp \frac{2\pi i}{3} |z|^{-2/3} \exp\left(-|z|^{-2/3}\right) . \quad (10.59)
$$

10.6   Type II Generalised Terminants

Now we consider the second type of generalised terminant, $S_{p,q}^{II}(N,z^\beta)$, as defined by Eq. (9.2). By introducing the integral representation for the gamma function into this series, we find that

$$
S_{p,q}^{II}(N,z^\beta) = \int_0^\infty dt \, e^{-t} t^{q-1} \sum_{k=N}^\infty \left(z^\beta t^p\right)^k . \qquad (10.60)
$$

From Ch. 4 we know that this form of the geometric series is divergent when $\Re z^\beta > 0$ and conditionally convergent when $\Re z^\beta < 0$. Therefore, in order to obtain a finite value for the second type of generalised terminant, we need to regularise the geometric series as we did in the case of the first type of generalised terminant.

According to Sec. 4.1, the regularised value is given by

$$S_{p,q}^{II}\left(N,z^{\beta}\right) \equiv \int_0^{\infty} dt \, \frac{e^{-t}\, t^{pN+q-1}}{1-z^{\beta}t^p} = -z^{\beta(N-1)}\, p^{-1}$$

$$\times \int_C ds \, \frac{s^{N+q/p-1}\, e^{-s^{1/p}}}{s-z^{-\beta}} \, , \tag{10.61}$$

where C is again the line contour along the positive real axis. We shall refer to the Cauchy integral in the above result as CI_2. The main difference between CI_2 and the Cauchy integral in the previous section is that the singularity occurs now at $s=z^{-\beta}$ rather than at $s=-z^{-\beta}$. Although there is no minus sign in the singularity as there was in the derivation of the regularised value of $S_{p,q}^{I}(N,z^{\beta})$, there is still ambiguity present since we can write the singularity in CI_2 as $s=\exp(-2ij\pi)z^{-\beta}$, where j can be any integer. Therefore, in deriving the regularised value of $S_{p,q}^{II}(N,z^{\beta})$ we shall adopt a similar approach to our study of $S^{I}(N,z^{\beta})$. In addition, because there is a difference in sign in the singularity of the above Cauchy integral, the Stokes sectors for the regularised value are now shifted by π to $2l\pi < \arg z^{-\beta} < 2(l+1)\pi$ compared with the first type of generalised terminant, where they were situated at $(2l-1)\pi < \arg z^{-\beta} < (2l+1)\pi$.

In Ch. 7 we studied the series, $S_1(N,z^{\beta}\exp(-2li\pi))$, as defined by Equivalence (7.52). The corresponding Type II generalised terminant, which will be the studied in this section, is, therefore, $S_{p,q}^{II}(N,z^{\beta}\exp(-2li\pi))$. If we assume that $0 < \arg z < 2\pi$, then the regularised value of $S_{p,q}^{II}(N,z^{\beta}\exp(-2li\pi))$ is effectively the regularised value of $S_{p,q}^{II}(N,z^{\beta})$ for $2l\pi/\beta < \arg z < (2l+2)\pi/\beta$. By applying Proposition 3 to the difference between consecutive values of l of this generalised terminant in the same manner as Equivalence (7.48), we arrive at

$$\Delta S_l^{II}\left(N,z^{\beta}e^{-2li\pi}\right) = S_{p,q}^{II}\left(N,z^{\beta}e^{-2li\pi}\right) - S_{p,q}^{II}\left(N,z^{\beta}e^{-2(l-1)i\pi}\right)$$

$$\equiv \int_{\substack{c-i\infty \\ \mathrm{Max}[N-1,-q/p]<c=\Re s<N}}^{c+i\infty} ds\, z^{\beta s}\, \Gamma(ps+q)e^{-2li\pi s} \, . \tag{10.62}$$

In obtaining this result we have interpreted $(-1)^s$ as $\exp(-i\pi s)$. According to the behaviour in Approximation (10.7), the above MB integral is only convergent when $(2l-p/2)\pi/\beta < \arg z < (2l+p/2)\pi/\beta$. By making the change of variable, $y=ps+q$, we find that Equivalence (10.62) becomes

$$\Delta S_l^{II}\left(N,z^{\beta}e^{-2li\pi}\right) \equiv \frac{z^{-\beta q/p}}{p}\, e^{2liq\pi/p} \int_{\substack{c-i\infty \\ c=\Re s>-q/p}}^{c+i\infty} dy\, \left(z^{-\beta}e^{2li\pi}\right)^{-y/p} \Gamma(y) \, . \tag{10.63}$$

As in the case of Equivalence (10.8) we treat the above integral as the inverse Mellin transform of $\exp(-x)$, where $x = z^{-\beta/p} \exp(2li\pi/p)$. Consequently, we find that

$$
S_{p,q}^{II}\left(N, z^{\beta} e^{-2li\pi}\right) - S_{p,q}^{II}\left(N, z^{\beta} e^{-2(l-1)i\pi}\right) \equiv \frac{2\pi i}{p} z^{-\beta q/p}
$$
$$
\times\; e^{2liq\pi/p} \exp\left(-z^{-\beta/p} e^{2li\pi/p}\right) = 2\pi i f(s)\Big|_{s=z^{-\beta}\exp(2li\pi)} \;, \tag{10.64}
$$

where $f(s)$ is again given by Eq. (10.4).

On the other hand, if we carry out Borel summation of $S_{p,q}^{II}(N, z^{\beta}\exp(-2li\pi))$, then we obtain

$$
S_{p,q}^{II}\left(N, z^{\beta} e^{-2li\pi}\right) \equiv -p^{-1} z^{\beta(N-1)} \int_C ds\; \frac{s^{N+q/p-1} e^{-s^{1/p}}}{s - z^{-\beta} e^{2li\pi}} \;, \tag{10.65}
$$

where, as before, C is the line contour along the positive real axis. Thus, we have another Cauchy integral, but now the singularity is situated at $s = z^{-\beta}\exp(2li\pi)$. Unfortunately, the rhs of the above result cannot possibly be the regularised value of the Type II generalised terminant because it yields the same form for all values of l, whereas from MB regularisation we have seen that the regularised value takes on a different form in the various domains of convergence. Therefore, we expect that the regularised value from the Borel-summed forms will be different in each Stokes sector. However, if we consider evaluating the residue in the above integral by (1) letting $s = z^{-\beta}\exp(2li\pi) + \varepsilon\exp(i\theta)$, (2) integrating over θ from γ_1 to γ_2, and (3) taking the limit as $\varepsilon \to 0$, then we find that

$$
I^{II} = -i p^{-1} z^{-\beta q/p} \lim_{\varepsilon \to 0} \int_{\gamma_1}^{\gamma_2} d\theta\; e^{2liq\pi/p} \exp\left(-z^{-\beta/p} e^{2li\pi/p}\right)
$$
$$
= -i\Delta\gamma\, f(s)\Big|_{s=z^{-\beta}\exp(2il\pi)} \;, \tag{10.66}
$$

where $\Delta\gamma = \gamma_2 - \gamma_1$, as before. Therefore, carrying out a complete revolution in a clockwise direction yields the rhs of Equivalence (10.63).

In summary, the regularised value of $\Delta S_{p,q}^{II}(N, z^{\beta}\exp(-2li\pi))$, which is defined for $(2l - p/2)\pi/\beta < \arg z < (2l + p/2)\pi/\beta$, is given by the full residue of the Cauchy integral on the rhs of Equivalence (10.65) for $s = z^{-\beta}\exp(2li\pi)$ taken in a clockwise direction. This is a surprising result because we have not at any stage employed the Cauchy residue theorem. Instead, we have arrived at the result by using the theory of Mellin transforms. A similar situation arose in the study of the Type I generalised terminant in Sec. 10.1. The result is even more intriguing since the residue is to be evaluated at $s = z^{-\beta} \exp(2li\pi)$ rather than at $s = z^{-\beta}$. As $\arg z^{-\beta}$ undergoes a clockwise rotation to the next Stokes sector, the factor of $\exp(2li\pi)$ appearing in the singularity of the Cauchy integral undergoes an anti-clockwise rotation and vice-versa. This means that the overall phase of $z^{-\beta} \exp(2li\pi)$ is conserved as z^{β} moves to different Stokes sectors. Although the residue of the Cauchy integral is altered when l is altered, the Cauchy integral itself does not change in value. Therefore, if there were no dependence on the residue of the Cauchy integral, then we would find that the regularised value of $\Delta S_{p,q}^{II}(N, z^{\beta} \exp(-2li\pi))$ would merely vanish and the regularised value of the Type II generalised terminant would be the same for all Stokes sectors, which is absurd.

At this stage we do not know what the regularised values of $S_{p,q}^{II}(N, z^{\beta} \exp(-2li\pi))$ and $S_{p,q}^{II}(N, z^{\beta} \exp(-2(l-1)i\pi))$ are. All we know is that they are different from each other according to Equivalence (10.66). From p. 412 of Ref. [24], we also know that the Cauchy integral develops a jump discontinuity when $\arg z^{-\beta}$ moves across the line contour or positive real axis. Furthermore, when an asymptotic series of the form of a Type II generalised terminant is derived, it is generally done so for positive real values of z^{β} or when $\arg z^{\beta} = 2j\pi$ and j is an integer. For these values of z^{β}, which represent the Stokes lines for the series, the Cauchy integral on the rhs of Equivalence (10.61) is simply undefined. If we remove the effect of the singularity at $s = z^{-\beta}$ by taking the principal value of the Cauchy integral, then the regularised value is real. However, this may only be valid for the particular Stokes line along which the asymptotic series was derived in the first place. Previously, we referred to such a Stokes line as the primary Stokes line. Without loss of generality, we chose it to be $\arg z^{\beta} = 0$. Thus, when an asymptotic series such as $S_{p,q}^{II}(N, z^{\beta})$ is derived, we shall assume that $\arg z^{\beta} = 0$, and not $2j\pi$, where j is a non-zero integer. The $j \neq 0$ case is left for the reader to consider. Consequently, for $\arg z = 0$, the regularised value of the Type II generalised terminant is given by

$$S_{p,q}^{II}\left(N, z^{\beta}\right) \equiv -z^{\beta(N-1)} p^{-1} P \int_{0}^{\infty} dt \, \frac{t^{N+q/p-1} e^{-t^{1/p}}}{t - z^{-\beta}} \, . \qquad (10.67)$$

The regularised value of the Type II generalised terminant is affected immediately by the slightest movement in the argument or phase of z, which is different from the first type of generalised terminant, where the regularised value remains uniform initially in a sector until $\arg z$ eventually encounters a Stokes line. Hence, the above regularised value develops jump discontinuities when $\arg z$ is no longer equal to zero. From the above analysis we expect that the jump discontinuities will be dependent upon the residue of the Cauchy integral evaluated at $s = z^{-\beta}$ since $l = 0$ corresponds to the primary Stokes line.

From Equivalence (10.66) we find that

$$S_{p,q}^{II}\left(N, z^{\beta}\right) - S_{p,q}^{II}\left(N, z^{\beta} e^{2i\pi}\right) \equiv \frac{2\pi i}{p} z^{-\beta q/p} \exp\left(-z^{-\beta/p}\right) , \qquad (10.68)$$

where $-p\pi/2\beta < \arg z < p\pi/2\beta$. Hence, the above result is the one that applies when the primary Stokes line is chosen as $\arg z = 0$. Above the primary Stokes line the regularised value is going to be the analytic continuation of the integral on the rhs of Equivalence (10.67) plus an extra jump discontinuity term, say Δ_1. Below the primary Stokes line the regularised value will be the analytic continuation of the integral on the rhs of Equivalence (10.67) plus another jump discontinuity term, say Δ_2. We know that when we average these two results across the primary Stokes line that we should obtain Equivalence (10.67). Hence, $\Delta_1 = -\Delta_2$. If, on the other hand, we subtract both regularised values from each other, then we should obtain the difference on the rhs of Equivalence (10.68). Therefore, we arrive at

$$\Delta_1 = -\frac{\pi i}{p} z^{-\beta q/p} \exp\left(-z^{-\beta/p}\right) . \qquad (10.69)$$

We are now in a position to give the regularised value of the Type II generalised terminant $S_{p,q}^{II}(N, z^{\beta})$ for the sectors of $0 < \arg z^{-\beta} < 2\pi$ and $-2\pi < \arg z^{-\beta} < 0$. It should be noted that although Type II generalised terminants are expressed in terms of z^{β}, we shall from here on discuss the Stokes sectors in terms of $\arg z^{-\beta}$ since it is the reciprocal of z^{β} that is evaluated in the residue of the Cauchy integral as indicated above. Therefore, for $0 < \arg z^{-\beta} < 2\pi$ or $-2\pi/\beta < \arg z < 0$, we obtain

$$S_{p,q}^{II}\left(N, z^{\beta}\right) \equiv -\frac{z^{\beta(N-1)}}{p} \int_C ds \, \frac{s^{N+q/p-1} e^{-s^{1/p}}}{s - z^{-\beta}}$$
$$+ \frac{\pi i}{p} z^{-\beta q/p} \exp\left(-z^{-\beta/p}\right) , \qquad (10.70)$$

while for $0 < \arg z < 2\pi/\beta$, the regularised value of the asymptotic series is given by

$$S_{p,q}^{II}\left(N, z^{\beta}\right) \equiv -\frac{z^{\beta(N-1)}}{p} \int_C ds \, \frac{s^{N+q/p-1} e^{-s^{1/p}}}{s - z^{-\beta}}$$
$$- \frac{\pi i}{p} z^{-\beta q/p} \exp\left(-z^{-\beta/p}\right) . \qquad (10.71)$$

Again, C represents the line contour along the positive real axis. The above results are consistent with the earlier statement that the difference between the regularised value for the upper Stokes sector given by Equivalence (10.70) and its immediate predecessor given by Equivalence (10.71) is $-2\pi i f(s)$. As expected, both results are complex conjugates of one another.

We can also apply the above results to Rule 8a in Ch. 3 and with the extra qualification that we only evaluate the principal value of the Cauchy integral, we arrive at the regularised value of $S_{p,q}^{II}(N, z^\beta)$ along the primary Stokes line given by Equivalence (10.67). Alternatively, this result can be written as

$$
S_{p,q}^{II}(N, z^\beta) \equiv -\frac{|z|^{\beta(N-1)}}{p} P \int_0^\infty dt \, \frac{t^{N+q/p-1} e^{-t^{1/p}}}{t - |z|^{-\beta}} \, .
\tag{10.72}
$$

On the other hand, if we had chosen the primary Stokes line to be $\arg z = 2j\pi/\beta$, then the above results would still apply, but in the sectors above and below the new line as well as on the new line. That is, the asymptotic forms for $S_{p,q}^{II}(N, z^\beta)$ given above would only be shifted to different Stokes sectors.

Now we consider the behaviour of the singularity at $s = z^{-\beta} \exp(2li\pi)$ in the Cauchy integral of Equivalence (10.65) as it moves from the Stokes sector of $(0, 2\pi)$ to the adjacent Stokes sector of $(2\pi, 4\pi)$. Just before $\arg z^{-\beta} = 2\pi$, the singularity lies below the line contour along the positive real axis and $l = 0$. When $\arg z^{-\beta}$ crosses over to the adjacent Stokes sector, the singularity lies above the line contour. According to Dingle's treatment of the singularity in his theory of terminants as discussed on p. 412 of Ref. [3], we indent the contour in a clockwise direction around the singularity, thereby ensuring that the singularity always lies below the line contour in the Cauchy integral. Because $\arg z^{-\beta}$ has undergone a 2π rotation in an anti-clockwise direction, the singularity in the Cauchy integral is now interpreted as being situated at $s = z^{-\beta} \exp(-2i\pi)$ or $l = -1$ in Eq. (10.66). Consequently, the regularised value has picked up the residue contribution of $-2\pi i$ times the residue of the Cauchy integral for $s = z^{-\beta} \exp(-2i\pi)$. Hence, for $-4\pi/\beta < \arg z < -2\pi/\beta$, we find that the regularised value of the Type II generalised terminant is given by

$$
S_{p,q}^{II}(N, z^\beta) \equiv -z^{\beta(N-1)} p^{-1} \int_C ds \, \frac{s^{N+q/p-1} e^{-s^{1/p}}}{s - z^{-\beta}} + \frac{\pi i}{p} z^{-\beta q/p}
$$
$$
\times \, \exp(-z^{-\beta/p}) + \frac{2\pi i}{p} z^{-\beta q/p} e^{-2iq\pi/p} \exp(-z^{-\beta/p} e^{-2\pi i/p}) \, .
\tag{10.73}
$$

To obtain the regularised value of this generalised terminant when z lies on the secondary Stokes line of $\arg z = -2\pi/\beta$, we must evaluate the principal value of the Cauchy integral in addition to including only the semi-residue contribution or half the final term in the above result. All this is in accordance with Rule 8a in Ch. 3. Then we obtain

$$
S_{p,q}^{II}(N, z^\beta) \equiv -|z|^{\beta(N-1)} p^{-1} P \int_0^\infty dt\, \frac{t^{N+q/p-1} e^{-t^{1/p}}}{t - |z|^{-\beta}} + \frac{\pi i}{p}\, z^{-\beta q/p}
$$

$$
\times\ \exp\left(-z^{-\beta/p}\right) + \frac{\pi i}{p}\, z^{-\beta q/p} e^{-2iq\pi/p} \exp\left(-z^{-\beta/p} e^{-2\pi i/p}\right). \quad (10.74)
$$

The last term in this result can be also be written as $\pi i |z|^{-\beta q/p} \exp(-|z|^{-\beta/p})/p$.

When $\arg z^{-\beta}$ moves to the next Stokes sector of $(4\pi, 6\pi)$, the singularity in the Cauchy integral is interpreted as being situated at $s = z^{-\beta} \exp(-4i\pi)$. That is, yet another rotation of 2π resulting from the movement of $\arg z^{-\beta}$ to the next Stokes sector has been countered by the introduction of an extra phase factor of $\exp(-2i\pi)$. So, we indent the line contour along the positive real axis in a clockwise direction, thereby ensuring that the singularity lies above the contour. This is basically following Dingle's insight again. As a consequence, the regularised value has collected an extra contribution of $-2\pi i$ times the residue of the Cauchy integral on the rhs of Equivalence (10.65) situated at $s = z^{-\beta} \exp(-4i\pi)$. For $-6\pi/\beta < \arg z < -4\pi/\beta$, the regularised value of $S_{p,q}^{II}(N, z^\beta)$ is given by

$$
S_{p,q}^{II}(N, z^\beta) \equiv -z^{\beta(N-1)} p^{-1} \int_C ds\, \frac{s^{N+q/p-1} e^{-s^{1/p}}}{s - z^{-\beta}} + \frac{\pi i}{p}\, z^{-\beta q/p}
$$

$$
\times\ \exp\left(-z^{-\beta/p}\right) + \frac{2\pi i}{p}\, z^{-\beta q/p} e^{-2i\pi q/p} \exp\left(-z^{-\beta/p} e^{-2i\pi/p}\right)
$$

$$
+ \frac{2\pi i}{p}\, z^{-\beta q/p} e^{-4i\pi q/p} \exp\left(-z^{-\beta/p} e^{-4i\pi/p}\right). \quad (10.75)
$$

For the case of the secondary Stokes line, where $\arg z = -4\pi/\beta$, we evaluate the principal value of the Cauchy integral and only take half of the final term in Equivalence (10.75) in accordance with Rule 8a in Ch. 3. Therefore, we find that

$$
S_{p,q}^{II}\left(N,z^{\beta}\right) \equiv -|z|^{\beta(N-1)} p^{-1} P \int_0^{\infty} dt \, \frac{t^{N+q/p-1} e^{-t^{1/p}}}{t - |z|^{-\beta}} + \frac{i\pi}{p} z^{-\beta q/p}
$$
$$
\times \; \exp\left(-z^{-\beta/p}\right) + \frac{2\pi i}{p} z^{-\beta q/p} e^{-2i\pi q/p} \exp\left(-z^{-\beta/p} e^{-2i\pi/p}\right)
$$
$$
+ \; \frac{\pi i}{p} z^{-\beta q/p} e^{-4i\pi q/p} \exp\left(-z^{-\beta/p} e^{-4i\pi/p}\right) \; . \tag{10.76}
$$

Again the last term can be written as $\pi i |z|^{-\beta/p} \exp(-|z|^{\beta/p})/p$.

As in the case of the Type I generalised terminant $S_{p,q}^{I}(N,z^{\beta})$, a similar pattern is developing that enables us to determine the regularised value of the Type II generalised terminant for any Stokes sector in the upper half of the complex plane. Hence, for $-2(M+1)\pi/\beta < \arg z < -2M\pi/\beta$, we find that the regularised value of $S_{p,q}^{II}(N,z^{\beta})$ becomes

$$
S_{p,q}^{II}\left(N,z^{\beta}\right) \equiv -z^{\beta(N-1)} p^{-1} \int_C ds \, \frac{s^{N+q/p-1} e^{-s^{1/p}}}{s - z^{-\beta}} + \frac{2i\pi}{p} z^{-\beta q/p}
$$
$$
\times \; \sum_{j=1}^{M} e^{-2ijq\pi/p} \exp\left(-z^{-\beta/p} e^{-2ij\pi/p}\right) + \frac{\pi i}{p} z^{-\beta q/p} \exp\left(-z^{-\beta/p}\right) \; . \tag{10.77}
$$

On the secondary Stokes line, where $\arg z = -2M\pi/\beta$, the regularised value of the series obeys Rule 8a in Ch. 3, which yields

$$
S_{p,q}^{II}\left(N,z^{\beta}\right) \equiv -\frac{|z|^{\beta(N-1)}}{p} P \int_0^{\infty} dt \, \frac{t^{N+q/p-1} e^{-t^{1/p}}}{t - |z|^{-\beta}} + \frac{i\pi}{p} z^{-\beta q/p}
$$
$$
\times \; \exp\left(-z^{-\beta/p}\right) + \frac{2i\pi}{p} z^{-\beta q/p} \sum_{j=1}^{M-1} e^{-2ijq\pi/p} \exp\left(-z^{-\beta/p} e^{-2ij\pi/p}\right)
$$
$$
+ \; \frac{i\pi}{p} z^{-\beta q/p} e^{-2Miq\pi/p} \exp\left(-z^{-\beta/p} e^{-2Mi\pi/p}\right) \; . \tag{10.78}
$$

As in the previous results for the secondary Stokes lines the last term in the above result can be written as $\pi i |z|^{-\beta q/p} \exp(-|z|^{-\beta/p})/p$.

10.7   Type II Special Values with Arg *z* < 0

Previously, we found that the finite sums in the regularised value of the Type I generalised terminant could be evaluated for special values of p. The same also applies to the regularised value of a Type II generalised terminant. For example, when p is the reciprocal of an integer, Equivalence (10.77) reduces to

$$
S_{p,q}^{II}\left(N,z^{\beta}\right) \equiv -\frac{z^{\beta(N-1)}}{p} \int_{C} ds\, \frac{s^{N+q/p-1}\, e^{-s^{1/p}}}{s-z^{-\beta}} + \frac{2i\pi}{p}\, z^{-\beta q/p} \exp\left(-z^{-\beta/p}\right)
$$
$$
\times\; e^{-(M+1)iq\pi/p}\, \frac{\sin(Mq\pi/p)}{\sin(q\pi/p)} + \frac{i\pi}{p}\, z^{-\beta q/p} \exp\left(-z^{-\beta/q}\right) . \quad (10.79)
$$

We can express the above result in terms of another multiplier, which possesses an extra factor of $\exp(\pm iq\pi/p)$ when compared with $Q_{I,\pm}^{BS}(M)$. In particular, Equivalence (10.79) can be written as

$$
S_{p,q}^{II}\left(N,z^{\beta}\right) \equiv -\frac{z^{\beta(N-1)}}{p} \int_{C} ds\, \frac{s^{N+q/p-1}\, e^{-s^{1/p}}}{s-z^{-\beta}} + \frac{2i\pi}{p}\, z^{-\beta q/p}
$$
$$
\times\; Q_{II,-}^{BS}(M)\exp\left(-z^{-\beta/p}\right) , \quad (10.80)
$$

where

$$
Q_{II,\pm}^{BS}(M) = e^{\pm iq\pi/p}\, Q_{I,\pm}^{BS}(M) + 1/2 , \quad (10.81)
$$

and $Q_{I,\pm}^{BS}(M)$ is given by Eq. (10.24). On the other hand, for the Stokes lines of $\arg z = -2M\pi/\beta$, we find that Equivalence (10.78) reduces to

$$
S_{p,q}^{II}\left(N,z^{\beta}\right) \equiv -\frac{|z|^{\beta(N-1)}}{p}\, P\int_{0}^{\infty} dt\, \frac{t^{N+q/p-1}\, e^{-t^{1/p}}}{t-|z|^{-\beta}} + \frac{2i\pi}{p}\, |z|^{-\beta q/p}
$$
$$
\times\; \exp\left(-|z|^{-\beta/p}\right) e^{-Miq\pi/p}\, \frac{\sin(Mq\pi/p)}{\sin(q\pi/p)}\, \cos(q\pi/p) , \quad (10.82)
$$

where, once again, we have used Eq. (9.23).

As was also the case for $S_{p,q}^{I}(N,z^{\beta})$, we can evaluate the sums in Equivalences (10.77) and (10.78) when $p=2$. For $M=2J$, where J is a positive integer, Equivalence (10.77) yields

$$
S_{2,q}^{II}\left(N, z^\beta\right) \equiv -\frac{z^{\beta(N-1)}}{2} \int_C ds \, \frac{s^{N+q/2-1} e^{-\sqrt{s}}}{s - z^{-\beta}} + 2\pi i \, z^{-\beta q/2} e^{-i(J+1/2)q\pi}
$$
$$
\times \, \cosh\left(z^{-\beta/2} - iq\pi/2\right) \frac{\sin(Jq\pi)}{\sin(q\pi)} + \frac{i\pi}{2} z^{-\beta q/2} \exp\left(-z^{-\beta/2}\right) .
$$
$$\tag{10.83}$$

For $M = 2J+1$, we obtain

$$
S_{2,q}^{II}\left(N, z^\beta\right) \equiv -\frac{z^{\beta(N-1)}}{2} \int_C ds \, \frac{s^{N+q/2-1} e^{-\sqrt{s}}}{s - z^{-\beta}} + i\pi \, z^{-\beta q/2}
$$
$$
\times \left(e^{-iq\pi/2} \cosh\left(z^{-\beta/2} - i\pi q/2\right) \left(1 + 2e^{-(J+1)iq\pi} \frac{\sin(Jq\pi)}{\sin(q\pi)}\right) \right.
$$
$$
\left. + \frac{1}{2} \exp\left(z^{-\beta/2} - iq\pi\right) \right) .
$$
$$\tag{10.84}$$

Unlike the result for when p is the reciprocal of an integer, viz. Equivalence (10.80), we cannot express the extra terms to the integrals in the above forms for the regularised value as a product of a multiplier and the same subdominant exponential factor that is applicable for all Stokes sectors. That is, the concept of a Stokes multiplier as described in the introduction in Ref. [9], irrespective of whether it experiences a rapid smoothing at a Stokes line or not, is only valid for specific values of p and q in a Type II generalised terminant. Therefore, in general, the conventional view of the Stokes phenomenon is not upheld.

To obtain the regularised value of the $p=2$ Type II generalised terminant for the Stokes lines of $\arg z = -2M\pi/\beta$, we require Equivalence (10.78). When $M = 2J$ and $p=2$, this equivalence reduces to

$$
S_{2,q}^{II}\left(N, z^\beta\right) \equiv -\frac{|z|^{\beta(N-1)}}{2} P \int_0^\infty dt \, \frac{t^{N+q/2-1} e^{-\sqrt{t}}}{t - |z|^{-\beta}} + \pi |z|^{-\beta q/2}
$$
$$
\times \, e^{Jiq\pi} \left(2i \, e^{-iq\pi/2} \cosh\left(|z|^{-\beta/2} + iq\pi/2\right) \frac{\sin(Jq\pi)}{\sin(q\pi)} \right.
$$
$$
\left. - \sin(Jq\pi) \exp\left(-|z|^{-\beta/2}\right) \right) ,
$$
$$\tag{10.85}$$

while for $M = 2J+1$, the regularised value is given by

$$S_{2,q}^{II}(N,z^\beta) \equiv -\frac{|z|^{\beta(N-1)}}{2} P \int_0^\infty dt \, \frac{t^{N+q/2-1} e^{-\sqrt{t}}}{t - |z|^{-\beta}} + i\pi |z|^{-\beta q/2}$$

$$\times \; e^{(J+1/2)iq\pi} \left(\frac{2\sin(Jq\pi)}{\sin(q\pi)} \cosh\left(|z|^{-\beta/2} - iq\pi/2\right) \right.$$

$$\left. + \; \cosh\left(|z|^{-\beta/2} + i(J+1/2)q\pi\right) \right) . \tag{10.86}$$

10.8   Type II Generalised Terminants for Arg z > 0

So far, we have been concerned with deriving the regularised value of the Type II generalised terminant for those Stokes sectors involving negative values of $\arg z$. Positive values for $\arg z$ ensue when we rotate $\arg z^{-\beta}$ continuously in a clockwise direction beginning with the sector of -2π to 0 or $(-2\pi, 0)$, which abuts the primary Stokes line.

We have already determined the regularised value of $S_{p,q}^{II}(N, z^\beta)$ when $0 < \arg z < 2\pi/\beta$, which is given by Equivalence (10.74). Now we consider moving $\arg z^{-\beta}$ to the adjacent sector of $(-4\pi, -2\pi)$. The pole in the Cauchy integral on the rhs of Equivalence (10.72) lies above the line contour or positive real axis before $\arg z^{-\beta}$ crosses to $(-4\pi, -2\pi)$. To ensure that the singularity remains above the line contour as $z^{-\beta}$ in crossing to the adjacent Stokes sector, we indent the line contour in an anti-clockwise direction. In addition, we interpret the singularity in the Cauchy integral as being situated at $s = z^{-\beta} \exp(2i\pi)$ or $l = 1$. We note again that the overall phase of $z^{-\beta} \exp(2il\pi)$ is being conserved by putting $l = 1$. Therefore, for $2\pi/\beta < \arg z < 4\pi/\beta$, we find that the regularised value of the Type II generalised terminant is given by

$$S_{p,q}^{II}(N,z^\beta) \equiv -z^{\beta(N-1)} p^{-1} \int_C ds \, \frac{s^{N+q/p-1} e^{-s^{1/p}}}{s - z^{-\beta}} - \frac{\pi i}{p} z^{-\beta q/p}$$

$$\times \; \exp\left(-z^{-\beta/p}\right) - \frac{2\pi i}{p} z^{-\beta q/p} e^{2\pi i q/p} \exp\left(-z^{-\beta/p} e^{2\pi i/p}\right) . \tag{10.87}$$

As expected, this is simply the complex conjugate of Equivalence (10.73). For the Stokes line of $\arg z = 2\pi/\beta$, in accordance with Rule 8a in Ch. 3 we evaluate the principal value of the Cauchy integral and take the semi-residue contribution at

$s = z^{-\beta} \exp(2i\pi)$, which means that we only need half the final term in the above result. Then the regularised value of $S_{p,q}^{II}(N, z^\beta)$ along this secondary Stokes line is found to be

$$
\begin{aligned}
S_{p,q}^{II}(N, z^\beta) &\equiv -|z|^{\beta(N-1)} p^{-1} P \int_0^\infty dt\, \frac{t^{N+q/p-1} e^{-t^{1/p}}}{t - |z|^{-\beta}} - \frac{\pi i}{p} z^{-\beta q/p} \\
&\times \exp\left(-z^{-\beta/p}\right) - \frac{\pi i}{p} z^{-\beta q/p} e^{2\pi i q/p} \exp\left(-z^{-\beta/p} e^{2\pi i/p}\right).
\end{aligned} \tag{10.88}
$$

The final term in the above result can be written as $-\pi i |z|^{-\beta/p} \exp(-|z|^{\beta/p})$, which means that except for the sign outside, it is identical to the final term obtained for the regularised value for each of the secondary Stokes lines when $\arg z < 0$.

As $\arg z^{-\beta}$ moves to the adjacent Stokes sector of $(-6\pi, -4\pi)$, once again we need to adjust the overall phase of the singularity in the Cauchy integral. As stated previously, each clockwise rotation of 2π in $\arg z^{-\beta}$ has to be counterbalanced by a factor of $\exp(2i\pi)$. Since two rotations have been made in getting to $(-6\pi, -4\pi)$, we put $l = 2$ in the Cauchy integral of Equivalence (10.74). This, of course, does not affect the evaluation of the Cauchy integral, but it does result in another residue term appearing in the regularised value of the Type II generalised terminant. Hence, to obtain the regularised value when $4\pi/\beta < \arg z < 6\pi/\beta$, we need to add to Equivalence (10.87) the full residue contribution or $-2\pi i$ times the residue at $s = z^{-\beta} \exp(4i\pi)$ of the Cauchy integral. On the Stokes line, where $\arg z = 4\pi/\beta$, we need only consider the semi-residue or half this contribution in addition to evaluating the principal value of the Cauchy integral. Therefore, for $4\pi/\beta < \arg z < 6\pi/\beta$, the regularised value of $S_{p,q}^{II}(N, z^\beta)$ is given by

$$
\begin{aligned}
S_{p,q}^{II}(N, z^\beta) &\equiv -z^{\beta(N-1)} p^{-1} \int_C ds\, \frac{s^{N+q/p-1} e^{-s^{1/p}}}{s - z^{-\beta}} - \frac{\pi i}{p} z^{-\beta q/p} \\
&\times \exp\left(-z^{-\beta/p}\right) - \frac{2\pi i}{p} z^{-\beta q/p} e^{2iq\pi/p} \exp\left(-z^{-\beta/p} e^{2i\pi/p}\right) \\
&- \frac{2\pi i}{p} z^{-\beta q/p} e^{4iq\pi/p} \exp\left(-z^{-\beta/p} e^{4i\pi/p}\right),
\end{aligned} \tag{10.89}
$$

while for $\arg z = 4\pi/\beta$, we find that it becomes

$$S_{p,q}^{II}\left(N,z^{\beta}\right) \equiv -z^{\beta(N-1)}p^{-1}P\int_0^{\infty}dt\,\frac{t^{N+q/p-1}|,e^{-t^{1/p}}}{t-z^{-\beta}} - \frac{\pi i}{p}\,z^{-\beta q/p}$$

$$\times\,\exp\left(-z^{-\beta/p}\right) - \frac{2\pi i}{p}\,z^{-\beta q/p}e^{2iq\pi/p}\exp\left(-z^{-\beta/p}e^{2i\pi/p}\right)$$

$$- \frac{\pi i}{p}\,z^{-\beta q/p}e^{4iq\pi/p}\exp\left(-z^{-\beta/p}e^{4i\pi/p}\right)\,.\tag{10.90}$$

Again, the last term can be written as $-\pi i|z|^{-\beta q/p}\exp(-|z|^{-\beta/p})/p$.

From these results we see that a similar pattern is developing for the regularised value of $S_{p,q}^{II}(N,z^{\beta})$ as we observed for $\arg z < 0$. That is, we can express all the full residue contributions in Equivalence (10.89) as finite sum from $j = 1$ to 2. Therefore, for $2M\pi/\beta < \arg z < 2(M+1)\pi/\beta$, all we need to do is replace the upper limit of 2 by M. Then we find that the regularised value of the Type II generalised terminant can be written as

$$S_{p,q}^{II}\left(N,z^{\beta}\right) \equiv -z^{\beta(N-1)}p^{-1}\int_C ds\,\frac{s^{N+q/p-1}e^{-s^{1/p}}}{s-z^{-\beta}} - \frac{\pi i}{p}\,z^{-\beta q/p}$$

$$\times\,\exp\left(-z^{-\beta/p}\right) - \frac{2\pi i}{p}\,z^{-\beta q/p}\sum_{j=1}^{M}e^{2ijq\pi/p}\exp\left(-z^{-\beta/p}e^{2ij\pi/p}\right)\,.\tag{10.91}$$

On the Stokes lines of $\arg z = 2M\pi/\beta$, we evaluate the principal value of the Cauchy integral and only evaluate half the last full residue contribution in Equivalence (10.91). Hence, we obtain a finite sum from $j = 1$ to $M-1$. Therefore, the regularised value of the Type II generalised terminant on the Stokes lines is given by

$$S_{p,q}^{II}\left(N,z^{\beta}\right) \equiv -\frac{|z|^{\beta(N-1)}}{p}\,P\int_0^{\infty}dt\,\frac{t^{N+q/p-1}e^{-t^{1/p}}}{t-|z|^{-\beta}} - \frac{2\pi i}{p}\,z^{-\beta q/p}$$

$$\times\,\sum_{j=1}^{M-1}e^{2ijq\pi/p}\exp\left(-z^{-\beta/p}e^{2ij\pi/p}\right) - \frac{\pi i}{p}\,z^{-\beta q/p}\exp\left(-z^{-\beta/p}\right)$$

$$- \frac{\pi i}{p}\,z^{-\beta q/p}e^{2Miq\pi/p}\exp\left(-z^{-\beta/p}e^{2Mi\pi/p}\right)\,,\tag{10.92}$$

where the last term can also be written as $-\pi i|z|^{-\beta q/p}\exp(-|z|^{-\beta/p})/p$ as is the case for all Stokes lines. As expected, Equivalences (10.91) and (10.92) are respectively the complex conjugates of Equivalences (10.77) and (10.78).

10.9 Type II Special Values with Arg z > 0

As we have done throughout this chapter, we can evaluate the general results for the regularised value of the Type II generalised terminant presented in the previous section for the special cases where p is either the reciprocal of a natural number or equal to 2. For the first case Equivalence (10.91) reduces to

$$
S^{II}_{p,q}\left(N,z^{\beta}\right) \equiv -\frac{z^{\beta(N-1)}}{p} \int_C ds \, \frac{s^{N+q/p-1} e^{-s^{1/p}}}{s-z^{-\beta}} - \frac{2\pi i}{p} z^{-\beta q/p}
$$
$$
\times \, Q^{BS}_{II,+}(M) \exp\left(-z^{-\beta/p}\right) , \tag{10.93}
$$

where $Q^{BS}_{1,-}(M)$ is given by Eq. (10.81). For the Stokes lines, which are given by $\arg z = 2M\pi/\beta$, the regularised value of the Type II generalised terminant can be derived via Equivalence (10.92). When p is the reciprocal of a natural number, we find that it yields

$$
S^{II}_{p,q}\left(N,z^{\beta}\right) \equiv -\frac{|z|^{\beta(N-1)}}{p} \, P\int_0^{\infty} dt \, \frac{t^{N+q/p-1} e^{-t^{1/p}}}{t-|z|^{-\beta}} - \frac{2\pi i}{p} |z|^{-\beta q/p}
$$
$$
\times \, \exp\left(-|z|^{-\beta/p}\right) e^{Miq\pi/p} \, \sin(Mq\pi/p) \cot(q\pi/p) . \tag{10.94}
$$

Equivalences (10.93) and (10.94) are respectively the complex conjugates of Equivalences (10.80) and (10.82).

For the second case, viz. $p=2$, we obtain different results depending upon whether M is an odd or even integer. When $M=2J+1$, one finds that Equivalence (10.91) becomes

$$
S^{II}_{2,q}\left(N,z^{\beta}\right) \equiv -\frac{z^{\beta(N-1)}}{2} \int_C ds \, \frac{s^{N+q/2-1} e^{-\sqrt{s}}}{s-z^{-\beta}} - i\pi z^{-\beta q/2}
$$
$$
\times \, \left(e^{iq\pi/2} \cosh\left(z^{-\beta/2}+i\pi q/2\right) \left(1+2e^{(J+1)iq\pi} \frac{\sin(Jq\pi)}{\sin(q\pi)}\right) \right.
$$
$$
\left. + \frac{1}{2} \exp\left(z^{-\beta/2}+iq\pi\right) \right) , \tag{10.95}
$$

whereas for the Stokes lines Equivalence (10.92) yields

$$
\begin{aligned}
S_{2,q}^{II}\left(N,z^{\beta}\right) &\equiv -\frac{|z|^{\beta(N-1)}}{2} \, P\int_{0}^{\infty} dt \, \frac{t^{N+q/2-1}\,e^{-\sqrt{t}}}{t-|z|^{-\beta}} - i\pi|z|^{-\beta q/2} \\
&\quad \times \; e^{-(J+1/2)iq\pi}\left(\frac{2\sin(Jq\pi)}{\sin(q\pi)}\cosh\left(|z|^{-\beta/2}+iq\pi/2\right)\right. \\
&\quad + \; \left.\cosh\left(|z|^{-\beta/2}-i(J+1/2)q\pi\right)\right) \; .
\end{aligned}
\tag{10.96}
$$

On the other hand, when $M=2J$, Equivalence (10.91) simplifies to

$$
\begin{aligned}
S_{2,q}^{II}\left(N,z^{\beta}\right) &\equiv -\frac{z^{\beta(N-1)}}{2}\int_{C} ds \, \frac{s^{N+q/2-1}\,e^{-\sqrt{s}}}{s-z^{-\beta}} - 2i\pi\,z^{-\beta q/2}e^{(J+1/2)iq\pi} \\
&\quad \times \; \cosh\left(z^{-\beta/2}+iq\pi/2\right)\frac{\sin(Jq\pi)}{\sin(q\pi)} - \frac{i\pi}{2}\,z^{-\beta q/2}\exp\left(-z^{-\beta/2}\right) \; ,
\end{aligned}
\tag{10.97}
$$

while from Equivalence (10.92) we find that

$$
\begin{aligned}
S_{2,q}^{II}\left(N,z^{\beta}\right) &\equiv -\frac{|z|^{\beta(N-1)}}{2} \, P\int_{0}^{\infty} dt \, \frac{t^{N+q/2-1}\,e^{-\sqrt{t}}}{t-|z|^{-\beta}} - \pi|z|^{-\beta q/2} \\
&\quad \times \; e^{-Jiq\pi}\left(2i\,e^{iq\pi/2}\cosh\left(|z|^{-\beta/2}-iq\pi/2\right)\frac{\sin(Jq\pi)}{\sin(q\pi)}\right. \\
&\quad + \; \left.\sin(Jq\pi)\exp\left(-|z|^{-\beta/2}\right)\right) \; .
\end{aligned}
\tag{10.98}
$$

Yet again, we see that the above forms for the regularised value cannot be expressed as the product of a Stokes multiplier and the same subdominant function. As we have seen throughout this chapter, the concept of a Stokes multiplier is only applicable to the cases where p is the reciprocal of a natural number including unity. Thus, the concept of a Stokes multiplier is nowhere near as general as is described in the introduction of Ref. [9].

10.10   Numerical Example

To re-inforce the material presented in this chapter before we proceed to the development of a more general theory in the following chapter, let us return to the example of the same Type II generalised terminant studied in the previous chapter, viz. $S_{1,1/5}^{II}(N,z^{6})$. As has already been stated, this series is basically a standard terminant as defined in Ch. 21 of Dingle's book [3] except that the power on z is no longer unity. In Ch. 7 we were able to compare the regularised values of the asymptotic series for the error and related functions with the actual values of the functions since Mathematica was able to evaluate their values using its intrinsic routines. Unfortunately, we do not know which special function this Type II generalised terminant represents. So, we were unable to ascertain whether the MB-regularised values presented at the end of the previous chapter are indeed the actual values of the unknown special function. Therefore, in order to demonstrate that the values in Table 9.1 and 9.2 are indeed correct, we now compare them with the values obtained from the corresponding Borel-summed forms derived in this chapter.

Unlike MB-regularised forms for the regularised value of Type II generalised terminants we now have to deal with the Stokes phenomenon. Therefore, we note that the Stokes lines for $S_{1,1/5}^{II}(N, z^6)$ occur at $\arg z = j\pi/3$, where j is any integer. In keeping with convention, we nominate the $j=0$ line as the primary Stokes line which was also the case when evaluating the MB-regularised forms for the regularised value. Consequently, as soon as z moves off this line, the regularised value of the series develops jump discontinuities in either direction. Furthermore, according to the Zwaan-Dingle principle the regularised value is real initially, but this will not always be the case when $\arg z$ encounters other (secondary) Stokes lines.

We now determine the Borel-summed regularised value of $S_{1,1/5}^{II}(N, z^6)$ for all Stokes sectors of z. By introducing $p=1$, $q=1/5$ and $\beta=6$ into Equivalences (10.77) and (10.91), we obtain

$$
\begin{aligned}
S_{1,1/5}^{II}\left(N, z^6\right) \equiv -z^{6(N-1)} \int_C ds\, \frac{s^{N-4/5}e^{-s}}{s - z^{-6}} \\
\mp\, 2\pi i\, S_\pm(M) z^{-6/5} \exp\left(-z^{-6}\right) ,
\end{aligned} \tag{10.99}
$$

where the multipliers $S_\pm(M)$ can be evaluated with the aid of Eq. (9.23). The upper sign version of Equivalence (10.99) applies to the positive Stokes sectors given by $M\pi/3 < \arg z < (M+1)\pi/3$, while the multiplier $S_+(M)$ is given by

$$
S_+(M) = e^{(M+1)i\pi/5}\, \frac{\sin(M\pi/5)}{\sin(\pi/5)} + \frac{1}{2} . \tag{10.100}
$$

On the other hand, the lower sign version of Equivalence (10.99) applies to the negative Stokes sectors defined by $-(M+1)\pi/3 < \arg z < -M\pi/3$. The multiplier

$S_-(M)$ is found to be

$$S_-(M) = e^{-(M+1)i\pi/5} \frac{\sin(M\pi/5)}{\sin(\pi/5)} + \frac{1}{2} , \qquad (10.101)$$

which, as expected, is the complex conjugate of Eq. (10.100).

For the case where $\arg z^6$ corresponds to a Stokes line, the Borel-summed regularised value of $S_{1,1/5}^{II}(N,z^6)$ is determined by putting $p=1$, $q=1/5$, and $\beta=6$ into Equivalences (10.78) and (10.92). These reduce to

$$S_{1,1/5}^{II}(N,z^6) \equiv -|z|^{6(N-1)} P \int_0^\infty dt \, \frac{t^{N-4/5} e^{-t}}{t - |z|^{-6}}$$
$$\mp \, 2\pi i S_\pm(M) |z|^{-6/5} \exp\left(-|z|^{-6}\right) , \qquad (10.102)$$

where

$$S_+(M) = 2e^{-iM\pi/5} \sin(M\pi/5) \cot(\pi/5) , \qquad (10.103)$$

for $\arg z = M\pi/3$, and

$$S_-(M) = 2e^{iM\pi/5} \sin(M\pi/5) \cot(\pi/5) , \qquad (10.104)$$

for $\arg z = -M\pi/3$.

We have been able to express the above results in terms of the multipliers $S_\pm(M)$ only because p is the reciprocal of an integer. For $M=0$, we see from Eqs. (10.100) and (10.101) that the multipliers with their outside signs cross from -1/2 to 1/2, which is the standard or conventional behaviour of a Stokes multiplier discussed in the introductory chapter. This is despite the fact that in this terminant $\beta=6$, not -1 as in the theory of terminants presented in Ref. [3]. In the above results the multipliers are complex whenever $(M+1) \bmod 5 \neq 0$ and $M \bmod 5 \neq 0$ and real for the other values of M. However, if we examine the values of $S_\pm(M)$ for consecutive values of M other than zero, then it is apparent that the multiplier does not experience jump discontinuities of unity, which contradicts the statement made by Berry in Ref. [9] that asymptotic expansions can be written in terms of Stokes multipliers, which take on a value of S_- on one side of a Stokes line and on crossing it they take on a value of $S_- + 1$ after a rapid smoothing. We have already seen in Ch. 6 that there is no rapid smoothing across a Stokes line. Now we see that the multiplier only experiences a jump discontinuity of unity across the primary Stokes line for p equal to a reciprocal of a natural number, but not for any other Stokes line. Therefore, the view that an asymptotic form can be written in terms of a multiplier that experiences jump discontinuities of unity across all Stokes lines is highly simplistic and only limited to special cases such as the asymptotic forms for the error function, which utilise standard terminants.

Table 10.1: Borel-summed regularised values of $S_{1,1/5}^{II}(0, z^6)$ with $|z|=7/3$

arg z	Borel-summed Regularised Value
$\pm 15\pi/17$	$-1.563\,354\,963\,772\,395\,1 \pm 0.324\,634\,838\,578\,440\,5\,i$
$\pm 13\pi/17$	$-1.522\,003\,005\,250\,218\,7 \mp 0.444\,176\,963\,902\,583\,65\,i$
$\pm 11\pi/17$	$-1.156\,473\,556\,261\,042\,2 \mp 0.989\,173\,042\,444\,334\,9\,i$
$\pm 9\pi/17$	$-0.683\,407\,869\,429\,492\,7 \mp 1.411\,677\,602\,015\,271\,5\,i$
$\pm 7\pi/17$	$0.041\,934\,793\,840\,419\,76 \mp 1.598\,302\,795\,235\,477\,i$
$\pm 5\pi/17$	$0.714\,519\,183\,875\,369\,7 \mp 1.372\,946\,978\,975\,310\,4\,i$
$\pm 3\pi/17$	$1.202\,040\,278\,119\,573\,4 \mp 0.966\,331\,167\,394\,293\,5\,i$
$\pm \pi/17$	$1.539\,575\,647\,279\,514\,8 \mp 0.366\,397\,084\,777\,978\,4\,i$

Table 10.1 presents a sample of the values obtained from Equivalence (10.99) for $N=0$ and the same values of z as in Table 9.1, i.e. $|z|=7/3$ with $\arg z$ ranging from $\pm \pi/17$ to $\pm 15\pi/17$. By comparing the corresponding results in both tables we see that the regularised values obtained via Borel summation are identical to those obtained via MB regularisation within the precision and accuracy goals set in the NIntegrate routine. Because two different approaches have been used to obtain these values, we can be confident that the results in Proposition 5 and Equivalences (10.77) and (10.91) give the regularised values of the Type II generalised terminant for all values of z. Consequently, we shall use the presentation in this chapter as a platform for developing expressions for the regularised value of more general asymptotic series than generalised terminants in the following chapter.

Now let us consider the regularised values obtained via Borel summation for $\arg z$ situated on all the Stokes lines within the principal branch of the complex plane. All the forms in Equivalence (10.102) require the evaluation of the principal value of the same Cauchy integral. In Mathematica 4.1 this means that we need to use the add-on package CauchyPrincipalValue to evaluate this integral, although in more recent versions such as Mathematica 6.0 and 7.0 one can use the NIntegrate routine by setting an option to evaluate the Cauchy principal value. In both cases, however, one must specify where the singularity occurs when the appropriate numerical integration routine is called. Hence, for $|z| = 7/3$ we must indicate that the singularity occurs at $(3/7)^6$. Along the primary Stokes line the regularised value via Borel summation is merely the Cauchy principal value, while for the secondary Stokes lines we need to include the jump discontinuities, which are expressed in terms of a multiplier and subdominant exponential as demonstrated in Equivalence (10.102).

Table 10.2: Borel-summed regularised values of $S_{1,1/5}^{II}(0, z^6)$ with $|z|$=7/3 and arg z lying on the Stokes lines within the principal branch

arg z	Borel-summed Regularised Value
0	$1.590\,061\,161\,675\,932\,8$
$\pm\pi/3$	$0.515\,841\,753\,698\,813\,8 \mp 1.478\,536\,171\,758\,954\,i$
$\pm 2\pi/3$	$-1.222\,281\,759\,782\,954\,7 \mp 0.913\,785\,607\,743\,186\,2\,i$
π	$-1.222\,281\,759\,782\,954\,7 + 0.913\,785\,607\,743\,186\,2\,i$

The regularised values of $S_{1,1/5}^{II}(0, z^6)$ obtained from the Borel-summed forms given by Equivalence (10.102) with $|z| = 7/3$ are presented in Table 10.2. Here, we see that for arg $z = 0$, corresponding to the primary Stokes line, the regularised value is real, while the value obtained from the respective MB-regularised form in Table 9.2 value gives a tiny imaginary term. As indicated previously, this term is effectively zero within the accuracy and precision goals, which were set to 14 in both modules containing the NIntegrate routine. As for the remaining values in Table 10.2 we see that they are also identical to their respective values in Table 9.2 within the accuracy and precision goals. Since the values produced by the routines effectively agree with each other, this confirms the main tenor of this work that the regularised value of a divergent series is a unique and distinct quantity. Therefore, one can use either the Borel-summed forms derived in this chapter or the MB-regularised forms of Propositions 4 and 5 to determine the regularised value of either type of generalised terminant.

CHAPTER 11

Extension of Borel Summation

Abstract. In order to demonstrate that it is regularisation and not Borel summation which is responsible for yielding meaningful values to asymptotic series, the gamma function in both types of generalised terminants is now replaced by the function $f(pk+q)$ in this chapter. Then the regularised values for both types of series are determined by assuming that $f(s)$ is a Mellin transform. These appear in Propositions 6 and 7. Although both types of series are different from generalised terminants, the proofs of the propositions are nonetheless based on the exposition in the preceding chapter. Consequently, the regularised values are referred to as extended Borel-summed forms. The chapter concludes by considering a complicated example of a Type II series, where the coefficients can be expressed as the Mellin transform of the product of the Bessel function $J_\nu(x)$ and the Macdonald function $K_\nu(x)$. As there is no special function equivalent to this asymptotic series, the MB-regularised forms for the regularised value are derived with the aid of the general theory in Ch. 7. Then a numerical study of both the extended Borel-summed and MB-regularised forms is carried out with the index ν set equal to 1/3 and -3/5 and for large and small values of $|z|$ over the principal branch. Once again, the Borel-summed and MB-regularised forms yield identical regularised values for the series.

In the two preceding chapters we derived the regularised values for generalisations of the two types of terminants introduced by Dingle in Ref. [3]. We were able to present various expressions for the regularised values of these asymptotic series by applying the techniques of MB regularisation and Borel summation. The latter expressions, which were presented in Ch. 10, represent generalisations of the results appearing in Ch. 3. Because the coefficients $f(k)$ in the both types of series, $S(N,z)$ and $S_1(N,z)$, as defined in the introduction to Ch. 7, were limited to the gamma function in the analysis presented in Ch. 10, we were able to introduce the integral representation for the gamma function and then interchange the order of the summation and integration in order to derive their Borel-summed regularised values. These, of course, represent the standard steps for carrying out Borel summation of a divergent series. However, it was found in Ch. 3 that the crucial step in Borel summation is regularisation, not the introduction of the integral representation for the gamma function. Because of this it should, therefore, be possible to

Victor Kowalenko

derive the regularised values for both types of asymptotic series, where the coefficients are no longer given by the gamma function. Hence, the issue addressed in this chapter is whether it is possible to derive the regularised values of even more general series than those defined by Eqs. (9.1) and (9.2) by extending Borel summation.

Since it is our aim to consider more general series than those studied in the previous two chapters, we shall return to the two types of series $S(N, z^\beta)$ and $S_1(N, z^\beta)$ of Ch. 7, which we now write as

$$S(N, z^\beta) = \sum_{k=N}^{\infty} f(pk + q)(-z^\beta)^k \ , \tag{11.1}$$

and

$$S_1(N, z^\beta) = \sum_{k=N}^{\infty} f(pk + q)(z^\beta)^k \ . \tag{11.2}$$

Because the coefficients are now given as $f(pk + q)$, the above series are even more general than those in Ch. 7. We shall assume that p and q are both real and positive. For the case where q is less than zero, we can evaluate the finite series to the first value of k, where $pk + q > 0$, and use this value to serve as the new value of N in the remaining series. Then we have a truncated or finite series plus a series that can be expressed in one of the above forms.

As we have seen previously, both types of series need to be analysed separately because their regularised values behave differently as $\arg z$ moves around the complex plane. The expressions for the regularised values of $S(N, z^\beta)$ vary for each Stokes sector. These are defined by $(2j-1)\pi < \arg z < (2j+1)\pi$, where j is an arbitrary integer. We refer to the primary Stokes sector as that sector for which there is no jump discontinuity term in the regularised value initially. As we have done previously, we shall set j equal to zero for this sector. One property of the primary sector is that the regularised values of $S(N, z^\beta)$ obtained for the same value of $|z|$, but with $\arg z$ equidistant from the central ray in the primary Stokes sector are complex-conjugates of each other.

In contrast, the regularised value of $S_1(N, z^\beta)$ is very much dependent upon isolating a primary Stokes line from the infinite number of Stokes lines given by $\arg z = j\pi/\beta$, where j is again an arbitrary integer. Generally, the primary Stokes line is taken to be the $j = 0$ line, but it can also be defined as the line along which the regularised value of $S_1(N, z^\beta)$ is real. Once the primary Stokes line is fixed, the regularised value of the second type of asymptotic series develops jump discontinuities as soon as $\arg z$ moves off it into the adjoining Stokes sectors. Then the regularised values for the same values of $|z|$ and $\arg z$ equidistant from the primary Stokes line are also complex conjugates, although in this instance different expressions are used to evaluate them. While Borel summation has been extremely useful in developing an understanding of the concept of regularisation, our aim in this chapter is to consider more general versions of both types of asymptotic series, $S(N, z^\beta)$ and $S_1(N, z^\beta)$, in particular where the coefficients $f(pk+q)$ are no longer Borel-summable. As we shall see, if the function $f(s)$ can be expressed in terms of a Mellin transform rather the gamma function, then we shall be able to derive expressions for the regularised values of both types of asymptotic series. Nevertheless, in deriving these expressions we shall require the basic properties of the Borel-summed regularised values of the previous chapter. Since this is basically adapting or extending Borel summation, the regularised values obtained in this chapter will be referred to as extended Borel-summed forms as opposed to the MB-regularised and Borel-summed forms of the previous chapters.

11.1   Type I Series

We begin by presenting the extended Borel-summed regularised forms for the regularised value of $S(N, z^\beta)$ in the following proposition.

Proposition 6. If (1) the coefficients of the series $S(N, z^\beta)$ can be expressed as a Mellin transform, i.e.

$$f(s) = \int_0^\infty dx\, x^{s-1} F(x) \,, \tag{11.3}$$

where $\Re s > s_0$, (2) $pN+q > s_0$, (3) for positively real x and $\varepsilon > 0$,

$$F(x) = \begin{cases} O\!\left(x^{-p(N-1)-q-\varepsilon}\right) \,, & x \to \infty \,, \\ O\!\left(x^{-pN-q+\varepsilon}\right) \,, & x \to 0 \,, \end{cases} \tag{11.4}$$

and (4) the primary Stokes sector is given by $|\arg z| < \pi/\beta$, then the regularised value of the series $S(N, z^\beta)$, as defined by Eq. (11.1), over the Stokes sector of

$(\pm 2M - 1)\pi/\beta < \arg z < (\pm 2M + 1)\pi/\beta$, where $M \geq 0$, is given by

$$S(N, z^\beta) \equiv (-1)^N z^{\beta(N-1)} p^{-1} \int_C ds \, \frac{s^{N + q/p - 1} F(s^{1/p})}{s + z^{-\beta}}$$

$$\mp \frac{2\pi i}{p} \sum_{j=1}^{M} e^{\pm(2j-1)iq\pi/p} z^{-\beta q/p} F\left(e^{\pm(2j-1)i\pi/p} z^{-\beta/p}\right) . \qquad (11.5)$$

In this result C represents the line contour along the positive real axis. For the Stokes lines, where $\arg z = \pm(2M+1)\pi/\beta$, the regularised value of the series is given by

$$S(N, z^\beta) \equiv -\frac{|z|^{\beta(N-1)}}{p} P \int_0^\infty dt \, \frac{t^{N + q/p - 1} F(s^{1/p})}{t - |z|^{-\beta}} \mp \frac{2\pi i}{p} |z|^{-\beta q/p}$$

$$\times \sum_{j=1}^{M} e^{\mp 2ijq\pi/p} F\left(|z|^{-\beta/p} e^{\mp 2ij\pi/p}\right) \mp \frac{\pi i}{p} |z|^{-\beta q/p} F\left(|z|^{-\beta/p}\right) . \qquad (11.6)$$

Proof. Much of the groundwork for the proof of this proposition has already been laid out in the previous chapter, especially when it is realised that the integral representation for the gamma function is itself a Mellin transform. Since $pN + q > s_0$, we can replace $f(pk + q)$ in Eq. (11.1) by its Mellin transform representation, which yields

$$S(N, z^\beta) = \sum_{k=N}^{\infty} (-z^\beta)^k \int_0^\infty dx \, x^{pk + q - 1} F(x) . \qquad (11.7)$$

Making the change of variable, $y = x^p$, and interchanging the order of the summation and integration, we obtain

$$S(N, z^\beta) = p^{-1} \int_0^\infty dy \, y^{q/p - 1} F(y^{1/p}) \sum_{k=N}^{\infty} \left(-z^\beta y\right)^k . \qquad (11.8)$$

The final series in Eq. (11.8) is the geometric series, which would not have arisen if the coefficients in Eq. (11.1) had been written as $f(g(k))$ with $g(k)$ being non-linear in k. Then we would not be able to use the material in Ch. 4 to regularise the resulting series, which is the next step in the proof.

From Ch. 4 we know that the geometric series is absolutely convergent when the magnitude of the variable, which is $-z^\beta y$ in Eq. (11.8), is less than unity.

Unfortunately, the series is not absolutely convergent for all values of y over the entire range of integration. However, if $\Re\left(-z^{\beta}y\right)$ is always less than or equal to unity, which occurs when $\Re z^{\beta} \geq 0$, then the series is, at worst, conditionally convergent. For these values of z^{β}, we can introduce the regularised value given by the rhs of Equivalence (4.6) with the equivalence symbol replaced by an equals sign. For all other values of z^{β} we require the equivalence symbol, but as the latter operator is less stringent, we shall use it for all values of z^{β}. Therefore, regularisation of Eq. (11.8) yields

$$
\begin{aligned}
S\left(N, z^{\beta}\right) &\equiv (-1)^N z^{\beta(N-1)} p^{-1} \int_0^{\infty} dy \, \frac{y^{N+q/p-1} F\left(y^{1/p}\right)}{y + z^{-\beta}} \\
&= (-1)^N z^{\beta(N-1)} p^{-1} \int_C ds \, \frac{s^{N+q/p-1} F\left(s^{1/p}\right)}{s + z^{-\beta}} ,
\end{aligned}
\tag{11.9}
$$

where C is again the line contour along the positive real axis. We shall denote the above Cauchy integral by CI$_3$. The third condition in the proposition is necessary for ensuring that the first representation of the integral in the above equivalence is convergent.

As discussed in the previous chapter, the rhs of Equivalence (11.9) cannot represent the regularised value for all values of z^{β} because we know from Ref. [24] that the Cauchy integral develops jump discontinuities when $-z^{-\beta}$ crosses the line contour or positive real axis. The resulting jump discontinuities need to be evaluated on each occasion that $z^{-\beta}$ moves across the axis into an adjacent Stokes sector, but before we can do so, we must nominate a primary Stokes sector over which the Cauchy integral on the rhs of the above equivalence represents the regularised value for $S\left(N, z^{\beta}\right)$. As was done in the previous chapter, we nominate $-\pi < \arg z^{-\beta} < \pi$ as the primary sector, which ensures that the complex conjugate of the regularised value of $S\left(N, z^{\beta}\right)$ corresponds to the complex conjugate of z^{β} when it is introduced into the Type I series. That is, if $f(z)$ is the regularised value of $S\left(N, z^{\beta}\right)$, then $f(z)^*$ is the regularised value of $S\left(N, (z^*)^{\beta}\right)$, where the asterisk superscript denotes that the complex conjugate is to be evaluated.

In Ch. 7 we studied $S\left(N, z^{\beta} e^{-2li\pi}\right)$ in order to derive the regularised value for the other Stokes sectors. We do likewise here. By performing the same steps that led to Equivalence (11.9), we obtain

$$
S\left(N, z^{\beta} e^{-2li\pi}\right) \equiv (-1)^N z^{\beta(N-1)} p^{-1} \int_C ds \, \frac{s^{N+q/p-1} F\left(s^{1/p}\right)}{s + z^{-\beta} e^{2li\pi}} ,
\tag{11.10}
$$

Comparing this result with Equivalence (10.5), we note that $\exp(-s^{1/p})$ has been replaced in the above equivalence by $F(s^{1/p})$. So the comments made after Equivalence (10.5) apply here as well. Hence, the above equivalence is only valid for one value of l, viz. the primary Stokes sector. To determine the regularised value for the other Stokes sectors, we need to apply Proposition 3 to the difference of the l and $l-1$ versions of Equivalence (11.10) just as we did in obtaining Equivalence (10.6). In fact, by letting

$$\Delta S_l(N, z^\beta) = S(N, z^\beta e^{-2li\pi}) - S(N, z^\beta e^{-2(l-1)i\pi}) \ , \tag{11.11}$$

we find that the application of Proposition 3 yields

$$\Delta S_l(N, z^\beta) \equiv \int_{\substack{c-i\infty \\ \text{Max}[N-1,-q/p]<c=\Re s<N}}^{c+i\infty} ds \, z^{\beta s} \, f(ps+q) \, e^{-(2l-1)i\pi s} \ . \tag{11.12}$$

If the substitution $y = ps + q$ is made, then we identify the integral as the inverse Mellin transform of $f(y)$. According to Eq. (11.3), the above equivalence reduces to

$$\Delta S_l(N, z^\beta) \equiv \frac{2\pi i}{p} z^{-\beta q/p} e^{(2l-1)iq\pi/p} F\left(z^{-\beta/p} e^{(2l-1)i\pi/p} \right) \ . \tag{11.13}$$

This represents the full residue of the Cauchy integral in Equivalence (11.10) for the singularity at $s = -z^{-\beta} \exp(2li\pi)$ taken in a clockwise direction. Here, -1 has been taken to equal $\exp(-i\pi)$.

From Equivalence (11.13) we see that the regularised value is different for each value of l. Yet, Equivalence (11.10) implies the opposite since its rhs yields the same regularised value for each of value of l. The only method of resolving this paradox is to let the rhs of Equivalence (11.10) become the regularised value for a particular value of l since the Cauchy integral CI_3 develops singularities once $z^{-\beta} \exp((2l-1)i\pi)$ moves to another Stokes sector as discussed on p. 412 of Ref. [24]. The regularised value in the other Stokes sectors can be related by using Equivalence (11.13). This particular value of l will, of course, become the primary Stokes sector. As we did previously for the Type I generalised terminant, we choose the primary sector to be $l=0$, although this is arbitrary. Therefore, for $-\pi/\beta < \arg z < \pi/\beta$, the regularised value of our Type I series is given by the rhs of Equivalence (11.10) with $l=0$. Hence, we arrive at

$$S(N, z^\beta) \equiv (-1)^N z^{\beta(N-1)} p^{-1} \int_C ds \, \frac{s^{N+q/p-1} F(s^{1/p})}{s + z^{-\beta}} \ . \tag{11.14}$$

This represents the $M=0$ form of the equivalence given in Proposition 6.

If we put $l=1$ in Equivalences (11.11) and (11.13), then we obtain

$$S\left(N,z^{\beta}e^{-2i\pi}\right) - S(N,z^{\beta}) \equiv \frac{2\pi i}{p}\, z^{-\beta q/p}e^{iq\pi/p}F\left(z^{-\beta/p}e^{i\pi/p}\right)\ . \qquad (11.15)$$

Since the above equivalence is valid only for $-\pi/\beta < \arg z < \pi/\beta$, we can replace $S(N,z^{\beta})$ by its regularised value since the Cauchy integral is also valid over this sector. Thus, Equivalence (11.15) becomes

$$S\left(N,z^{\beta}e^{-2i\pi}\right) \equiv (-1)^{N}p^{-1}z^{\beta(N-1)}\int_{C} ds\, \frac{s^{N+q/p-1}\,F\left(s^{1/p}\right)}{s-(-z^{-\beta})}$$
$$+ \frac{2\pi i}{p}\, z^{-\beta q/p}e^{iq\pi/p}F\left(z^{\beta/p}e^{i\pi/p}\right)\ . \qquad (11.16)$$

Next we replace $z\exp(-2i\pi/\beta)$ by z, while z on the rhs is replaced by z_1, where $z_1 = z\exp(2i\pi/\beta)$. Then we find that

$$S\left(N,z^{\beta}\right) \equiv (-1)^{N}p^{-1}z_1^{\beta(N-1)}\int_{C} ds\, \frac{s^{N+q/p-1}\,F\left(s^{1/p}\right)}{s+z_1^{-\beta}}$$
$$+ \frac{2\pi i}{p}\, z_1^{-\beta q/p}e^{iq\pi/p}F\left(z_1^{-\beta/p}e^{i\pi/p}\right)\ , \qquad (11.17)$$

which is now valid for $-3\pi/\beta < \arg z < -\pi/\beta$.

As we did in Sec. 10.1, we obtain the regularised value of $S(N,z^{\beta})$ on the Stokes lines by applying Rule 8a in Ch. 3. For $\arg z = -\pi/\beta$ this means averaging Equivalences (11.14) and (11.17) after z_1 in the latter results is replaced by $z\exp(2i\pi/\beta)$. In addition, the contour integral must be modified so that only the Cauchy principal value is evaluated. Consequently, the regularised value of $S(N,z^{\beta})$ along the Stokes line of $\arg z = -\pi/\beta$ is given by

$$S(N,z^{\beta}) \equiv -p^{-1}|z|^{\beta(N-1)}P\int_{0}^{\infty} dt\, \frac{t^{N+q/p-1}\,F\left(t^{1/p}\right)}{t-|z|^{-\beta}}$$
$$+ \frac{\pi i}{p}\, |z|^{-\beta q/p}F\left(|z|^{-\beta/p}\right)\ . \qquad (11.18)$$

Thus, the regularised value of $S(N,z^{\beta})$ for the Stokes line of $\arg z = -\pi/\beta$ is composed of the regularised value in the primary Stokes sector except that the Cauchy principal value must be evaluated and half the $l = 1$ or semi-residue contribution from the rhs of Equivalence (11.13).

If we put $l = 2$ in Equivalences (11.11) and (11.13), then we obtain

$$S\left(N, z^\beta e^{-4i\pi}\right) - S\left(N, z^\beta e^{-2i\pi}\right) \equiv \frac{2\pi i}{p} z^{-\beta q/p}$$
$$\times e^{3iq\pi/p} F\left(z^{-\beta/p} e^{3iq\pi/p}\right) . \tag{11.19}$$

We can remove $S(N, z^\beta \exp(-2i\pi))$ in the above equivalence by introducing Equivalence (11.16). Then we find that

$$S\left(N, z^\beta e^{-4i\pi}\right) \equiv (-1)^N p^{-1} z^{\beta(N-1)} \int_C ds \, \frac{s^{N+q/p-1} F\left(s^{1/p}\right)}{s - (-z^{-\beta})} + \frac{2\pi i}{p} z^{-\beta q/p}$$
$$\times e^{3iq\pi/p} F\left(z^{-\beta/p} e^{3iq\pi/p}\right) + \frac{2\pi i}{p} z^{-\beta q/p} e^{iq\pi/p} F\left(z^{-\beta/p} e^{i\pi/p}\right) . \tag{11.20}$$

In this instance we replace $z \exp(-4i\pi/\beta)$ by z on the lhs of the above equivalence, while on the rhs we substitute z by z_2, which equals $z \exp(4i\pi/\beta)$. Hence, the regularised value of the Type I series becomes

$$S\left(N, z^\beta\right) \equiv (-1)^N p^{-1} z_2^{\beta(N-1)} \int_C ds \, \frac{s^{N+q/p-1} F\left(s^{1/p}\right)}{s + z_2^{-\beta}} + \frac{2\pi i}{p} z_2^{-\beta q/p}$$
$$\times \sum_{j=1}^{2} e^{i(2j-1)q\pi/p} F\left(z_2^{-\beta/p} e^{i(2j-1)\pi/p}\right) , \tag{11.21}$$

where $-(2(2)+1)\pi/\beta < \arg z < -(2(2)-1)\pi/\beta$ or $-5\pi/\beta < \arg z < -3\pi/\beta$. When z lies on the Stokes line that borders the $l = 1$ and $l = 2$ sectors, i.e. $\arg z = -3\pi/\beta$, all we need to do is take the average of Equivalences (11.17) and (11.21) and make sure that the principal value of the Cauchy integral is evaluated. Before we can average the two equivalences we must replace z_1 in Equivalence (11.17) by $z \exp(2i\pi/\beta)$ and z_2 in Equivalence (11.21) by $z \exp(4i\pi/\beta)$ so that z is the same variable in both equivalences. After taking the average of the two modified equivalences and setting $\arg z$ equal to $-3\pi/\beta$, we find that the regularised value of our general Type I series for $\arg z = -3\pi/\beta$ is given by

$$S\left(N, z^\beta\right) \equiv -p^{-1} |z|^{\beta(N-1)} P \int_0^\infty dt \, \frac{t^{N+q/p-1} F\left(t^{1/p}\right)}{t - |z|^{-\beta}} + \frac{2\pi i}{p} |z|^{-\beta q/p}$$
$$\times e^{2iq\pi/p} F\left(|z|^{-\beta/p} e^{2i\pi/p}\right) + \frac{\pi i}{p} |z|^{-\beta q/p} F\left(|z|^{-\beta/p}\right) . \tag{11.22}$$

As we found when studying Type I generalised terminants, a pattern is emerging that allows us to derive the regularised value of the general Type I series for the arbitrary $l = M$ Stokes sector. To accomplish this, all that we need to do is to replace the upper limit of 2 by M in the sum on the rhs of Equivalence (11.21). As a consequence, we find that the regularised value of $S(N, z^\beta)$ can be expressed more generally as

$$
S(N, z^\beta) \equiv (-1)^N z_M^{\beta(N-1)} p^{-1} \int_C ds \, \frac{s^{N+q/p-1} F(s^{1/p})}{s + z_M^{-\beta}} + \frac{2\pi i}{p} \, z_M^{-\beta q/p}
$$
$$
\times \sum_{j=1}^{M} e^{i(2j-1)q\pi/p} F\left(z_M^{-\beta/p} e^{i(2j-1)\pi/p} \right) , \tag{11.23}
$$

which is now valid for $-(2M+1)\pi/\beta < \arg z < -(2M-1)\pi/\beta$. In the above result $z_M = z \exp(2Mi\pi/\beta)$. When the latter quantity is introduced into the above equivalence, one obtains the lower sign version of Equivalence (11.5). On the other hand, for the Stokes line of $\arg z = -(2M+1)\pi/\beta$, the regularised value is obtained by taking the average of the regularised values for the two adjacent Stokes sectors bordered by the line, *viz.* $l = M$ and $l = M+1$, and evaluating the principal value of the resultant Cauchy integral. After a little algebra one finds that the regularised value of $S(N, z^\beta)$ reduces to

$$
S(N, z^\beta) \equiv -|z|^{\beta(N-1)} p^{-1} P \int_0^\infty dt \, \frac{t^{N+q/p-1} F(t^{1/p})}{t - |z|^{-\beta}} + \frac{2\pi i}{p} \, |z|^{-\beta q/p}
$$
$$
\times \sum_{j=1}^{M} e^{2ijq\pi/p} F\left(|z|^{-\beta/p} e^{2ij\pi/p} \right) + \frac{\pi i}{p} \, |z|^{-\beta q/p} F\left(|z|^{-\beta/p} \right) . \tag{11.24}
$$

The conditions on $F(x)$ in the proposition are necessary for ensuring that the principal value integral is always convergent. Equivalence (11.24) represents the lower sign version of Equivalence (11.6).

To determine the regularised value of $S(N, z^\beta)$ when $\pi/\beta < \arg z < 3\pi/\beta$, we now put $l = 0$ in Equivalences (11.11) and (11.13). This yields

$$
S\left(N, z^\beta e^{2i\pi} \right) - S(N, z^\beta) \equiv - \frac{2\pi i}{p} \, z^{-\beta q/p} e^{-iq\pi/p} F\left(z^{-\beta/p} e^{-i\pi/p} \right) . \tag{11.25}
$$

Since $-\pi/\beta < \arg z < \pi/\beta$, we can replace $S(N, z^\beta)$ in Equivalence (11.25) by introducing the rhs of Equivalence (11.18). Thus, we find that the regularised value of $S(N, z^\beta \exp(2i\pi))$ can be written as

$$
S\left(N, z^\beta e^{2i\pi}\right) \equiv -p^{-1} z^{\beta(N-1)} \int_C ds \, \frac{s^{N+q/p-1} F\left(s^{1/p}\right)}{s + z^{-\beta}}
$$
$$
- \frac{2\pi i}{p} z^{-\beta q/p} e^{-iq\pi/p} F\left(z^{-\beta/p} e^{-i\pi/p}\right) . \qquad (11.26)
$$

If $z \exp(2i\pi/\beta)$ is replaced by z on the lhs and z on the rhs by z_{-1}, then the above result becomes

$$
S(N, z^\beta) \equiv -p^{-1} z^{\beta(N-1)} P \int_C dt \, \frac{s^{N+q/p-1} F\left(s^{1/p}\right)}{s + z^{-\beta}}
$$
$$
- \frac{2\pi i}{p} z_{-1}^{-\beta q/p} e^{-iq\pi/p} F\left(z_{-1}^{-\beta/p} e^{-i\pi/p}\right) , \qquad (11.27)
$$

where $z_{-1} = z \exp(-2i\pi/\beta)$. To obtain the regularised value for the Stokes line of $\arg z = \pi/\beta$, we average Equivalences (11.14) and (11.26) and consider only the principal value of the Cauchy integral in accordance with Rule 8a in Ch. 3. Thus, the regularised value of the Type I series is given by

$$
S(N, z^\beta) \equiv -p^{-1} |z|^{\beta(N-1)} P \int_0^\infty dt \, \frac{t^{N+q/p-1} F\left(t^{1/p}\right)}{t - |z|^{-\beta}}
$$
$$
- \frac{\pi i}{p} |z|^{-\beta q/p} F\left(|z|^{-\beta/p}\right) . \qquad (11.28)
$$

As expected, Equivalences (11.27) and (11.28) are respectively the complex conjugates of Equivalences (11.17) and (11.18).

To determine the regularised value of $S(N, z^\beta)$ for $3\pi/\beta < \arg z < 5\pi/\beta$, we consider $l = -1$ in Equivalences (11.11) and (11.13). This gives

$$
S\left(N, z^\beta e^{4i\pi}\right) - S\left(N, z^\beta e^{2i\pi}\right) \equiv -\frac{2\pi i}{p} z^{-\beta q/p}
$$
$$
\times \, e^{-3iq\pi/p} F\left(z^{-\beta/p} e^{-3i\pi/p}\right) . \qquad (11.29)
$$

We can remove $S(N, z^\beta \exp(2i\pi))$ in the above equivalence by introducing Equivalence (11.25). Consequently, Equivalence (11.29) becomes

$$
S\left(N, z^\beta e^{4i\pi}\right) - S(N, z^\beta) \equiv -\frac{2\pi i}{p} z^{-\beta q/p} \sum_{j=1}^{2} e^{-i(2j-1)q\pi/p}
$$
$$
\times \, F\left(z^{-\beta/p} e^{-i(2j-1)\pi/p}\right) . \qquad (11.30)
$$

Now we introduce Equivalence (11.14) into the above result. After this is done, we set z equal to $z\exp(4i\pi/\beta)$ on the lhs and substitute z on the rhs by z_{-2}. Eventually we arrive at

$$S(N,z^\beta) \equiv (-1)^N p^{-1} z_{-2}^{\beta(N-1)} \int_C ds \, \frac{s^{N+q/p-1} F\left(-s^{1/p}\right)}{s+z_{-2}^{-\beta}} - \frac{2\pi i}{p} z_{-2}^{-\beta q/p}$$

$$\times \sum_{j=1}^{2} e^{-i(2j-1)q\pi/p} F\left(z_{-2}^{-\beta/p} e^{-i(2j-1)q\pi/p}\right) , \qquad (11.31)$$

where $(2(2)-1)\pi/\beta < \arg z < (2(2)+1)\pi/\beta$ or $3\pi/\beta < \arg z < 5\pi/\beta$. Equivalence (11.31) is the complex conjugate of Equivalence (11.21). For the Stokes line of $\arg z = 3\pi/\beta$, we average the rhs's of Equivalences (11.26) and (11.30) after Equivalence (11.14) has been introduced into the latter. Then in accordance with Rule 8a in Ch. 3, only the principal value of the Cauchy integral is to be evaluated. Alternatively, we can simply determine the complex conjugate of Equivalence (11.22) in order to obtain the regularised value of the Type I series on this Stokes line. Therefore, we find that

$$S(N,z^\beta) \equiv -p^{-1} |z|^{\beta(N-1)} P \int_0^\infty dt \, \frac{t^{N+q/p-1} F\left(t^{1/p}\right)}{t - |z|^{-\beta}} - \frac{2\pi i}{p} |z|^{-\beta q/p}$$

$$\times \, e^{-2iq\pi/p} F\left(|z|^{-\beta/p} e^{-2i\pi/p}\right) - \frac{\pi i}{p} |z|^{-\beta q/p} F\left(|z|^{-\beta/p}\right) . \qquad (11.32)$$

Again, a similar pattern is emerging as was observed in the case of the anti-clockwise rotations of $z^{-\beta}$. Previously, we were able to derive a general result for the regularised value of $S(N,z^\beta)$ for the arbitrary $l=M$ Stokes sector by replacing the index and upper limit of 2 on the rhs of Equivalence (11.21) by M. We do likewise in Equivalence (11.31). This results in

$$S(N,z^\beta) \equiv (-1)^N p^{-1} z_{-M}^{\beta(N-1)} \int_C ds \, \frac{s^{N+q/p-1} F\left(s^{1/p}\right)}{s+z_{-M}^{-\beta}} - \frac{2\pi i}{p} z_{-M}^{-\beta q/p}$$

$$\times \sum_{j=1}^{M} e^{-i(2j-1)q\pi/p} F\left(z_{-M}^{-\beta/p} e^{-i(2j-1)\pi/p}\right) , \qquad (11.33)$$

which is valid for $(2M-1)\pi/\beta < \arg z < (2M+1)\pi/\beta$ and $M > 0$. Since $z_{-M} = z\exp(-2iM\pi/\beta)$, we notice that Equivalence (11.33) represents the complex conjugate of Equivalence (11.23). If we make the substitution $j = M - j + 1$ in the above result, then we obtain the regularised value given in Proposition 6, viz. the upper sign version of Equivalence (11.5). On the other hand, for the arbitrary Stokes line of $\arg z = (2M+1)\pi/\beta$, all we need to do in the above equivalence is: (1) evaluate the principal value of the Cauchy integral, (2) replace M by $M-1$ in the sum and (3) evaluate only half of the final or $j=M$ term in the sum. After a little algebra the regularised value of $S(N, z^\beta)$ is found to be

$$
S(N, z^\beta) \equiv -|z|^{\beta(N-1)} p^{-1} P \int_0^\infty dt \, \frac{t^{N+q/p-1} F\left(t^{1/p}\right)}{t - |z|^{-\beta}} - \frac{2\pi i}{p} |z|^{-\beta q/p}
$$
$$
\times \sum_{j=1}^{M-1} e^{-2ijq\pi/p} F\left(|z|^{-\beta/p} e^{-2ij\pi/p}\right) - \frac{\pi i}{p} |z|^{-\beta q/p} F\left(|z|^{-\beta/p}\right) , \quad (11.34)
$$

which is the complex conjugate of Equivalence (11.24). In this case if we let $j = M - j + 1$, then the above result reduces to the upper sign version of Equivalence (11.6). This completes the proof of Proposition 6.

Since $F(t^{1/p})$ is arbitrary, it is apparent that the jump discontinuous terms that arise when z moves to other Stokes sectors need no longer be subdominant. In fact, it is even possible that they may dominate the Cauchy integral from the outset, which is contrary to the conventional view of the Stokes phenomenon, where they are regarded as being subdominant initially. Only with the movement of $\arg z$ within the Stokes sector do they ever become dominant. Although the regularised values of both generalised and standard terminants can be classed as exhibiting this conventional behaviour, it is in reality fortuitous due to the fact that their coefficients, i.e. $f(pk+q)$ in Eq. (11.1), represent the Mellin transform of a decaying or subdominant exponential.

11.2 Type II Series

In this section we study the regularisation of the second type of series, $S_1(N, z^\beta)$, as defined by Eq. (11.2). This type of series is different from the first type studied in the previous section in that the Stokes lines now occur whenever $\arg z = 2j\pi/\beta$ rather than at $\arg z = (2j+1)\pi/\beta$. That is, the Stokes lines coincide with the line contour in the Cauchy integral that arises out of the process of regularisation. Consequently, we need to define a primary Stokes line now rather than a primary Stokes sector as we did when evaluating the regularised value of $S(N, z^\beta)$

in the previous section. As soon as $z^{-\beta}$ moves off this primary Stokes line, the regularised value of the Type II series immediately acquires jump discontinuities, which are equal in magnitude, but opposite in sign in both directions from the primary line. Because of this, the derivation of the regularised value of $S_1(N, z^\beta)$ as presented in the following proposition turns out to be rather different from that for a Type I series.

Proposition 7. Given that (1) the coefficients of the series $S_1(N, z^\beta)$ can be expressed as the Mellin transform given by Eq. (11.3), (2) $pN + q > s_0$, (3) $F(x)$ possesses the same behaviour as in the previous proposition and (4) the primary Stokes line is given by $\arg z = 0$, then the regularised value of $S_1(N, z^\beta)$ over the Stokes sector of $-2(M+1)\pi/\beta < \arg z < -2M\pi/\beta$, where $M \geq 0$, is given by

$$S_1\left(N, z^\beta\right) \equiv -\frac{z^{\beta(N-1)}}{p} \int_C ds \, \frac{s^{N+q/p-1}F\left(s^{1/p}\right)}{s - z^{-\beta}} + \frac{\pi i}{p} z^{-\beta q/p} F\left(z^{-\beta/p}\right)$$
$$+ \frac{2\pi i}{p} \sum_{j=1}^{M} e^{-2jiq\pi/p} z^{-\beta q/p} F\left(z^{-\beta/p} e^{-2ji\pi/p}\right) . \tag{11.35}$$

Along the primary Stokes line, the regularised value of $S_1(N, z^\beta)$ is given by

$$S_1\left(N, z^\beta\right) \equiv -\frac{|z|^{\beta(N-1)}}{p} P\int_0^\infty dt \, \frac{t^{N+q/p-1}F\left(t^{1/p}\right)}{t - |z|^{-\beta}} , \tag{11.36}$$

while for the secondary Stokes lines of $\arg z = -2M\pi/\beta$, it is given by

$$S_1\left(N, z^\beta\right) \equiv -\frac{|z|^{\beta(N-1)}}{p} P\int_0^\infty dt \, \frac{t^{N+q/p-1}F\left(t^{1/p}\right)}{t - |z|^{-\beta}} + \frac{\pi i}{p} |z|^{-\beta q/p}$$
$$\times \, e^{2Miq\pi/p} F\left(|z|^{-\beta/p} e^{2Mi\pi/p}\right) + \frac{\pi i}{p} |z|^{-\beta q/p} F\left(|z|^{-\beta/p}\right)$$
$$+ \frac{2\pi i}{p} |z|^{-\beta q/p} \sum_{j=1}^{M-1} e^{2jiq\pi/p} F\left(|z|^{-\beta/p} e^{2ji\pi/p}\right) . \tag{11.37}$$

For the Stokes sector of $2M\pi/\beta < \arg z < 2(M+1)\pi/\beta$, where $M \geq 1$, the regularised value of the Type II series is

$$S_1\left(N, z^\beta\right) \equiv -\frac{z^{\beta(N-1)}}{p} \int_C ds \, \frac{s^{N+q/p-1}F\left(s^{1/p}\right)}{s - z^{-\beta}} - \frac{\pi i}{p} z^{-\beta q/p} F\left(z^{-\beta/p}\right)$$
$$- \frac{2\pi i}{p} \sum_{j=1}^{M} e^{2jiq\pi/p} z^{-\beta q/p} F\left(z^{-\beta/p} e^{2ji\pi/p}\right) , \tag{11.38}$$

while along the secondary Stokes lines of $\arg z = 2M\pi/\beta$ $(M \geq 1)$ it is given by

$$
\begin{aligned}
S_1\left(N, z^\beta\right) &\equiv -\frac{|z|^{\beta(N-1)}}{p} P \int_0^\infty dt\, \frac{t^{N+q/p-1}F\left(t^{1/p}\right)}{t - |z|^{-\beta}} - \frac{\pi i}{p}|z|^{-\beta q/p} \\
&\times e^{-2Miq\pi/p}F\left(|z|^{-\beta/p}e^{-2Mi\pi/p}\right) - \frac{\pi i}{p}|z|^{-\beta q/p}F\left(|z|^{-\beta/p}\right) \\
&- \frac{2\pi i}{p}|z|^{-\beta q/p}\sum_{j=1}^{M-1}e^{-2jiq\pi/p}F\left(|z|^{-\beta/p}e^{-2ji\pi/p}\right) .
\end{aligned}
\tag{11.39}
$$

In these results C again represents the line contour along the positive real axis.

Proof. Since $pN + q > s_0$, we can introduce the Mellin transform for $f(s)$ into Eq. (11.2). On making the change of variable, $y = x^p$, we obtain

$$
S_1\left(N, z^\beta\right) = p^{-1}\int_0^\infty dy\, y^{q/p-1}F\left(y^{1/p}\right)\sum_{k=N}^\infty \left(z^\beta y\right)^k .
\tag{11.40}
$$

Since y ranges from zero to infinity, the geometric series in the above equation is conditionally convergent for $\Re z \leq 0$, while for $\Re z > 0$, it is divergent. Nevertheless, because the regularised value of the geometric series is identical to its conditionally convergent limit, we can write the above as

$$
\begin{aligned}
S\left(N, z^\beta\right) &\equiv -z^{-\beta(N-1)}p^{-1}\int_0^\infty dy\, \frac{y^{N+q/p-1}\,F\left(y^{1/p}\right)}{y - z^{-\beta}} \\
&= -z^{-\beta(N-1)}p^{-1}\int_C ds\, \frac{s^{N+q/p-1}\,F\left(s^{1/p}\right)}{s - z^{-\beta}} ,
\end{aligned}
\tag{11.41}
$$

where C is line contour along the positive real axis. If $z^\beta \exp(-2il\pi)$ is replaced by z^β, then the above result can be written more generally as

$$
S\left(N, z^\beta e^{-2li\pi}\right) \equiv -z^{-\beta(N-1)}p^{-1}\int_C ds\, \frac{s^{N+q/p-1}\,F\left(s^{1/p}\right)}{s - z^{-\beta}e^{2li\pi}} .
\tag{11.42}
$$

We shall refer to the Cauchy integral in the two preceding results as CI$_4$. Again, it should be noted that the conditions on $F(x)$ ensure that CI$_4$ is convergent.

Aside from the phase factor of $(-1)^N$, Equivalence (11.42) is basically the same as Equivalence (11.10) except that the pole is situated at $s = z^{-\beta}\exp(2li\pi)$. Therefore, for the same reasons concerned with Equivalence (11.9) the rhs of the above equivalence cannot be the regularised value of $S_1(N, z^\beta)$ for all values of z, or more specifically for all values of $\arg z$, in the complex plane. It can only be the regularised value of the Type II series over a particular Stokes sector or along a definite Stokes line. For the latter case the line contour of integration has to be modified anyway since CI_4 is undefined.

Because of the difference in sign between the singularity in Equivalence (11.42) and that in Equivalence (11.10), the Stokes lines for $S_1(N, z^\beta)$ occur at different locations than those for $S(N, z^\beta)$. From p. 412 of Ref. [24] we know that CI_4 develops jump discontinuities whenever $s = z^{-\beta} \exp(2li\pi)$ for l, an arbitrary integer. This means that the Stokes lines for $S_1(N, z^\beta)$ are defined by $\arg z^{-\beta} = -2l\pi$, instead of $-(2l+1)\pi$ as for the Type I series of the previous section. More of a problem, however, is the fact that they now coincide with the line contour of integration. Therefore, we not only have to contend with the pole of a Cauchy integral lying on a Stokes line initially, but as soon as its argument moves above or below the positive real axis, we need to determine the jump discontinuities that form the regularised value of $S_1(N, z^\beta)$. At best, the rhs of Equivalence (11.41) can only be valid for one Stokes line, but even then the Cauchy integral or CI_4 will be undefined there. For Type I series $-z^{-\beta}$ was situated initially in a Stokes sector and consequently, we needed to consider a primary Stokes sector. For Type II series, however, we must consider a primary Stokes line, which we define as the particular Stokes line where the regularised value of $S_1(N, z^\beta)$ is real. This, of course, is the new interpretation of the Zwaan-Dingle principle. As stated previously, the choice of the primary Stokes line is arbitrary. If $\arg z^{-\beta} = -2l\pi$ is selected as the primary Stokes line, then the complex conjugate of the regularised value of $S_1(N, z^\beta)$ will equal the regularised value of $S_1(N, z_l^\beta)$, where $z_l = |z| \exp(i(2l\pi/\beta - \arg z))$. To derive the results in the proposition, however, we choose the $l = 0$ Stokes line as the primary Stokes line. Then the complex conjugate of the regularised value of $S_1(N, z^\beta)$ will equal the regularised value of the series when z is replaced by its complex conjugate, viz. $S_1(N, (z^*)^\beta)$.

For $z^{-\beta}$ situated on the primary Stokes line the Cauchy integral is undefined. Therefore, it has to be modified, perhaps by indenting the line contour around the pole. This will require evaluating the semi-residue contribution due to the singularity in the integral. The residue of CI_4 is found to be

$$\mathrm{Res}\left\{CI_4\right\}_{s=z^{-\beta}\exp(-2li\pi)} = -p^{-1}e^{-2liq\pi/p}z^{-\beta q/p}F\left(z^{-\beta/p}e^{-2li\pi/p}\right) . \quad (11.43)$$

As we have chosen the primary Stokes line to correspond with $l = 0$, the semi-residue contributions are simply $\pm \pi i$ times the above result with $l=0$. Whichever sign is adopted depends upon whether a clockwise or anti-clockwise rotation is taken around the pole. If either contribution is to be included in the modification to Equivalence (11.41), then the regularised value becomes complex. However, we have stated that the regularised value of $S_1(N, z^\beta)$ must be real along the primary Stokes line. This can only be achieved by neglecting the singularity in CI$_4$. Therefore, we must evaluate the principal value of the Cauchy integral in order to ensure that the regularised value is real. Hence, the regularised value of $S_1(N, z^\beta)$ for $\arg z = 0$ is given by

$$S_1\left(N, z^\beta\right) \equiv -z^{-\beta(N-1)} p^{-1} P \int_0^\infty dt \, \frac{t^{N+q/p-1} \, F\left(t^{1/p}\right)}{t - z^{-\beta}} \, . \qquad (11.44)$$

This is essentially Equivalence (11.36) in Proposition 7. Comparing this result with its analogue for Type II generalised terminants, viz. Equivalence (10.67), we see that $\exp(-t^{1/p})$ in the principal value integral has been replaced by $F(t^{1/p})$ in the above result.

Let us assume that $\arg z^{-\beta}$ is infinitesimally greater than zero, but moves down to the positive real axis. This means that the singularity in the Cauchy integral is initially above the line contour. In order to ensure that the singularity remains above the line contour when $z^{-\beta}$ lands on the positive real axis, we indent the contour in an anti-clockwise direction around the singularity. Thus, the Cauchy integral for $\arg z^{-\beta} = 0$ is composed of the principal value integral plus the semi-residue contribution at $s = z^{-\beta}$ taken in an anti-clockwise direction. However, in obtaining the regularised value of $S_1(N, z^\beta)$ for $\arg z^{-\beta} = 0$ given by Equivalence (11.44) we removed the semi-residue contribution, leaving only the principal value integral. Therefore, although the singularity in the Cauchy integral is no longer situated on the line contour when $\arg z^{-\beta} > 0$, to be consistent with the behaviour for $\arg z = 0$, we need to remove the semi-residue contribution from the Cauchy integral. This line of reasoning is basically Dingle's argument on p. 412 of Ref. [3]. Hence, the regularised value of $S_1(N, z^\beta)$ for $-2\pi/\beta < \arg z < 0$ becomes

$$\begin{aligned} S_1\left(N, z^\beta\right) &\equiv -z^{-\beta(N-1)} p^{-1} \int_C ds \, \frac{s^{N+q/p-1} \, F\left(s^{1/p}\right)}{s - z^{-\beta}} - \pi i \\ &\times \left(-p^{-1}\right) z^{-\beta q/p} F\left(z^{-\beta/p}\right) \, . \end{aligned} \qquad (11.45)$$

Therefore, the Cauchy integral has effectively acquired the semi-residue contribution at $s = z^{-\beta}$ taken in a clockwise direction. Note also that the P notation has been dropped since the contour integral is no longer affected by the singularity at $s=z^{-\beta}$. In comparing this result with the corresponding regularised value of the Type II generalised terminant, viz. Equivalence (10.70), we see that $\exp(-s^{1/p})$ and $\exp(-z^{-\beta/p})$ have been replaced by $F(s^{1/p})$ and $F(z^{-\beta/p})$ respectively.

Now we consider $\arg z^{-\beta}$ infinitesimally below zero and moving upwards to the positive real axis. In order to ensure that the singularity in the Cauchy integral remains below the line contour when $z^{-\beta}$ reaches the positive real axis, we indent the contour in a clockwise direction around the singularity. Therefore, the Cauchy integral for $\arg z = 0$ is again composed of the principal value integral and the semi-residue contribution at $s=z^{-\beta}$ except on this occasion the semi-residue contribution is evaluated in a clockwise direction. This semi-residue contribution also does not appear in the regularised value of $S_1(N,z^\beta)$ when $\arg z^{-\beta} = 0$. Therefore, in order to obtain the regularised value of $S_1(N,z^\beta)$ when $\arg z > 0$, we must remove the semi-residue contribution from the Cauchy integral for the same reason used in the derivation of Equivalence (11.45). Hence, we find that

$$
S_1\left(N,z^\beta\right) \equiv -z^{-\beta(N-1)}\, p^{-1} \int_C ds\, \frac{s^{N+q/p-1}\, F\left(s^{1/p}\right)}{s-z^{-\beta}} - \pi i
$$
$$
\times\; p^{-1} z^{-\beta q/p} F\left(z^{-\beta/p}\right)\,, \tag{11.46}
$$

for $0 < \arg z < 2\pi/\beta$. Thus, the Cauchy integral has effectively acquired the semi-residue contribution at $s=z^{-\beta}$ taken in an anti-clockwise direction. Furthermore, the above result represents the complex conjugate of the regularised value of Equivalence (11.45) when the complex conjugate of z is introduced into it. By comparing the above result with its analogue for a Type II generalised terminant or Equivalence (10.71), we see that the exponential terms in the latter have again been replaced by $F(z)$.

To determine the regularised value of $S_1(N,z^\beta)$ for the Stokes sector of $2\pi < \arg z^{-\beta} < 4\pi$, we note that $z^{-\beta}$ has undergone an extra anti-clockwise rotation of 2π. In order to conserve the overall phase of the singularity, we now interpret the pole in the Cauchy integral as occurring at $s=z^{-\beta}\exp(-2i\pi)$ rather than at $s=z^{-\beta}$ as we did for the first Stokes sector above the primary Stokes line. Just before $z^{-\beta}$ reaches 2π, the singularity in the Cauchy integral lies below the line contour. In order to maintain this behaviour as it crosses the secondary Stokes line, we indent the contour in a clockwise direction. Therefore, for the Stokes sector of $-4\pi/\beta < \arg z < -2\pi/\beta$, the Cauchy integral has acquired a full residue contribution of $-2\pi i$ times the rhs of Eq. (11.43) evaluated with $s=z^{-\beta}\exp(-2i\pi)$. Previously, we only acquired a semi-residue term in deriving Equivalence (11.45). This was because we began with the regularised value of $S_1(N, z^\beta)$ for $z^{-\beta}$ on a Stokes line. Now we are moving from one Stokes sector to another, which involves a jump discontinuity to the secondary Stokes line and then another from the secondary Stokes line to the new Stokes sector. Thus, the regularised value of $S_1(N, z^\beta)$ for $-4\pi/\beta < \arg z < -2\pi/\beta$ becomes

$$S_1\left(N, z^\beta\right) \equiv -\frac{z^{-\beta(N-1)}}{p} \int_C ds \, \frac{s^{N+q/p-1} \, F\left(s^{1/p}\right)}{s - z^{-\beta}} + \frac{\pi i}{p} z^{-\beta q/p}$$
$$\times \ F\left(z^{-\beta/p}\right) + \frac{2\pi i}{p} z^{-\beta q/p} e^{-2iq\pi/p} F\left(z^{-\beta/p} e^{-2i\pi/p}\right) \ . \quad (11.47)$$

To obtain the regularised value along the secondary Stokes line of $\arg z = -2\pi/\beta$, we simply average Equivalences (11.45) and (11.47) and replace the Cauchy integral by its principal value in accordance with Rule 8a in Ch. 3. This yields

$$S_1\left(N, z^\beta\right) \equiv -\frac{|z|^{-\beta(N-1)}}{p} \, P \int_0^\infty dt \, \frac{t^{N+q/p-1} \, F\left(t^{1/p}\right)}{t - |z|^{-\beta}} + \frac{\pi i}{p} z^{-\beta q/p}$$
$$\times \ F\left(z^{-\beta/p}\right) + \frac{\pi i}{p} z^{-\beta q/p} e^{-2iq\pi/p} F\left(z^{-\beta/p} e^{-2i\pi/p}\right) \ . \quad (11.48)$$

From these results we see that a pattern is again developing as $z^{-\beta}$ moves to higher Stokes sectors. For $2M\pi < \arg z^{-\beta} < 2(M+1)\pi$, the regularised value is the sum of the regularised value for the previous Stokes sector plus the full residue contribution taken at $s = z^{-\beta} \exp(-2Mi\pi)$ in a clockwise direction. All the full residue contributions in the regularised value can then be combined into a finite sum. As a consequence, we arrive at Equivalence (11.35).

For the secondary Stokes line of $\arg z = -2M\pi/\beta$, all we need to do is: (1) replace the Cauchy integral by its principal value, (2) evaluate only half of the final term in the sum in Equivalence (11.35) and (3) replace M by $M-1$ in the sum. Then the regularised value can be written as

$$S_1\left(N, z^\beta\right) \equiv -\frac{|z|^{\beta(N-1)}}{p} \, P \int_0^\infty dt \, \frac{t^{N+q/p-1} \, F(t^{1/p})}{t - |z|^{-\beta}} + \frac{\pi i}{p} z^{-\beta q/p}$$
$$\times \ F\left(z^{-\beta/p}\right) + \frac{2\pi i}{p} \sum_{j=1}^{M-1} e^{-2jiq\pi/p} z^{-\beta q/p} F\left(e^{-2ji\pi/p} z^{-\beta/p}\right)$$
$$+ \ \frac{\pi i}{p} z^{-\beta q/p} e^{-2Miq\pi/p} F\left(z^{-\beta/p} e^{-2Mi\pi/p}\right) \ . \quad (11.49)$$

By putting $z = |z| \exp(-2Mi\pi/\beta)$ into the above result and changing j to $M-j$ in the sum, one eventually obtains Equivalence (11.37) in the proposition.

To evaluate the regularised value of $S_1(N, z^\beta)$ for $-4\pi < \arg z^{-\beta} < -2\pi$, we begin with Equivalence (11.46), which, as has already been stated, is valid over $-2\pi < \arg z^{-\beta} < 0$. At the lower end of this Stokes sector we have effectively carried out a 2π rotation in a clockwise direction, which means that we interpret the singularity in the Cauchy integral as occurring at $s = z^{-\beta} \exp(2i\pi)$. That is, the overall phase of the singularity is conserved as it was when we were dealing with anti-clockwise rotations of $z^{-\beta}$. Prior to reaching the secondary Stokes line of $\arg z^{-\beta} = -2\pi$, the singularity was situated above the line contour in the Cauchy integral or CI$_4$. To be consistent with this behaviour, as $z^{-\beta}$ crosses into the adjacent Stokes sector, we indent the line contour in an anti-clockwise direction. For $-4\pi < \arg z^{-\beta} < -2\pi$, this means that the Cauchy integral has now acquired a full residue contribution of $2\pi i$ times the rhs of Eq. (11.43) evaluated at $s = z^{-\beta} \exp(2i\pi)$. Therefore, the regularised value of $S_1(N, z^\beta)$ for $2\pi/\beta < \arg z < 4\pi/\beta$ is given by

$$S_1\left(N, z^\beta\right) \equiv -\frac{z^{-\beta(N-1)}}{p} \int_C ds \, \frac{s^{N+q/p-1} \, F\left(s^{1/p}\right)}{s - z^{-\beta}} - \frac{\pi i}{p} z^{-\beta q/p}$$
$$\times \, F\left(z^{-\beta/p}\right) - \frac{2\pi i}{p} z^{-\beta q/p} e^{2iq\pi/p} F\left(z^{-\beta/p} e^{2i\pi/p}\right) \, . \qquad (11.50)$$

This is simply the complex conjugate of Equivalence (11.47) when the complex conjugate of $z^{-\beta}$ is introduced into it. For the Stokes line of $\arg z = 2\pi/\beta$, we replace the Cauchy integral by its principal value and evaluate only the semi-residue contribution in an anti-clockwise direction. This results in half the last term in Equivalence (11.50) appearing in the regularised value. Then we obtain

$$S_1\left(N, z^\beta\right) \equiv -\frac{z^{-\beta(N-1)}}{p} P \int_0^\infty dt \, \frac{t^{N+q/p-1} \, F\left(t^{1/p}\right)}{t - z^{-\beta}} - \frac{\pi i}{p} z^{-\beta q/p}$$
$$\times \, F\left(z^{-\beta/p}\right) - \frac{\pi i}{p} z^{-\beta q/p} e^{2iq\pi/p} F\left(z^{-\beta/p} e^{2i\pi/p}\right) \, . \qquad (11.51)$$

A similar pattern is evolving as for the case of the anti-clockwise rotations of $z^{-\beta}$. For each rotation in a clockwise direction we encounter another secondary Stokes line with the singularity gaining an extra factor of $\exp(2i\pi)$. As a result, the overall phase of the singularity is conserved. For the Stokes sector of $-2(M+1)\pi < \arg z^{-\beta} < -2M\pi$, the regularised value of $S_1(N, z^\beta)$ is the sum of the regularised value for the previous Stokes sector given by $-2M\pi < \arg z^{-\beta} < -2(M-1)\pi$ and the full residue contribution, which is obtained by taking $2\pi i$ times the residue of the Cauchy integral with the singularity at $s = z^{-\beta}\exp(2Mi\pi)$. Therefore, by recursion we find that the regularised value of $S_1(N, z^{\beta})$ for the Stokes sector of $2M\pi/\beta < \arg z < 2(M+1)\pi/\beta$ is the sum of the rhs of Equivalence (11.46) and the finite sum of the full residue contributions, i.e. $2\pi i$ times the rhs of Eq. (11.43), from $l = 1$ to M. This gives Equivalence (11.38).

For the secondary Stokes line of $\arg z = 2M\pi/\beta$ all we need to do is to replace the Cauchy integral in Equivalence (11.38) by its principal value and only consider half of the $j = M$ term in the finite sum. As a consequence, we obtain the complex conjugate of Equivalence (11.49) when z is replaced by its complex conjugate. Furthermore, if we substitute z by $|z|\exp(2Mi\pi/\beta)$ and change j to $M - j$ in the sum, then we finally arrive at Equivalence (11.39) in the proposition. This completes the proof of the proposition.

11.3   Type II Example

To complete this chapter, let us consider an example that makes use of the general theory in Ch. 7 in addition to the material in Proposition 7. Suppose for positive real values of z we derive the following asymptotic series:

$$P(z) = \sum_{k=1}^{\infty} \frac{\Gamma(2k)\,\Gamma(k+\nu/2)}{\Gamma(\nu/2-k+1)}\, z^{7k/3} \quad , \tag{11.52}$$

where $|\Re\nu| < 1$. The above series represents the solution that can be obtained by employing iteration as discussed in Ch. 2 to the ordinary differential equation given by

$$r^3 \frac{d^4}{dr^4}Y(r) + \frac{11r^2}{2}\frac{d^3}{dr^3}Y(r) - \left(\frac{\nu^2}{4} - \frac{11}{2}\right) r\frac{d^2}{dr^2}Y(r)$$

$$- \left(\frac{\nu^2}{8} - \frac{1}{2}\right)\frac{d}{dr}Y(r) + r^{-1}Y(r) - \frac{\nu}{8r} = 0 \quad , \tag{11.53}$$

where $r = 1/4z^{7/3}$. Specifically, this means that in order to obtain Eq. (11.52) we set $Y(r) \equiv \sum_{k=1}^{\infty} f(k)r^{-k}$ in the above equation.

The asymptotic series $P(z)$ is an example of the general Type II series defined by Eq. (11.2), in which $N=1$, $\beta=7/3$ and the coefficients are given by

$$f(pk+q) = \frac{\Gamma(2k)\,\Gamma(k+v/2)}{\Gamma(v/2-k+1)} \ . \tag{11.54}$$

In order to apply the results in Proposition 7, we need to find a Mellin transform for the above quotient of gamma functions. According to No. 2.16.21.1 in Ref. [53], we have

$$\int_0^\infty dx\, x^{pk+q-1} J_v(ax)\, K_v(ax) = 2^{pk+q-3} a^{-pk-q}\, \Gamma\big((pk+q)/2\big)$$

$$\times\ \frac{\Gamma\big((pk+q)/4+v/2\big)}{\Gamma\big(1-(pk+q)/4+v/2\big)} \ . \tag{11.55}$$

By putting $p=4$, $q=0$ and $a=2$ in Eq. (11.55), we obtain the rhs of Eq. (11.54) except for a numerical factor of $1/8$. Hence, we find that $F(x)$ in Proposition 7 is given by

$$F(x) = 8\, J_v(2x)\, K_v(2x) \ . \tag{11.56}$$

Since we are dealing with a Type II series, we nominate $\arg z=0$ as the primary Stokes line. Then the regularised value of the series for z lying on this line can be obtained from Equivalence (11.36). This reduces to

$$P(z) \equiv -2P \int_0^\infty dt\, \frac{J_v\big(2t^{1/4}\big) K_v\big(2t^{1/4}\big)}{t-z^{-7/3}} \ . \tag{11.57}$$

Because $\beta=7/3$, four Stokes sectors are projected onto the principal branch of the complex plane. However, we shall consider all the Stokes sectors encompassing $|\arg z| < 2\pi$ in order to gain a better appreciation of how the regularised value acquires jump discontinuities *via* Proposition 7. When we conduct a numerical comparison with the corresponding MB-regularised values later, only those results pertaining to the principal branch of the complex plane will be considered. The regularised value of this Type II series for the Stokes sector of $0<\arg z<6\pi/7$ can be obtained from Equivalence (11.38) with $M=0$. Then we find that

$$P(z) \equiv -2 \int_C ds\, \frac{J_v\big(2s^{1/4}\big) K_v\big(2s^{1/4}\big)}{s-z^{-7/3}} - 2\pi i$$

$$\times\ J_v\big(2z^{-7/12}\big) K_v\big(2z^{-7/12}\big) \ . \tag{11.58}$$

For the secondary Stokes line of $\arg z = 6\pi/7$, we put M equal to unity in Equivalence (11.39). Hence, the regularised value of $P(z)$ reduces to

$$P(z) \equiv -2P \int_0^\infty dt\, \frac{J_v\left(2t^{1/4}\right) K_v\left(2t^{1/4}\right)}{t - |z|^{-7/3}} - 2\pi i J_v\left(2|z|^{-7/12}\right)$$

$$\times\ K_v\left(2|z|^{-7/12}\right) - 2\pi i J_v\left(-2i|z|^{-7/12}\right) K_v\left(-2i|z|^{-7/12}\right). \quad (11.59)$$

To obtain the regularised value for the Stokes sector of $6\pi/7 < \arg z < 12\pi/7$, we use Equivalence (11.38) again, but in this instance M is set equal to unity. Consequently, we arrive at

$$P(z) \equiv -2\int_C ds\, \frac{J_v\left(2s^{1/4}\right) K_v\left(2s^{1/4}\right)}{s - z^{-7/3}} - 2\pi i J_v\left(2z^{-7/12}\right)$$

$$\times\ K_v\left(2z^{-7/12}\right) - 4\pi i J_v\left(2iz^{-7/12}\right) K_v\left(2iz^{-7/12}\right). \quad (11.60)$$

For the secondary Stokes line of $\arg z = 12\pi/7$, once again, we use Equivalence (11.39), but now $M = 2$. Then the regularised value of the Type II series becomes

$$P(z) \equiv -2P \int_0^\infty dt\, \frac{J_v\left(2t^{1/4}\right) K_v\left(2t^{1/4}\right)}{t - |z|^{-7/3}} - 2\pi i J_v\left(2|z|^{-7/12}\right)$$

$$\times\ K_v\left(2|z|^{-7/12}\right) - 4\pi i J_v\left(-2i|z|^{-7/12}\right) K_v\left(-2i|z|^{-7/12}\right)$$

$$-\ 2\pi i J_v\left(-2|z|^{-7/12}\right) K_v\left(-2|z|^{-7/12}\right). \quad (11.61)$$

For the Stokes sector of $12\pi/7 < \arg z < 18\pi/7$, the regularised value of $P(z)$ is determined yet again by using Equivalence (11.38), but now we set $M = 2$. Therefore, we obtain

$$P(z) \equiv -2\int_C ds\, \frac{J_v\left(2s^{1/4}\right) K_v\left(2s^{1/4}\right)}{s - z^{-7/3}} - 2\pi i J_v\left(2z^{-7/12}\right) K_v\left(2z^{-7/12}\right) - 4\pi i$$

$$\times\ J_v\left(2iz^{-7/12}\right) K_v\left(2iz^{-7/12}\right) - 4\pi i J_v\left(-2z^{-7/12}\right) K_v\left(-2z^{-7/12}\right). \quad (11.62)$$

Now we consider anti-clockwise rotations of $z^{-\beta}$. For the first Stokes sector above the primary Stokes line, which means that $-6\pi/7 < \arg z < 0$, the regularised value of $P(z)$ can be obtained from Equivalence (11.35) with M set equal to zero. This gives

$$P(z) \equiv -2\int_C ds\, \frac{J_v\left(2s^{1/4}\right) K_v\left(2s^{1/4}\right)}{s - z^{-7/3}}$$

$$+\ 2\pi i J_v\left(2z^{-7/12}\right) K_v\left(2z^{-7/12}\right). \quad (11.63)$$

For the secondary Stokes line of $\arg z = -6\pi/7$, we put M equal to unity in Equivalence (11.37). Then the regularised value of $P(z)$ becomes

$$
P(z) \equiv -2P \int_0^\infty dt \, \frac{J_v\left(2t^{1/4}\right) K_v\left(2t^{1/4}\right)}{t - |z|^{-7/3}} + 2\pi i J_v\left(2|z|^{-7/12}\right)
$$
$$
\times \; K_v\left(2|z|^{-7/12}\right) + 2\pi i J_v\left(2i|z|^{-7/12}\right) K_v\left(2i|z|^{-7/12}\right) \; . \tag{11.64}
$$

For the next anti-clockwise rotation of $z^{-\beta}$, which means that $-12\pi/7 < \arg z < -6\pi/7$, we use Equivalence (11.35) again, but on this occasion $M = 1$. Consequently, we find that

$$
P(z) \equiv -2 \int_C ds \, \frac{J_v\left(2s^{1/4}\right) K_v\left(2s^{1/4}\right)}{s - z^{-7/3}} + 2\pi i J_v\left(2z^{-7/12}\right)
$$
$$
\times \; K_v\left(2z^{-7/12}\right) + 4\pi i J_v\left(-2iz^{-7/12}\right) K_v\left(-2iz^{-7/12}\right) \; . \tag{11.65}
$$

For the secondary Stokes line of $\arg z = -12\pi/7$, we put $M = 2$ in Equivalence (11.37), thereby obtaining

$$
P(z) \equiv -2P \int_0^\infty dt \, \frac{J_v\left(2t^{1/4}\right) K_v\left(2t^{1/4}\right)}{t - |z|^{-7/3}} + 2\pi i J_v\left(2|z|^{-7/12}\right)
$$
$$
\times \; K_v\left(2|z|^{-7/12}\right) + 4\pi i J_v\left(2i|z|^{-7/12}\right) K_v\left(2i|z|^{-7/12}\right)
$$
$$
+ \; 2\pi i J_v\left(-2|z|^{-7/12}\right) K_v\left(-2|z|^{-7/12}\right) \; . \tag{11.66}
$$

For the third anti-clockwise rotation of $z^{-\beta}$, whereby $-18\pi/7 < \arg z < -12\pi/7$, the regularised value of $P(z)$ is determined once again via Equivalence (11.38), but with $M = 2$. Hence, we arrive at

$$
P(z) \equiv -2 \int_C ds \, \frac{J_v\left(2s^{1/4}\right) K_v\left(2s^{1/4}\right)}{s - z^{-7/3}} + 2\pi i J_v\left(2z^{-7/12}\right)
$$
$$
\times \; K_v\left(2z^{-7/12}\right) + 4\pi i J_v\left(-2iz^{-7/12}\right) K_v\left(-2iz^{-7/12}\right)
$$
$$
+ \; 4\pi i J_v\left(-2z^{-7/12}\right) K_v\left(-2z^{-7/12}\right) \; . \tag{11.67}
$$

Since we are unable to express the regularised value of $P(z)$ directly in terms of known special functions by solving the ordinary differential equation for $Y(r)$, we cannot be sure that the extended Borel-summed forms for the regularised value of this Type II series given above are indeed correct. We can, however, derive the MB-regularised forms for the regularised value of the Type II series and then carry out a numerical study for specific values of v and various values of z over the principal branch of the complex plane. If the values agree with those obtained from the corresponding extended Borel-summed counterparts, then we can be re-assured that the above results are indeed valid.

In order to determine the MB-regularised values for the $P(z)$, we cannot use the material in Proposition 5 since $P(z)$ is not a generalised terminant. Instead, we must use the general theory introduced in Ch. 7, where initially the regularised value of $P(z)$ can be expressed in terms of the general forms appearing in Equivalence (7.60). These results are expressed in terms of the MB integrals denoted by $I_l^*(z^\beta)$ and $\Delta_{j,j-1}^*(z^\beta)$, which are defined by Eqs. (7.51) and (7.53) respectively. In the integrands of these integrals $f(s)$ is given by

$$f(s) = \frac{\Gamma(2s)\,\Gamma(s+v/2)}{\Gamma(v/2-s+1)} \;. \tag{11.68}$$

The limits in both MB integrals are based on the assumption that there exists a real value c between zero and unity such that the poles of $f(s)\Gamma(s)$ lie to the left of c and that those of $f(s)\Gamma(1-s)$ lie to the right of c. With $f(s)$ given above, we notice that c can take any value between zero and unity when $0<\Re v<1$, but for $-1<\Re v<0$ and $N=1$, $c>-\Re v/2$. Hence, we must introduce the extra condition that $c>\mathrm{Max}[N-1,-\Re v/2]$ below the lower limits of the MB integrals. A similar condition was introduced in Ch. 8 when we were evaluating the MB-regularised values for the asymptotic forms of $u(a)$ and related functions.

In order to derive the MB-regularised values for $P(z)$, we need to determine the domains of convergence for the MB integrals, which, in turn, means determining A and B in Approximation (7.39). In particular, we need to establish that both values are negative. If we introduce Stirling's approximation for the gamma function or Approximation (8.5) into Eq. (11.68), then we find that $A=B=-\pi$. Therefore, according to Equivalences (7.56) and (7.60), we find that the MB-regularised value of $P(z)$ can be written as

$$P(z) \equiv \begin{cases} I_0^*(z^{7/3}) - \frac{1}{2}\,\Delta I_{0,-1}^*(z^{7/3}) & , \quad -3\pi/7 < \theta < 9\pi/7 \quad , \\ I_{-1}^*(z^{7/3}) + \frac{1}{2}\,\Delta I_{0,-1}^*(z^{7/3}) & , \quad -9\pi/7 < \theta < 3\pi/7 \quad . \end{cases} \tag{11.69}$$

In the above result $\theta = \arg z$ as before. In addition, the MB integral $I_0^*(z^{7/3})$ is given by

$$
I_0^*\!\left(z^{7/3}\right) = \int_{\substack{c-i\infty \\ \mathrm{Max}[0,-\Re v/2]<c=\Re s<1}}^{c+i\infty} ds\, \frac{z^{7s/3}\,e^{-i\pi s}}{\left(e^{-i\pi s}-e^{i\pi s}\right)}\, \frac{\Gamma(2s)\,\Gamma(s+v/2)}{\Gamma(v/2-s+1)}\ , \qquad (11.70)
$$

while $I_{-1}^*\!\left(z^{7/3}\right)$ is given by

$$
I_{-1}^*\!\left(z^{7/3}\right) = \int_{\substack{c-i\infty \\ \mathrm{Max}[0,-\Re v/2]<c=\Re s<1}}^{c+i\infty} ds\, \frac{z^{7s/3}\,e^{i\pi s}}{\left(e^{-i\pi s}-e^{i\pi s}\right)}\, \frac{\Gamma(2s)\,\Gamma(s+v/2)}{\Gamma(v/2-s+1)}\ . \qquad (11.71)
$$

The jump discontinuity term, which arises as soon as z moves off the primary Stokes line, can be determined from Eq. (7.59), which is found to be

$$
\Delta I_{0,-1}^*\!\left(z^{7/3}\right) = \int_{\substack{c-i\infty \\ \mathrm{Max}[0,-\Re v/2]<c=\Re s<1}}^{c+i\infty} ds\, z^{7s/3}\, \frac{\Gamma(2s)\,\Gamma(s+v/2)}{\Gamma(v/2-s+1)}\ . \qquad (11.72)
$$

We can regard Eq. (11.72) as an inverse Mellin transform. Hence, according to Eq. (11.55) we arrive at

$$
\Delta I_{0,-1}^*\!\left(z^{7/3}\right) = \pi i\, J_v\!\left(2z^{-7/12}\right) K_v\!\left(2z^{-7/12}\right)\ . \qquad (11.73)
$$

This means that Equivalence (11.69) can be simplified further. For $-3\pi/7 < \arg z < 9\pi/7$, the MB-regularised value of $P(z)$ becomes

$$
\begin{aligned}
P(z) \equiv &\int_{\substack{c-i\infty \\ \mathrm{Max}[0,-\Re v/2]<c=\Re s<1}}^{c+i\infty} ds\, \frac{z^{7s/3}\,e^{-i\pi s}}{\left(e^{-i\pi s}-e^{i\pi s}\right)}\, \frac{\Gamma(2s)\,\Gamma(s+v/2)}{\Gamma(v/2-s+1)} \\
&- 2\pi i\, J_v\!\left(2z^{-7/12}\right) K_v\!\left(2z^{-7/12}\right)\ ,
\end{aligned} \qquad (11.74)
$$

while for $-9\pi/7 < \arg z < 3\pi/7$, it can be written as

$$
\begin{aligned}
P(z) \equiv &\int_{\substack{c-i\infty \\ \mathrm{Max}[0,-\Re v/2]<c=\Re s<1}}^{c+i\infty} ds\, \frac{z^{7s/3}\,e^{i\pi s}}{\left(e^{-i\pi s}-e^{i\pi s}\right)}\, \frac{\Gamma(2s)\,\Gamma(s+v/2)}{\Gamma(v/2-s+1)} \\
&+ 2\pi i\, J_v\!\left(2z^{-7/12}\right) K_v\!\left(2z^{-7/12}\right)\ .
\end{aligned} \qquad (11.75)
$$

The interesting feature about the above results is that there are only two different forms required for evaluating the regularised value of $P(z)$ over the entire principal branch of the complex plane, whereas there are seven results obtained from Proposition 7. Therefore, the evaluation of the regularised value is more complex using the extended Borel-summed results than by using the MB-regularised results. In addition, the MB-regularised results can be checked against each other in the overlapping sectors of their domains of convergence, while those obtained from Proposition 7 only apply to distinct or non-intersecting sectors and lines in the complex plane. Furthermore, the MB-regularised forms do not involve the integration of complex integrands, only the exponentially-decaying gamma function. As a consequence, MB-regularised values can often be computed to a much greater degree of accuracy more expediently, a property that was first observed in Ref. [13].

To convince the reader that the MB-regularised forms given by Equivalences (11.74) and (11.75) yield the same regularised values of $P(z)$ as the extended Borel-summed forms over the principal branch of the complex plane for z, we present a numerical example for two specific values of v. For the first case we choose v to equal $1/3$, which means that the offset c in the MB integrals of Equivalences (11.74) and (11.75) can be any value between zero and unity. For the second case we choose v to equal $-3/5$, which means that c must be greater than $3/10$ and less than unity in the MB integrals. Consequently, in carrying out our numerical study we shall set c equal to $1/2$ for both values of v when calling the NIntegrate routine.

As in all previous numerical studies we shall select one small value of $|z|$, viz. $|z| = 2/11$, and one "relatively large" value for which there is no optimal point of truncation, viz. $|z| = 11/3$. Then we shall vary the argument or phase of z over the principal branch of the complex plane. Equivalence (11.74) will be referred to as Form 1 for the regularised value of $P(z)$, while Equivalence (11.75) will be referred to as Form 2.

Table 11.1 displays a sample of the MB-regularised values of $P(z)$ obtained by varying $\arg z$ over the principal branch for $v = 1/3$ and $|z| = 2/11$. The first feature that the reader should notice about these results is all the values are presented to more than 16 decimal places, which means they have been obtained by running Mathematica 6.0/7.0 on a Dell workstation rather than using Mathematica 4.1 on a Pentium computer. Consequently, the results are no longer limited by machine precision. Because of the greater precision of the alternative computing system, both AccuracyGoal and PrecisionGoal options were set to 30 when the NIntegrate routine was called to evaluate the MB integrals in Equivalences (11.74) and (11.75), while WorkingPrecision was set equal to 60. Although the results were accurate to more than 30 decimal places as is evidenced by the small complex er-

Table 11.1: MB-regularised values of $P(z)$ as defined by Eq. (11.52) with $\nu=1/3$ and $|z|=2/11$

arg z	Form	MB-regularised Value
$-6\pi/7$	2	$-77.048\,740\,792\,436\,604\,763 + 106.198\,600\,785\,131\,808\,919\,i$
$-3\pi/4$	2	$5.097\,694\,959\,416\,571\,318\,3 + 40.759\,458\,001\,116\,859\,958\,0\,i$
$-5\pi/7$	2	$8.503\,397\,719\,960\,710\,291\,0 + 25.449\,207\,893\,776\,492\,136\,7\,i$
$-5\pi/8$	2	$5.514\,183\,552\,829\,511\,629\,3 + 6.662\,148\,355\,288\,387\,158\,6\,i$
$-\pi/2$	2	$1.220\,887\,721\,329\,726\,943\,8 + 1.011\,191\,935\,557\,866\,179\,i$
$-3\pi/7$	2	$0.421\,428\,110\,758\,552\,762\,7 + 0.396\,000\,929\,709\,204\,309\,7\,i$
$-\pi/7$	1	$-0.006\,137\,007\,923\,491\,624\,3 + 0.007\,408\,380\,289\,360\,609\,8\,i$
$-\pi/7$	2	$-0.006\,137\,007\,923\,491\,624\,3 + 0.007\,408\,380\,289\,360\,609\,8\,i$
0	1	$-0.001\,250\,905\,901\,612\,879\,12 + 3. \times 10^{-38}\,i$
0	2	$-0.001\,250\,905\,901\,612\,879\,12 + 3. \times 10^{-38}\,i$
$\pi/4$	1	$0.004\,082\,433\,809\,964\,800\,9 - 0.045\,927\,494\,820\,684\,432\,7\,i$
$\pi/4$	2	$0.004\,082\,433\,809\,964\,800\,9 - 0.045\,927\,494\,820\,684\,432\,7\,i$
$3\pi/8$	1	$0.170\,405\,593\,959\,868\,354\,3 - 0.210\,543\,817\,167\,501\,517\,4\,i$
$3\pi/8$	2	$0.170\,405\,593\,959\,868\,354\,3 - 0.210\,543\,817\,167\,501\,517\,4\,i$
$\pi/2$	1	$1.220\,887\,721\,329\,726\,943\,8 - 1.001\,119\,193\,555\,786\,6179\,i$
$6\pi/7$	1	$-77.048\,740\,792\,436\,604\,763 - 106.198\,600\,785\,131\,808\,919\,i$
$7\pi/8$	1	$-109.213\,890\,318\,121\,779\,84 - 111.677\,628\,657\,187\,033\,11\,i$
π	1	$412.670\,234\,922\,360\,852\,82 + 162.182\,107\,796\,385\,428\,04\,i$

ror when arg $z=0$, they have been presented to slightly less than 20 decimal places due to limited space. Consequently, where both equivalences are used to evaluate the regularised value for the same value of arg z, there is no discrepancy between them due to round-off error.

The most important feature of the table, however, is that the regularised values of $P(z)$ are identical over the common region of the domains of convergence for the MB integrals in Equivalences (11.74) and (11.75), namely when $|\arg z| < 3\pi/7$. This is a necessary condition for establishing the validity of the concept of regularisation as explained in Ch. 4. That is, whatever method or form is used to evaluate the regularised value of a divergent series, it should always yield the same value. Another interesting feature in the table is that for arg $z=0$, Equivalences (11.74) and (11.75) yield the same real values, which we expect to obtain from the corresponding extended Borel-summed form given by Equivalence (11.57) since these values of arg z are situated on the primary Stokes line.

Table 11.2: MB-regularised values of $P(z)$ as defined by Eq. (11.52) with ν=-3/5 and $|z|=11/3$

arg z	Form	MB-regularised Value
$-7\pi/8$	2	$5.368\,417\,231\,571\,883\,993\,9 - 7.682\,934\,978\,043\,766\,500\,0\,i$
$-6\pi/7$	2	$5.189\,170\,191\,958\,708\,759\,4 - 7.353\,534\,476\,343\,461\,809\,1\,i$
$-4\pi/7$	2	$2.975\,244\,895\,339\,302\,860\,5 - 2.980\,641\,852\,648\,480\,419\,4\,i$
$-\pi/2$	2	$2.587\,311\,339\,571\,093\,990\,3 - 2.235\,894\,984\,514\,044\,424\,8\,i$
$-3\pi/8$	1	$2.030\,087\,955\,176\,440\,505\,5 - 1.264\,204\,212\,214\,207\,239\,9\,i$
$-3\pi/8$	2	$2.030\,087\,955\,176\,440\,505\,5 - 1.264\,204\,212\,214\,207\,239\,9\,i$
$-\pi/4$	1	$1.620\,105\,304\,335\,512\,406\,3 - 0.641\,712\,337\,906\,040\,937\,4\,i$
$-\pi/4$	2	$1.620\,105\,304\,335\,512\,406\,3 - 0.641\,712\,337\,906\,040\,937\,4\,i$
$-\pi/7$	1	$1.391\,750\,495\,710\,218\,139\,3 - 0.304\,204\,567\,244\,082\,444\,7\,i$
$-\pi/7$	2	$1.391\,750\,495\,710\,218\,139\,3 - 0.304\,204\,567\,244\,082\,444\,7\,i$
0	1	$1.278\,611\,241\,990\,052\,662\,48 + 0.\times 10^{-35}\,i$
0	2	$1.278\,611\,241\,990\,052\,662\,48 + 0.\times 10^{-35}\,i$
$\pi/8$	1	$1.365\,386\,040\,616\,151\,304\,2 + 0.259\,949\,294\,529\,798\,312\,9\,i$
$\pi/8$	2	$1.365\,386\,040\,616\,151\,304\,2 + 0.259\,949\,294\,529\,798\,312\,9\,i$
$2\pi/7$	1	$1.721\,968\,580\,406\,993\,907\,4 + 0.789\,960\,668\,046\,559\,402\,2\,i$
$2\pi/7$	2	$1.721\,968\,580\,406\,993\,907\,4 + 0.789\,960\,668\,046\,559\,402\,2\,i$
$3\pi/7$	1	$2.250\,788\,167\,704\,792\,795\,5 + 1.632\,279\,928\,689\,053\,013\,2\,i$
$\pi/2$	1	$2.587\,311\,339\,571\,093\,990\,3 + 2.235\,894\,984\,514\,044\,424\,8\,i$
$5\pi/7$	1	$3.933\,305\,223\,432\,828\,001\,4 + 4.911\,977\,097\,617\,071\,943\,3\,i$
$3\pi/4$	1	$4.217\,324\,665\,434\,752\,732\,8 + 5.481\,801\,151\,539\,858\,689\,0\,i$
$6\pi/7$	1	$5.189\,170\,191\,958\,708\,759\,4 + 7.353\,534\,476\,343\,461\,809\,1\,i$
π	1	$6.729\,271\,541\,605\,415\,377\,9 + 10.029\,666\,159\,221\,635\,563\,2\,i$

Table 11.2 presents a sample of the MB-regularised values of $P(z)$ for $\nu = -3/5$ and $|z| = 11/3$. These results have also been obtained by running Mathematica 6.0/7.0 on a Dell workstation. Hence, they are not limited by the machine precision of the computing system as many of the examples presented in previous chapters. As in Table 11.1 the regularised values have been presented to slightly less than 20 decimal places due to limited space. In this case, however, the complex error term occurring along the primary Stokes line is slightly larger than for the first value of ν, but nevertheless, is well within the accuracy and precision

goals set in the NIntegrate routine. Aside from this, the results display the same general features as those in Table 11.1. In particular, the regularised values obtained from both forms are identical when the same value of $\arg z$ is used in the common region of the domains of convergence for both MB integrals.

Table 11.3: Regularised values of $P(z)$ using the extended Borel-summed forms with $\nu=1/3$ and $|z|=2/11$

$\arg z$	Regularised Value
$-6\pi/7$	$-77.048\,740\,792\,436\,25 + 106.198\,600\,785\,130\,85\,i$
$-3\pi/4$	$5.097\,694\,959\,416\,122 + 40.759\,458\,001\,116\,84\,i$
$-5\pi/7$	$8.503\,397\,719\,960\,07 + 25.449\,207\,893\,7765\,i$
$-5\pi/8$	$5.514\,183\,552\,829\,374 + 6.662\,148\,355\,288\,888\,i$
$-\pi/2$	$1.220\,887\,721\,329\,726 + 1.001\,119\,193\,555\,7842\,i$
$-3\pi/7$	$0.421\,428\,110\,758\,552\,97 + 0.396\,000\,929\,709\,2044\,i$
$-\pi/7$	$-0.006\,137\,007\,923\,491\,69 + 0.007\,408\,380\,289\,360\,651\,i$
0	$-0.001\,250\,905\,901\,614\,753$
$\pi/4$	$0.004\,082\,433\,809\,964\,793 - 0.045\,927\,494\,820\,6845\,i$
$3\pi/8$	$0.170\,405\,593\,959\,868\,73 - 0.210\,543\,817\,167\,501\,42\,i$
$\pi/2$	$1.220\,887\,721\,329\,7327 - 1.001\,119\,193\,555\,7848\,i$
$6\pi/7$	$-77.048\,740\,792\,436\,25 - 106.198\,600\,785\,130\,85\,i$
$7\pi/8$	$-109.213\,890\,318\,1219 - 111.677\,628\,657\,187\,06\,i$
π	$-412.670\,234\,922\,361 + 162.182\,107\,796\,3845\,i$

Now we turn our attention to evaluating the regularised values of $P(z)$ using the extended Borel-summed forms derived earlier in this chapter, i.e. Equivalences (11.57)-(11.60) and (11.63)-(11.65). Table 11.3 displays a sample of the results obtained by putting $v = 1/3$, $|z| = 2/11$ and then varying $\arg z$ in the principal branch of the complex plane. On the other hand, Table 11.4 presents the regularised values of $P(z)$ for $v = -3/5$, $|z| = 11/3$ and various values of $\arg z$ in the principal branch. In obtaining these results, it was necessary to carry out a change of variable, viz. $y=t^{1/4}$, in the Cauchy integral of the various equivalences or else convergence problems arose. Therefore, the following integral was actually computed by using the NIntegrate and CauchyPrincipalValue routines in Mathematica 4.1:

$$I = 8z^{7/3} \int_0^\infty dy \, \frac{y^3 J_v(2y) K_v(2y)}{1 - z^{7/3} y^4} \; . \tag{11.76}$$

The results in Tables 11.3 and 11.4 are nowhere near as accurate as the MB-regularised results in Tables 11.1 and 11.2 since they have been evaluated by using our standard computing system of Mathematica 4.1 on a Pentium computer. As a result, AccuracyGoal and PrecisionGoal were set to 14, while WorkingPrecision was set to 16 in the NIntegrate routine. When the numerical study was first performed, the latest version of Mathematica at that time was Version 6.0. Surprisingly, convergence problems appeared when the NIntegrate routine was used to evaluate the integral in Eq. (11.76). This means that Version 6.0 is not 100 percent upwards compatible from earlier versions of the software. In private communication with Wolfram Research the bug was attributed to the product of Bessel functions specified in the NIntegrate routine. It has since been fixed in the latest version of the software, Version 7.0, although strangely, the problem did not arise when using the NIntegrate and CauchyPrincipalValue routines in Version 4.1 as evidenced by the results in Tables **11.3** and **11.4**. With the advent of Mathematica 7.0, it has been confirmed that the accuracy of the results in Tables 11.3 and 11.4 is no longer limited by machine precision. That is, the latest version of the software is able to calculate the extended Borel-summed regularised values for $P(z)$ to the level of accuracy as displayed in Tables **11.1** and **11.2**.

It should also be mentioned that there is no complex error term in the regularised value along the primary Stokes line since they have been evaluated by using the CauchyPrincipalValue routine in Mathematica 4.1. This routine was also used when evaluating the extended Borel-summed forms along the secondary Stokes lines of $\arg z = \pm 6\pi/7$ and was much slower than all the other modules used to evaluate the regularised values of $P(z)$. Interestingly, this add-on package does not exist in Versions 6.0 and 7.0. Instead, one introduces PrincipalValue as a method in the NIntegrate routine. Despite this, however, the regularised values in Tables **11.3** and **11.4** agree with the corresponding values in Tables **11.1** and **11.2** to the specified accuracy and precision, which is the primary aim of this study. We also observe that the regularised value is definitely complex along the secondary Stokes line even though all the terms in $P(z)$ are all of the same sign and homogeneous in phase. Only along the primary Stokes line of $\arg z = 0$ is the regularised value real.

In summary, we have seen that the extended Borel-summed forms, which are given by Equivalences (11.57)-(11.60) and (11.63)-(11.65) yield identical values to those obtained via the MB-regularised forms that are given by Equivalences (11.74) and (11.75). The extended Borel-summed results presented in this chapter have not been derived with the introduction of the integral representation of the

Table 11.4: Regularised values of $P(z)$ using the extended Borel-summed forms with ν=-3/5 and $|z|$=11/3

arg z	Regularised Value
$-7\pi/8$	$5.368\,417\,231\,571\,886 - 7.682\,934\,978\,043\,763i$
$-6\pi/7$	$5.189\,170\,191\,958\,712 - 7.353\,534\,476\,343\,465i$
$-4\pi/7$	$2.975\,244\,895\,339\,310\,2 - 2.980\,641\,852\,648\,481i$
$-\pi/2$	$2.587\,311\,339\,571\,092\,3 - 2.235\,894\,984\,514\,045\,4i$
$-3\pi/8$	$2.030\,087\,955\,176\,439\,3 - 1.264\,204\,212\,214\,208\,8i$
$-\pi/4$	$1.620\,105\,304\,335\,515\,4 - 0.641\,712\,337\,906\,04i$
$-\pi/7$	$1.391\,750\,495\,710\,220\,2 - 0.304\,204\,567\,244\,079\,6i$
0	$1.278\,611\,241\,990\,055\,3$
$\pi/8$	$1.365\,386\,040\,616\,149 + 0.259\,949\,294\,529\,801\,2i$
$2\pi/7$	$1.721\,968\,580\,406\,994\,2 + 0.789\,960\,668\,046\,557\,3i$
$3\pi/7$	$2.250\,788\,167\,704\,794 + 1.632\,279\,928\,689\,052\,6i$
$\pi/2$	$2.587\,311\,339\,571\,087\,4 + 2.235\,894\,984\,514\,047\,2i$
$5\pi/7$	$3.933\,305\,223\,432\,828 + 4.911\,977\,097\,617\,072i$
$3\pi/4$	$4.217\,324\,665\,434\,761 + 5.481\,801\,151\,539\,861i$
$6\pi/7$	$5.189\,170\,191\,958\,712 + 7.353\,534\,476\,343\,465i$
π	$6.729\,271\,541\,605\,42 + 10.029\,666\,159\,221\,64i$

gamma function, which is one of the necessary steps in Borel summation. Instead, they have been derived by being able to express the coefficients of the two types of series defined by Eqs. (11.1) and (11.2) as a Mellin transform. Therefore, Borel summation is not the reason why we have been able to obtain meaningful results or limits to the divergent series appearing in the preceding chapters. Rather it has been the ability to regularise the geometric series as discussed in Ch. 4 that has enabled them to be evaluated.

CHAPTER 12

Conclusion

Abstract. In the final chapter a summary of the major issues surrounding this work is presented together with a discussion of the main results. The conclusion begins by relating the discovery of the Stokes phenomenon in the terms of the concepts appearing in this book plus the subsequent developments over the next sesquicentenary or so. It then proceeds to explain why these developments have been unsuccessful in providing a satisfactory explanation of the phenomenon. This is attributed to the fact that they have not employed the important concepts of equivalence and regularisation, which are crucial for obtaining meaningful values from each component series in a complete asymptotic expansion. Such values have been referred to as regularised values throughout this work. Then the two main techniques of regularising a divergent series, viz. Borel summation and MB-regularisation, are discussed and compared with each other. It is pointed out that the former is responsible for the behaviour of an asymptotic expansion as it experiences the Stokes phenomenon, while the latter produces broader sectors over which an asymptotic expansion is valid. Not only do the broader sectors overlap one another, there is no evidence of discontinuities across specific rays as there is with Borel-summed forms for the regularised value of an asymptotic series. Furthermore, the concept of regularisation allows us to go beyond Borel summation and consider asymptotic series whose coefficients are not dependent upon the gamma function. Finally, the ramifications of this work on the subject of asymptotics are addressed.

This work has presented an in-depth study of the Stokes phenomenon, which refers to the emergence of jump discontinuities in asymptotic expansions as the variable of the original function or integral experiences changes in its phase or argument in the complex plane. Consequently, an asymptotic expansion remains uniform in a sector of the complex plane known as a Stokes sector before acquiring a jump discontinuity when the argument of the variable crosses to another Stokes sector. At the borders of the Stokes sectors, which are called Stokes lines, only half the jump discontinuity applies. The jump discontinuity may itself be another asymptotic series and hence, it may be divergent or it may simply be a constant as witnessed in Stokes's first example of the phenomenon that now bears his name [2].

We began our study of the Stokes phenomenon with this first example, which was denoted by the function $u(a)$ as defined by Eq. (2.1) in Ch. 2. By deriving the asymptotic forms for $u(a)$ over the entire complex plane using the same method as used by Stokes, we were able to present the asymptotic forms for the error function and its variants since they, too, are related to $u(a)$. In this work an asymptotic form has been defined as the complete asymptotic expansion for a function over a Stokes sector, while a complete asymptotic expansion has been defined as an expansion that not only contains all the terms of a dominant asymptotic series, but also all the terms of any subdominant asymptotic series, if they should exist. This means that neither the dominant nor the subdominant asymptotic series in a complete expansion has been truncated at any stage, despite the fact that their remainders can be divergent. Because of their decaying exponential nature when compared with the dominant asymptotic series, subdominant asymptotic series are said to lie beyond all orders [12, 13]. We have veered away from referring to the Stokes discontinuities as being subdominant because although they are initially subdominant, as the argument of the variable continues to progress over a Stokes sector, these subdominant terms eventually become the dominant terms in a complete asymptotic expansion and vice-versa.

As explained in Ch. 2 Stokes applied the well-known method of iteration to the differential equation for $u(a)$ in order to derive the asymptotic forms over the sectors of $-\pi < \arg a < 0$, $0 < \arg a < \pi$, and $\pi < \arg a < 2\pi$. He referred to the resulting series, which is today called an asymptotic series, as the descending series since it was an expansion in powers of a^{-2}. More importantly, he was aware that the series was divergent and possessed an optimal point of truncation. However, whilst he was aware of the divergence occurring in the remainder of the asymptotic series, he was not aware that it possessed an infinity, which arose from the asymptotic method he had used to derive it. This means that like all other asymptotic methods such as the method of steepest descent or Laplace's method, the method of iteration is improper. Because of the infinity that resides in the remainder when an asymptotic method is used, the lhs, which is equal to $u(a)$ and, therefore, finite, cannot always be equal to the asymptotic forms that were derived by Stokes on the rhs. Consequently, the less stringent equivalence symbol has been introduced in order to relate a finite function to its asymptotic form. That is, it is simply incorrect to refer to the original mathematical statement as an equation as occurs quite frequently in text-books on mathematical analysis. Consequently, we have referred to the modified mathematical statements relating asymptotic forms to finite functions or integrals as equivalences throughout this work. For example, the asymptotic forms derived by Stokes for $u(a)$ become

Equivalence (2.11) in Ch. 2.

For the next century or so after the discovery of the Stokes phenomenon, it seems that mathematicians were content to determine asymptotic expansions without the need for developing a general theory, which had the effect of making the phenomenon even more arcane. This state of affairs changed in 1957 when Heading found a recognizable pattern for the Stokes phenomenon in his study of the asymptotic solutions to certain n-th order differential equations [4, 5]. From this work Dingle was able to present a set of general rules describing how the phenomenon affects asymptotic expansions. These are discussed at length in Ch. 1 of Ref. [3], whilst a summary appears in Ch. 3 of this book. Moreover, Dingle claimed that with ingenuity one could determine all the asymptotic forms for a function or integral, especially when there are only two component asymptotic series. This was demonstrated in his derivation of the various asymptotic forms for the parabolic cylinder function, but as discussed also in Ch. 3 here, his heuristic approach was fortuitous in that he was able to relate the multipliers of the subdominant terms to the reflection formula for the gamma function. What is really required are general formulae for deriving such asymptotic forms arising out of a genuine understanding of the Stokes phenomenon, instead of relying on external mathematical relations to calculate the factors appearing in subdominant asymptotic series. In addition, Dingle was unable to provide the reader with a means of evaluating the limits of both series in the asymptotic forms other than by invoking truncation. Therefore, whilst he was able to present the various asymptotic forms of the parabolic cylinder function over many Stokes sectors, he was not able to obtain exact values for the function from any of its asymptotic forms.

The first of Dingle's eight rules, as far the present work is concerned, is perhaps the most important rule of all since it indicates where the Stokes discontinuities occur in the complex plane. It simply states that Stokes lines occur for those arguments of the variable whereby all the terms in an asymptotic series are homogeneous in phase and are of the same sign. Rules 2 to 5 discuss the behaviour of the dominant and subdominant asymptotic series in a complete asymptotic expansion as the phase/argument of the variable varies over a Stokes sector. As mentioned above, we have tended to avoid studying which part of an asymptotic expansion is dominant and which is subdominant. Instead, we have concentrated on how to obtain exact values of the original function from all component asymptotic series. Consequently, Rules 2 to 5 are deemed to be redundant at least as far this work is concerned. On the other hand, Rule 6 follows naturally when one realises that the jump discontinuity is related to the residue contribution from the singularity in a Cauchy integral. Hence, it is not necessary for developing general asymptotic forms such as those presented in Chs. 10 and 11. Technically, Rule 7 is correct, but it should be noted that when the second type of asymptotic series has been derived where all its terms homogeneous in phase and of the same sign, the variable is initially situated on what is referred to here as the primary Stokes line. As soon as the variable moves above or below this line, jump discontinuities ensue which are complex conjugates of one another. This is not the same situation as when the argument of the variable in the second type of asymptotic series approaches a secondary Stokes line or when the variable approaches a Stokes line for the first type of asymptotic series studied in this work, where all the terms initially alternate in sign. Then the asymptotic form acquires half the discontinuity on reaching the line and another half on leaving the line, which is the true meaning of Rule 7. While the behaviour at a primary Stokes line is different from the behaviour of all the other Stokes lines, it is nevertheless consistent with Rule 7. Although Rule 8 is true, it has been adapted here to become Rule 8a. This new version of the final rule states that averaging the asymptotic forms across two adjacent Stokes sectors gives the asymptotic form for the Stokes line that represents their common border provided the principal value is evaluated when the asymptotic form possesses a Cauchy integral.

Although Dingle claims that with ingenuity one can determine all the asymptotic forms for a function using these rules, he still sees a need for presenting a theory of terminants, which appears in Ch. 22 of Ref. [3]. Terminants represent power series whose coefficients a_k are expressed in terms of the gamma function, i.e. $\Gamma(k + \alpha + 1)$. From the discussion of asymptotic series in the previous paragraph, it follows that two distinct types exist: one that applies over a primary Stokes sector in which the terms alternate in sign and the other applying to a primary Stokes line in which the terms are of the same sign and phase. The primary reason Dingle introduces these asymptotic series is that he claims rather boldly that all asymptotic series for sufficiently large values of k, say greater than N, behave essentially as one type of terminant. Presumably, for $k < N$, this implies that one should use the true values of the a_k.

Unlike the situation for the asymptotic forms of the parabolic cylinder function, Dingle is able to give limit values for both types of terminant. These values, which are given by "Eqs." (3.3) and (3.5), are expressed in terms of the functions $\Lambda_{N+\alpha}(z)$ and $\bar{\Lambda}_{N+\alpha}(-z)$, whose integral representations are given by Eqs. (3.4) and (3.6). The reason for the quotation marks is that depending upon the phase of z, both terminants possess an infinity according to Ch. 2 and therefore, cannot possibly be equal to a finite quantity on the rhs. That is, they should be equivalences, not equations. In this work we have referred to the finite quantity or limit in an equivalence statement as the regularised value since it can only be obtained by employing a regularisation technique. Therefore, the integral representations of the Λ functions in Ch. 3 represent regularised values because they have been determined via Borel summation, which, as we show in the following chapter, is a regularisation technique.

The second type of terminant is the more confronting of the two types of terminant. When it is evaluated as an asymptotic expansion, it is generally done so, initially, for real values of the variable and then analytically continued into the complex plane. This analytic continuation is wrought with danger because the positive real axis of the variable corresponds to a Stokes line for the terminant. Hence, as soon as the phase of the variable moves off the positive real axis in either direction, jump discontinuities emerge as described earlier, although they are initially subdominant. With remarkable insight Dingle relates the jump discontinuities to the singularity in the Cauchy integral representation for $\bar{\Lambda}_{N+\alpha}(-z)$. As a consequence, he is able to determine the regularised value for this terminant in the adjacent Stokes sectors bordered by the positive real axis when it is represented as $\arg z = 0$. These results are combined into one form given here by Eq. (3.8). Again, this equation should be an equivalence statement. The Cauchy integral representation for the regularised value is also undefined when the variable lies on the positive real axis. Therefore, in order to make sense, the integral is modified so that its principal value becomes the regularised value of the second type of terminant along the positive real axis.

In order to clarify the confusion in Dingle's results for both terminants, it has been found necessary to stipulate that the Λ-notation be used specifically to refer to the series themselves. Then we introduce the equivalence symbol when relating the asymptotic series to the finite integrals on the rhs. Hence, Dingle's theory of terminants has been modified at the end of Ch. 3 with the presentation of Equivalences (3.12) and (3.13).

Ch. 4 is devoted to an exposition of the fundamental concept upon which the present work is based — regularisation. Simply put, this concept deals with the removal of the infinity in the remainder of a divergent series so as to make it summable. Such an infinity can arise when part of the integrand is expanded outside of its radius of absolute convergence within the limits of an integral. On its own, regularisation represents a mathematical abstraction, but in asymptotics it is required in order to correct the impropriety of the method used to obtain the asymptotic expansion in the first place. It does not, however, affect the divergent behaviour when an asymptotic series is truncated beyond the optimal point of truncation. To compensate for this behaviour, regularisation ensures that the remainder also diverges, but in the opposite sense to the truncated series.

To enable the reader to gain an understanding of the concept, various examples of regularisation of a divergent series are presented in Ch. 4. The first, and perhaps most important of these, is the regularisation of the geometric series, which is known to be absolutely convergent within the unit circle of the complex plane. The reason why this example is so important is that besides being used to determine the regularised values of the more complicated series in Ch. 4, it is one of the steps employed in the Borel summation of asymptotic series including the terminants of the previous chapter. When an asymptotic series is expressed in terms of the geometric series, it is simply regularised by introducing the regularised value for the geometric series. Not only does this apply to Borel summation of Dingle's terminants, but it also occurs when one regularises the generalisation of the geometric series, the binomial series, which is denoted as $_1\mathcal{F}_0(\rho; z)$ in the second example of regularisation discussed in Ch. 4. This notation should not be confused with the standard hypergeometric function notation of $_1F_0(\rho; z)$, which can only be applied when the series is absolutely convergent, i.e. for $|z| < 1$.

The final example in Ch. 4 is the regularisation of the series denoted by $_2\mathcal{F}_1(a+1, b+1; a+b+2-x; 1)$. In the first two examples the regularised value was found to be the same value as the limit value for the corresponding series when they were either conditionally convergent or absolutely convergent. For example, the geometric series is conditionally convergent for $|z| > 1$ and $\Re z < 1$ and equals $1/(1-z)$ as it does when $|z| < 1$. Furthermore, regularisation of the series when $\Re z > 1$ yields the same limit value of $1/(1-z)$. With $_2\mathcal{F}_1(a+1, b+1; a+b+2-x; 1)$ the situation is different because the regularised value of this series when it is divergent, i.e. $\Re x < 0$, is not the same as the limit value when the series is convergent, *viz.* for $\Re x > 0$. Although this example demonstrates that the regularised value of the divergent form of a power series may be different from its limit value when the power series is convergent, this difference does not manifest itself for the various series studied later in this work since they only involve the geometric series and not this type of series whose coefficients consist of a complicated quotient of gamma functions. In fact, if this behaviour did not occur in the case of the geometric series, then the results in Chs. 10 and 11 would simply not be valid and it would have been necessary to introduce an equals sign for those values when the series were conditionally convergent and an equivalence symbol for the regularised values only when the series were divergent as in Proposition 2. That is, as far as the results in Chs. 10 and 11 are concerned, we have been able to replace the equals sign in them by an equivalence symbol only because their regularised values are equal to the limit values obtained when the series are conditionally convergent.

In Ch. 5 Dingle's theory of terminants is applied to the asymptotic forms for the more familiar error function, $\text{erf}(z)$, rather than its relative, $u(a)$, considered by Stokes. The asymptotic forms for the latter function, however, are presented later in the chapter once the asymptotic forms for the error function of imaginary argument, $\text{erfi}(z)$, are derived. The asymptotic forms for the error function are given by Equivalence (5.1), which is written in terms of a multiplicative factor S. This factor, which is referred to as a Stokes multiplier in Refs. [9, 11], oscillates between -1/2 and 1/2 for the Stokes sectors and is equal to zero whenever z lies on a Stokes line. According to Equivalence (5.1), the Stokes discontinuity for the error function equals -1, 0 or 1 depending upon the value of $\arg z$ and is, therefore, very simple.

Because the regularised value of the geometric series has been introduced in this analysis, it follows that Borel summation of the asymptotic series in Equivalence (5.1) is a method of regularising a divergent series. By using this fact we present the regularised value of the series for all the Stokes sectors covering the principal branch of the complex plane in Equivalence (5.16). Then the asymptotic forms for the error function of imaginary argument and $u(a)$ are derived. The results for the error function of imaginary argument are verified by carrying out an exhaustive numerical analysis for large and small values of $|z|$ and various values of $\arg z$. Of particular interest are the regularised value where $\arg z$ is close to zero, i.e. in the vicinity of the Stokes line. In all instances, we find that the asymptotic forms yield the exact values of the original function within the numerical accuracy of the computing system. The values for small $|z|$ are particularly noteworthy because truncating the asymptotic form is unable to provide an accurate approximation to the original function. In standard asymptotics jargon the expansion is said to lie outside its range of validity for small values of $|z|$, which is due to the fact that there is no longer an optimal point of truncation. Hence, the asymptotic expansion is simply referred to as a large $|z|$ expansion, but by regularising the entire series we have been able to demonstrate that a complete asymptotic expansion is merely another of representation of the original function regardless of the size of $|z|$.

As a consequence of the numerical analysis of the asymptotic forms for the error function of imaginary argument, we derive an exact representation for the Stokes multiplier, which is given by Eq. (5.27). We show numerically for both small and large values of $|z|$ in the neighbourhood of the Stokes line at $\arg z = 0$ that the Stokes multiplier experiences a discontinuous jump from $-1/2$ to $1/2$, while on the line itself, it equals zero. This contradicts the radical view espoused by Berry and Olver respectively in Refs. [9] and [11] that the Stokes multiplier undergoes a rapid smoothing in the vicinity of a Stokes line. In the following chapter the origin of the purported smoothing is traced back to the truncation of the asymptotic expansion that these authors derive for the Stokes multiplier. In fact, their result for the Stokes multiplier corresponds to the leading order term of Eq. (5.27) and thus, can only be regarded as an approximation.

In addition to Stokes smoothing, Ch. 6 presents an alternative view to the Stokes phenomenon based on the theory of resurgent functions. The proponents of this theory claim that it is a natural extension of Stokes's seminal work, but here we show that their interpretation bears little semblance to Stokes's conception of the phenomenon. Furthermore, this approach has little practical benefit, a fact conceded by its proponents. Their solution to deducing numerical computations from purely "formal" calculations of asymptotic series is to use the inexact hyperasymptotic method of Berry and Howls [47], which can be incorporated into resurgence analysis or theory since the results from hyperasymptotic resummations can be classified as Borel-Laplace transforms.

In Ch. 7 we present an alternative method for regularising a divergent series, which is known as Mellin-Barnes (MB) regularisation [13, 14]. In this method a power series is expressed as the sum of a convergent MB integral and a contour integral along the great arc. The latter, which is zero for a convergent series and infinite for a divergent series, is discarded in accordance with the process of regularisation as discussed in the proof of Proposition 3. Initially, we consider the general type of series defined as $S(N, z)$ in Proposition 3, but later we extend the study to $S(N, z^\beta \exp(-2\pi i l))$, where l is an arbitrary integer. The regularised value of the latter series is given by the MB integral, $I_l(z^\beta)$ in Equivalence (7.18) and reduces to the regularised value in Proposition 3 when $l = 0$ and $\beta = 1$. Furthermore, the MB integral has a domain of convergence given by $(2l - 1 - \varepsilon_1)\pi/\beta < \arg z < (2l + 1 + \varepsilon_2)\pi/\beta$, where the ε_1 and ε_2 of Proposition 3 are greater than zero. Since the domains of convergence for successive values of l overlap, we can relate successive values of $I_l(z^\beta)$ by evaluating their difference as in Equivalence (7.20). If this difference between successive values of $I_l(z^\beta)$ can be analytically continued outside the common region of the overlapping domains of convergence, then one can derive the regularised value of $S(N, z^\beta)$ for all values of z in the complex plane, not just a branch. Thus, the two forms for the regularised value encompassing the Stokes sectors of $(2M-1)\pi < \arg z < (2M+1)\pi$ and $(-2M-1)\pi < \arg z < (-2M+1)\pi$ are given by Equivalences (7.27) and (7.35) respectively.

Proposition 3 is then applied to another type of general series, $S_1(N, z^\beta)$, as defined in Equivalence (7.48). The terms in this series are homogeneous in sign and phase whenever $\arg z = 2\pi j/\beta$ for j an integer, which represent the Stokes lines for the series. Because of the extra multi-valued factor of $(-1)^s$ in the integrand of the MB integral, which can be interpreted in an infinite number of ways, the regularised value becomes ambiguous. This ambiguity is resolved by extending the analysis of the series to $S_1(N, z^\beta \exp(-2\pi ilk))$ as in Equivalence (7.52). The regularised value of this version of the original series is now written in terms of the MB integral of $I_l^*(z^\beta)$ that appears in Equivalence (7.51). As in the case of the MB-regularised forms for the regularised value of $S(N, z^\beta)$, the domains of convergence for $I_l^*(z^\beta)$ overlap for successive values of l, but, unfortunately, yield different values over their common sectors of the complex plane. This contradicts the principal tenet of regularisation that a regularised value should be unique. However, as result of the study into the second type of terminant in Ch. 3, when an asymptotic expansion is derived where all its terms are homogeneous in sign and phase initially, i.e. of the type of $S_1(N, z^\beta)$, it means that z automatically lies on a Stokes line of discontinuity. As the variable moves off this primary Stokes line, the regularised value immediately acquires jump discontinuities, which cannot be determined from Proposition 3 since the proposition is only applicable to the variable lying within a Stokes sector. By taking this into consideration, we are able to determine the missing jump discontinuities from the difference between successive $I_l^*(z^\beta)$. This allows us to obtain the proper regularised value of $S_1(N, z^\beta)$ over the domain of convergence for each $I_l^*(z^\beta)$, which is given by Equivalence (7.68) for positive values of l and by Equivalence (7.72) for negative values of l.

Ch. 8 begins with the replacement of the series in the asymptotic forms for $u(a)$ by their respective MB-regularised forms determined from the material in Ch. 7. Then a description of how these MB-regularised forms for the regularised value can be evaluated by using a numerical integration routine appears. This is necessary because as discussed in Ref. [13], MB-regularised asymptotic forms are often capable of yielding the regularised value far more rapidly and with much greater precision than the corresponding Borel-summed forms. Specifically, the description deals with how the NIntegrate routine in Mathematica [19] can be used to evaluate the regularised value of the MB-regularised asymptotic forms of $u(a)$, which are given by Equivalences (8.7) and (8.8). These can be combined into one form by introducing the Stokes multiplier S. As in previous numerical examples, both a small and large value of $|a|$ are selected with the argument or phase of a varying over the principal branch of the complex plane. The results confirm that the multiplier is indeed equal to $1/2$ for $-\pi < \arg z < 0$ and $-1/2$ for $0 < \arg a < \pi$. Again, no rapid smoothing is observed in the vicinity of the positive real axis. Furthermore, both asymptotic forms give the correct regularised value outside the Stokes sectors because the domains of convergence for the resultant MB integrals are broader. Specifically, Equivalence (8.7) is valid over $-5\pi/4 < \arg a < \pi/4$, while Equivalence (8.8) is valid over $-\pi/4 < \arg a < 5\pi/4$. This also means that either form gives the regularised value in the common sector of $|\arg a| < \pi/4$, a property that cannot occur with Borel-summed forms for the regularised value. Consequently, we do not require a separate asymptotic form when a lies on the Stokes line at $\arg a = 0$, although by averaging the two asymptotic forms, we are able to present another asymptotic form for the regularised value there. This result given by Equivalence (8.9), however, is not restricted to the Stokes line, but is valid over the entire common sector. From these results we see that MB-regularised forms for the regularised value do not experience the Stokes phenomenon as their Borel-summed counterparts and thus, MB regularisation is a completely different method of regularisation from Borel summation.

The asymptotic forms for $u(a)$ are then used to determine MB-regularised asymptotic forms for the error function, whose forms can also be combined into one form with the introduction of a multiplier. The main difference between the results for the error function and those for $u(a)$ is that the Stokes sectors have been shifted. Hence, the Stokes lines for the asymptotic forms of $u(a)$ now occur along the imaginary axis of the complex plane. Because of the multi-valued factor of z^{-2s} in the integrand of the MB integral, care must be exercised when evaluating the MB integrals via the NIntegrate routine in Mathematica when $|\arg z| > \pi/2$. To avoid this problem, we introduce the factor of $\exp(2i\pi jk)$ into the asymptotic series for the error function. As a result, a more general asymptotic form for $\mathrm{erf}(z)$ is derived, which is given by Equivalence (8.17). This gives the regularised value for all Stokes sectors in the complex plane. It also means that the multiplier is dependent upon each Stokes sector and is, therefore, written as $S(j)$ as in Eq. (8.21). With these general results for the error function we then determine the regularised value outside the Stokes sector of $|\arg z| < \pi/2$. Once again, the numerical results in the vicinity of the Stokes lines, i.e. near the positive and negative imaginary axes, confirm that the Stokes multiplier is discontinuous and does not, by any means, experience a rapid smoothing.

The greatest problem with using a software package such as Mathematica to evaluate the regularised value is that it is limited by the machine precision of the computing system at one's disposal. For most of the numerical work reported here our computing system was composed of Mathematica 4.1 and a Pentium computer, which meant that most of the numerical results involving numerical integration were, at best, accurate to only 16 figures. In general, this does not represent a problem because one seldom requires such precision. However, in situations where truncating an asymptotic form does not yield an accurate approximation, problems will arise. As the truncated series diverges, the MB integral also diverges, but in the opposite sense. Combining the two results still yields the correct regularised value, but it will come at the cost of accuracy. Thus, it is recommended that the truncation parameter N be kept below or around the optimal point of truncation for each asymptotic series and for the cases where there is no optimal point of truncation, N should be as close as possible to zero. It should also be pointed out that with the advent of Versions 6.0 and 7.0 Mathematica is no longer restricted by the machine precision of a computing system, although as the precision is increased, the longer it will take to perform the calculations.

In Ch. 9 we set the coefficients of the series studied in Ch. 7 equal to $\Gamma(pk+q)$. This results in generalisations of Dingle's terminants in Ch. 3, which are defined by Eqs. (9.1) and (9.2). Because the coefficients are now more specific, we are able to present in Propositions 4 and 5 the MB-regularised forms for the regularised values of both types of series over the entire complex plane. As shown later in the chapter, these results simplify drastically when p is equal to the reciprocal of an integer including unity or when $p=2$. In particular, for the first case the extra terms to the MB integrals can be expressed in terms of a multiplier, but this is often not possible with the $p=2$ regularised values except for trivial values of q. Hence, the conventional view of the Stokes phenomenon where the subdominant terms of an asymptotic expansion can be expressed in terms of a multiplier by the same function for all sectors of the complex plane is only valid for special cases and is, therefore, simplistic.

Ch. 10 describes how one can obtain the regularised values of the generalised terminants via Borel summation over the entire complex plane. In obtaining these results contact is made with Dingle's results ($p = 1$) given in Ch. 3, though the latter are largely restricted to the Stokes sectors encompassing the principal branch of the complex plane. That is, the results in this chapter are far more general than Dingle's results even when $p = 1$. Although the general asymptotic forms can be expressed in terms of a Stokes multiplier for specific values of p, as exemplified by the asymptotic forms for $u(a)$ and related functions in earlier chapters, this cannot be done generally for all values of p as was the case with the MB-regularised asymptotic forms in Ch. 9. Thus, the concept of a Stokes multiplier is not valid.

It is also found that the major difference between the MB-regularised and Borel-summed asymptotic forms is that the latter experience jump discontinuities across Stokes lines, whereas the former only possess domains of convergence which extend beyond the Stokes sectors. As we found in the numerical studies in the earlier chapters, the Borel-summed asymptotic forms yield regularised values that vary below, on and above a Stokes line, whereas the MB-regularised forms yield regularised values that are equal to one another over the common sectors where the domains of convergence for the MB integrals overlap. Nevertheless, regardless of the form that is used to evaluate the regularised value, one always obtains the values of the original function.

When determining the regularised value for the second type of generalised terminant denoted by $S_1(N, z^\beta)$ via Borel summation, one needs to differentiate between a primary Stokes line where the regularised value is real and a secondary Stokes line, which may yield a complex regularised value as a result of acquiring a jump discontinuity emanating from the residue of the Cauchy integral representation for the series. Although the choice of the primary Stokes line is arbitrary, by convention it has been chosen to be $\arg z = 0$ throughout this work. A different primary Stokes line would result in different asymptotic forms, which, although not presented, here can be obtained by developing the theory in Ch. 10 based on the premise that the regularised value is real along the alternative primary Stokes line.

Ch. 11 considers more general types of divergent series as given by Eqs. (11.1) and (11.2) than the two types of generalised terminants of the previous chapter. Because the coefficients in these series are no longer equal to the gamma function, i.e. they are now set equal to $f(pk + q)$, they cannot be Borel-summed since there is no gamma function that can be replaced by its integral representation. However, based on the material in Ch. 4, we have seen that if a divergent series can be regularised as opposed to being Borel-summable, then one can obtain a finite limit value for the series. Therefore, with the aid of the material in the latter part of the previous chapter, we are able to determine the regularised values for these more general types of series whose regularised values are presented in Propositions 6 and 7. Nevertheless, conditions need to be placed on the coefficients $f(pk+q)$, of which the most important is that $f(s)$ must be expressible as a Mellin transform. This is actually not very surprising since Borel summation employs the integral representation for the gamma function, which is basically the Mellin transform of the exponential function.

From the general results and numerous examples presented in this work, we have seen that when regularised, a complete asymptotic expansion is merely an alternative representation for the function from which it is derived. In the past the subject of asymptotics has suffered a bad name mainly because asymptotic expansions have been limited in accuracy and range of applicability. The limitation in accuracy stems from the fact that asymptotic series are invariably truncated after the first few terms, while the range of applicability is dependent upon the magnitude of the variable in the asymptotic series being of the appropriate size so that the optimal point of truncation is not small. By introducing the concept of regularisation we are no longer impeded by these drawbacks or limitations and as a consequence, asymptotics has now been elevated to a true mathematical discipline yielding precise answers.

On a final note, the reader may well ask if there are other methods or techniques of regularising divergent series than MB regularisation and Borel summation which have formed the basis for the new explanation of the Stokes phenomenon presented here. The answer to the question is affirmative since the Euler-Maclaurin summation formula is another method of regularising a divergent series. This, however, represents the subject of another work.

APPENDIX

In this appendix we derive exact formulae for the polynomials $g_k(A)$ introduced in Ch. 6. These were defined according to Equivalence (6.8), but can also be written as

$$\sum_{k=0}^{\infty} g_k(A)t^k \equiv (1+t)^A e^{-At} e^{At^2/2} \ . \tag{A.1}$$

We now introduce into the rhs of Equivalence (A.1) the power series expansions for each of the exponential factors and the binomial series as given in Proposition 1. Then Equivalence (A.1) becomes

$$\sum_{k=0}^{\infty} g_k(A)t^k \equiv \sum_{k=0}^{\infty} \frac{\Gamma(k-A)}{\Gamma(-A)\,k!}\,(-t)^k \sum_{k=0}^{\infty} \frac{(-At)^k}{k!} \sum_{k=0}^{\infty} \frac{(At^2/2)^k}{k!} \ . \tag{A.2}$$

The last two series on the rhs of the above result are absolutely convergent for all values of t, but the first series from Proposition 1 is only absolutely convergent for $|t| < 1$. This means that we can replace the $\equiv$ symbol by an equals sign for these values of t. Then we multiply out the series, thereby obtaining

$$\sum_{k=0}^{\infty} g_{2k}(A)t^{2k} + \sum_{k=0}^{\infty} g_{2k+1}(A)t^{2k+1} = \sum_{k=0}^{\infty} t^{2k} \sum_{l=0}^{k} \frac{\Gamma(2k-2l-A)}{\Gamma(-A)(2k-2l)!}$$

$$\times \sum_{j=0}^{l} \frac{2^{-j}A^{2l-j}}{(2l-2j)!\,j!} + \sum_{k=0}^{\infty} t^{2k+2} \sum_{l=0}^{k} \frac{\Gamma(2k-2l+1-A)}{\Gamma(-A)(2k-2l)!} \sum_{j=0}^{l} \frac{2^{-j}A^{2l-j+1}}{(2l-2j+1)!\,j!}$$

$$- \sum_{k=0}^{\infty} t^{2k+1} \sum_{l=0}^{k} \frac{\Gamma(2k-2l+1-A)}{\Gamma(-A)(2k-2l+1)!} \sum_{j=0}^{l} \frac{2^{-j}A^{2l-j}}{(2l-2j)!\,j!} - \sum_{k=0}^{\infty} t^{2k+1}$$

$$\times \sum_{l=0}^{k} \frac{\Gamma(2k-2l-A)}{\Gamma(-A)(2k-2l)!} \sum_{j=0}^{l} \frac{2^{-j}A^{2l-j+1}}{(2l-2j+1)!\,j!} \ . \tag{A.3}$$

Although t lies within the unit circle, it is, nevertheless, arbitrary. This means that for the two sides of the above equation to equal one another, the coefficients of the powers of t on both sides must be equal to another. Thus, by equating the coefficients, we find that $g_0(A) = 1$, $g_1(A) = g_2(A) = 0$ and $g_3(A) = A/3$, which agree with the results obtained via the recursion relation in Ch. 6. More generally, we find that

$$
g_{2k+1}(A) = -\sum_{l=0}^{k} \frac{\Gamma(2k-2l+1-A)}{\Gamma(-A)(2k-2l+1)!} \sum_{j=0}^{l} \frac{2^{-j}A^{2l-j}}{(2l-2j)!\,j!}
$$

$$
- \sum_{l=0}^{k} \frac{\Gamma(2k-2l-A)}{\Gamma(-A)(2k-2l)!} \sum_{j=0}^{l} \frac{2^{-j}A^{2l+1-j}}{(2l-2j+1)!\,j!} \ , \tag{A.4}
$$

while for $k \geq 1$, the even numbered polynomials are given by

$$
g_{2k}(A) = \sum_{l=0}^{k} \frac{\Gamma(2k-2l-A)}{\Gamma(-A)(2k-2l)!} \sum_{j=0}^{l} \frac{2^{-j}A^{2l-j}}{(2l-2j)!\,j!}
$$

$$
+ \sum_{l=0}^{k-1} \frac{\Gamma(2k-2l-1-A)}{\Gamma(-A)(2k-2l-2)!} \sum_{j=0}^{l} \frac{2^{-j}A^{2l+1-j}}{(2l-2j+1)!\,j!} \ . \tag{A.5}
$$

REFERENCES

[1] Morse PM and Feshbach H. Methods of Theoretical Physics Part 1. New York: McGraw-Hill; 1953: 609.

[2] Stokes GG. On the discontinuity of arbitrary constants which appear in divergent developments. In: Collected Mathematical and Physical Papers Vol 4. Cambridge: Cambridge University Press; 1904: 77-109.

[3] Dingle RB. Asymptotic Expansions: Their Derivation and Interpretation. London: Academic Press; 1973.

[4] Heading J. The Stokes phenomenon and certain nth-order differential equations I. Preliminary investigation of the equations. Proc Camb Phil Soc 1957; 53: 399-418.

[5] Heading J. The Stokes phenomenon and certain nth-order differential equations II. The Stokes phenomenon. Proc Camb Phil Soc 1957; 53: 419-441.

[6] Whittaker ET and Watson GN. A Course in Modern Analysis. Cambridge: Cambridge University Press; 1973.

[7] Copson ET. An Introduction to the Theory of Functions of a Complex Variable. Oxford: Clarendon Press; 1976.

[8] Heading J. An Introduction to Phase-integral Methods. London: Methuen; 1962.

[9] Berry MV. Uniform asymptotic smoothing of Stokes's discontinuities. Proc R Soc Lond 1989; A422: 7-21.

[10] Paris RB, Kaminski D. Asymptotics and Mellin-Barnes Integrals. Cambridge: Cambridge University Press; 2001.

[11] Olver FWJ. On Stokes' phenomenon and converging factors. In: Proc Internat Conf on Asymptotic and Computational Analysis, Winnipeg, Canada: June 5-7, 1989; R Wong, Ed.; New York: Marcel Dekker, 1990: pp. 329-355.

[12] Segur H, Tanveer S, Levine H (Eds.). Asymptotics beyond All Orders. New York: Plenum Press; 1991.

[13] Kowalenko V, Frankel N E, Glasser M, Taucher T. Generalised Euler-Jacobi Inversion Formula and Asymptotics beyond All Orders. London Mathematical Society Lecture Note 214. Cambridge: Cambridge University Press; 1995.

[14] Kowalenko V. Towards a theory of divergent series and its importance to asymptotics. In: Recent Research Developments in Physics Vol 2. Trivandrum India: Transworld Research Network; 2001: pp. 17-68.

[15] Kowalenko V. Exactification of the asymptotics for Bessel and Hankel Functions. Appl Math Comput 2002; 133: 487-518.

[16] Lighthill MJ. Introduction to Fourier analysis and generalised functions. Cambridge: Cambridge University Press; 1975: Ch 3.

[17] Kowalenko V, Rawlinson AA. Mellin-Barnes regularization, Borel summation and the Bender-Wu asymptotics for the anharmonic oscillator. J Phys A 1998: 31; L663-670.

[18] Kowalenko V. Landau damping, Stokes phenomenon and the weakly magnetized Maxwellian plasma. Phys Letts A 2005: 337; 408-418.

[19] Wolfram S. Mathematica- A System for Doing Mathematics by Computer. Reading Mass: Addison-Wesley; 1992.

[20] Prudnikov AP, Brychkov YuA, Marichev OI. Integrals and Series: Vol 1 Elementary Functions. New York: Gordon and Breach; 1986.

[21] Gradshteyn IS/Ryzhik, IM Jeffrey A (Ed.). Table of Integrals, Series and Products Fifth Ed. London: Academic Press; 1994.

[22] Munkhammar J. Integrating Factor. Available from: MathWorld-A Wolfram Web Resource. http://mathworld.wolfram.com//IntegratingFactor.html.

[23] Zwaan A. Intensitäten in Ca-Funkenspektrum. Arch. Néerlandaises des Sciences Exactes 1929: 12; 1-76.

[24] Carrier GF, Krook M, Pearson CE. Functions of a Complex Variable- Theory and Technique. New York: McGraw-Hill; 1966.

[25] Gel'fand IM, Shilov GE. Generalized Functions: Vol 1- Properties and Operations. New York: Academic Press; 1964.

[26] Farassat F. Introduction to Generalized Functions with Applications in Aerodynamics and Aeroacoustics. NASA Technical Paper 3428. Hampton Virginia: Langley Research Center; 1994.

[27] Boyd JP. Hyperasymptotics and the Linear Boundary Layer Problem: Why Asymptotic Series Diverge. SIAM Review 2005; 47(3): 553-575.

[28] Prudnikov AP, Brychkov YuA, Marichev OI. Integrals and Series: Vol 3 More Special Functions. New York: Gordon and Breach; 1990.

[29] Ninham BW. Generalised Functions and Divergent Integrals. Numerische Mathematik 1966: 8; 444-457.

[30] Abramowitz M, Stegun IA. Handbook of Mathematical Functions. New York: Dover; 1964.

[31] Paris RB. Smoothing of the Stokes Phenomenon using Mellin-Barnes integrals. J Comp Appl Math 1992: 41; 391-426.

[32] Olver FWJ. Asymptotic expansions of the coefficients in asymptotic series in asymptotic series solutions of linear differential equations. Methods and Applic of Analysis 1994: 1; 1-13.

[33] Olver FWJ. On an Asymptotic Expansion of a Ratio of Gamma Functions. Proc Roy Irish Acad 1995: 95A; 5-9.

[34] Havil J. Gamma- Exploring Euler's Constant. Princeton: Princeton University Press; 2003.

[35] Kowalenko V. Properties and Applications of the Reciprocal Logarithm Numbers. Acta Appl Math 2009: DOI:10.1007/s10440-008-9325-0.

[36] Apelblat A. Volterra Functions. New York: Nova Science Publishers; 2008.

[37] Erdelyi A et al. Higher Transcendental Functions Vol II. New York: McGraw Hill; 1955.

[38] Kowalenko V, Frankel NE. Asymptotics of the Bose Kummer Function. J Math Phys 1994: 35; 6179-7009.

[39] Kowalenko V. The Non-relativistic Charged Bose Gas in a Magnetic Field II Quantum Properties. Ann Phys (NY) 1999: 274; 165-250.

[40] Kowalenko V. Generalizing the Reciprocal Logarithm Numbers by Adapting the Partition Method for a Power Series Expansion. Acta Appl Math 2009: 106; 369-420, DOI:10.1007/s10440-008-9304-5.

[41] Ecalle J. Les Fonctions Resurgentes Tome I: Les algèbres de fonctions résurgentes. Publications Mathematiques D'Orsay Université de Paris-Sud 81-05.

[42] Ecalle J. Les Fonctions Resurgentes Tome II: Les fonctions résurgentes appliquées à l'itération. Publications Mathematiques D'Orsay Université de Paris-Sud 81-06.

[43] Ecalle J. Les Fonctions Resurgentes Tome III: L'équation du pont et la classification analytique des objets locaux. Publications Mathematiques D'Orsay Université de Paris-Sud 85-05.

[44] Paris RB, Wood AD. Stokes phenomenon demystified. IMA Bulletin 1995: 31; 21-28.

[45] Sternin BYu, Shatalov VE. Borel-Laplace Transform and Asymptotic Theory. Boca Raton Fl:CRC Press; 1996.

[46] Delabaere E, Dillinger H, Pham F. Exact semiclassical expansions for one-dimensional quantum oscillators. J Math Phys 1997: 38; 6126-6184.

[47] Berry MV, Howls CJ. Hyperasymptotics. Proc R Soc Lond A 1990: 443; 653-668.

[48] Boyd JP. Weakly Nonlocal Solitary Waves and Beyond-All-Orders Asymptotics: Generalized Solitons and Hyperasymptotic Perturbation Theory. Amsterdam: Kluwer; 1998.

[49] Boyd JP. The Devil's Invention: Asymptotic, Superasymptotic and Hyperasymptotic Series. Acta Appl Math 1999; 56(1): 1-98.

[50] Fedoryuk MV. Asymptotic Analysis. Berlin: Springer-Verlag; 1993.

[51] Murray JD. Asymptotic Analysis. Oxford: Clarendon Press; 1974.

[52] Oberhettinger F. Tables of Mellin Transforms. Berlin: Springer-Verlag; 1974.

[53] Prudnikov AP, Brychkov YuA, Marichev OI. Integrals and Series Vol 2: Special Functions. New York: Gordon and Breach; 1986.

[54] Watson GN. A Treatise on the Theory of Bessel Functions. Cambridge: Cambridge University Press; 1995.

AUTHOR INDEX

SUBJECT INDEX